Third Edition

Practical Guide to the Packaging of Electronics

Thermal and Mechanical Design and Analysis

Third Edition

Practical Guide to the Packaging of Electronics

Thermal and Mechanical Design and Analysis

Ali Jamnia

CRC Press is an imprint of the
Taylor & Francis Group, an **informa** business

CRC Press
Taylor & Francis Group
6000 Broken Sound Parkway NW, Suite 300
Boca Raton, FL 33487-2742

First issued in paperback 2021

Version Date: 20160502

ISBN-13: 978-1-03-209782-4 (pbk)
ISBN-13: 978-1-4987-5395-1 (hbk)

Publisher's Note
The publisher has gone to great lengths to ensure the quality of this reprint but points out that some imperfections in the original copies may be apparent.

Library of Congress Cataloging-in-Publication Data

Names: Jamnia, Ali, 1961- author.
Title: Practical guide to the packaging of electronics : thermal and mechanical design and analysis / Ali Jamnia.
Description: Third edition. | Boca Raton : Taylor & Francis, CRC Press, 2016. | Includes bibliographical references and index.
Identifiers: LCCN 2015049165 | ISBN 9781498753951 (alk. paper)
Subjects: LCSH: Electronic packaging.
Classification: LCC TK7870.15 .J36 2016 | DDC 621.381/046--dc23
LC record available at https://lccn.loc.gov/2015049165

Visit the Taylor & Francis Web site at
http://www.taylorandfrancis.com

and the CRC Press Web site at
http://www.crcpress.com

To the memory of Dr. Javad Nurbakhsh.

Out of regret for your loss, I say:

"You have gone, and I don't know where you are."

Contents

List of Figures

List of Tables

Preface to the Third Edition

In retrospect, I might have named this book "Practical Guide to *Reliable* Packaging of Electronics." When I look back, in the past nearly 20 years with the start of my Society for Automotive Engineers (SAE) workshops, my goal has been to relay a certain sense of identifying areas where failures occur and means of overcoming their sources. In my opinion, developing a reliable product means to understand failures and anticipate the conditions under which they occur.

For this reason, in this edition, in addition to providing some information on advanced materials and technologies in the heat transfer and vibration isolation areas—to provide more information to the reader—I have added two more chapters; The first is on acoustics. This is an area that is hardly discussed in books on packaging of electronics; yet, it is an important subject, because some systems become noisier as they degrade. We need to develop an understanding of acoustics fundamentals. The other added chapter provides a fuller treatment of reliability and its fundamentals. To do this, I have also changed the chapter on electrical reliability in the second edition. Now, this chapter is more focused on electrical failures and causes—much more similar to the chapter on mechanical reliability.

In the past, my goal has been to develop a book that can be read and comprehended over a short period of time; in other words, express my thoughts in as few pages as possible. As I am adding more material, I am more focused on providing material that may be readily comprehendible and usable in a work environment. I hope I have managed that aspect of the book as well.

Ali Jamnia

Preface to the Second Edition

It has been six years since the first edition of this book was published. My goal has always been to write a book that would enable the reader to pick up what he or she needs to evaluate an electronics system and have a set of tools for back of the envelop calculations. Unfortunately, in my first attempt, I had not paid enough attention to details and had assumed that the reader would make the leaps across developing formulas and derivations with me. I realized I was mistaken during one of my workshops; one of the participants brought this to my attention.

I can only hope that I have been more successful this time. I am still trying to have a simple enough book that is not inundated with lots of equations and too many details. A few of my colleagues still believe that there should not be any equations at all—just design tips. Although this approach may work for some—particularly for those who are concerned only with design—many of us need to make sure that our designs are functional particularly if we are working on a new product. This assurance only comes first through calculations and then through testing.

The arrangement of the chapters in this edition is somewhat different than the first. There are 15 chapters now as opposed to 11. Also, there are seven appendices. Hopefully, the information flows more naturally in this configuration.

Ali Jamnia

Preface to the First Edition

Here, I have a chance to talk to you—the reader—face to face and give you a brief history of how this book came into existence. In 1993–1994, I developed an interest in the issues of electronics packaging. By 1995, I could easily simulate an electronic system using state-of-the-art computer programs and calculate its thermal and vibration characteristics. It became apparent to me, however, that without these sophisticated tools, I did not have a simple way for estimating the same characteristics and hence could not do back of the envelope calculations. I noticed that there are many good books and references on electronics packaging on the market, but the majority seem to make the assumption that the reader is already familiar with the basic approaches and how to make back of the envelop calculations.

Later on, I discovered—much to my surprise—that there are not many engineers who have this set of tools. It was at that time that I embarked on developing a basic understanding of the engineering involved in electronics packaging and eventually presenting them throughout this book.

Herein, I have not set forth to bring together the latest and most accurate techniques, or to cover all aspects of electronics packaging. My goal has been to develop a book that can be read (and comprehended) either in a week's time or over a few weekends. And it would provide the basics that an engineer—mechanical, biomedical, or electrical—needs to keep in mind when designing a new system or troubleshooting a current one. Furthermore, this book will serve program and engineering managers as well as quality assurance directors to refresh their basics every once in a while. I hope that this book will be of service to them as well.

This book is also based on my seminar notes that the Society of Automotive Engineers has sponsored.

Ali Jamnia

Acknowledgments

In my career as a research engineer (and now as an engineering specialist), I have been blessed with meeting some very brilliant people who have left their imprint on me. Two persons have played key technical roles in that they have helped me make substantial changes in the direction that my career has taken. The first of these two was Mr. Robert E. Walter who helped me bridge the gap between the world of research and concepts and the world of "real" engineering and manufacturing. The second person who helped me take even a more important leap is Dr. Jack Chen who at one point of time accepted the role of being my mentor. Through Dr. Chen's guidance, I have brought the world of research and engineering together in order to develop an understanding of what it means to be an innovator.

Although having a technical mentor is important, it is just as essential to have a guide in developing the so-called soft skills; the art of developing emotional intelligence and understanding that to complete projects one needs to have developed the art of interacting with and influencing people. In this regard, I like to express my appreciation to Dr. Rex Kuriger and Mr. Kurt Steinbrenner who helped me understand what emotional intelligence means and how to utilize it in the work environment.

Writing this book (in all of its editions) has not been easy. The third edition took just about as much time as the first two. It meant time spent away from my daughter Naseem, my son Seena, and my wife Mojdeh. They have been wonderful and supportive, and this is the time for me to say thanks for their support.

Author

Dr. Ali Jamnia has published a large number of engineering papers and presentations as well as nontechnical articles. He also has a number of patents and patent applications. He has a demonstrable understanding of the engineering principles, particularly in the areas of fluid flow, heat transfer, solid mechanics and reliability.

He has focused on the issues of electronics packaging since the early 1990s. And since 1995, he has been involved with the development of innovative electronics system to aid individuals with either physical or cognitive disabilities. In fact, his prime achievement has been the development of a specialized computer systems called the Learning Station™ to be used as a teaching tool for individuals with cognitive disabilities.

His primary expertise lies in electromechanical systems design and development. In addition, he enjoys conducting analysis of various engineering problems using numerical approximations and computer simulations.

Dr. Jamnia enjoys teaching and mentoring junior engineers. In fact, the Society for Automotive Engineers (SAE) had sponsored him for nearly 10 years to teach workshops on electronics packaging from a thermal and mechanical design and analysis point of view.

1

Introduction

Issues in Electronics Packaging Design

Let us assume that you have the responsibility of developing a new electronics system. Let us also assume that your budget allows you to bring a team of experts together. Where do you begin? Whom do you hire?

It does make sense to hire a team of electrical engineers to design the electronics and people to lay out the boards, and maybe even people who will eventually manufacture them. Also, you have been advised that overheating may be a problem, so you consider hiring a thermal engineer, but one of your team members points out that he has a few tricks up his sleeve and it is better to spend the money elsewhere.

In the last leg of your project you hire a junior designer to develop your enclosure and you send the product to the market ahead of schedule. Everyone is happy, but ...

In a few months, you have a problem. Your field units fail too often. The majority seem to have an overheating problem. There is a fan to cool the system, but it is not enough; you decide to add another one, but to no avail.

Well, your patience runs out and you decide to hire the thermal engineer after all. His initial reaction is to point out that thermal considerations have not been built into the system design, but after a few weeks he manages to find a solution; however, it is expensive and cumbersome. Well, you have no other choice; you accept his recommendations and all of the systems are retrofitted.

Before you have a chance to take a sigh of relief, you have another problem facing you. The field units fail again but for different reasons. Some fail at the printed circuit board assembly level, others fail on the surface of the enclosure, and still others fail for no apparent reason.

What have you overlooked?

What knowledge base do you need to have to answer this question? This guide was developed precisely to help you answer this question. The objectives are

- To develop a fundamental grasp of the engineering issues involved in electronics packaging
- To develop the ability to define guidelines for system's design when the design criteria and components are not fully known
- To identify reliability issues and concerns
- To develop the ability to conduct more complete analyses for the final design

Technical Management Issues

Let us review the technical issues that require engineering management. These issues are briefly discussed here.

Electronics Design

An electronics engineer is generally concerned with designing the electronics to complete a particular task or choosing a commercial-off-the-shelf board accomplishing the same task. In other applications, an integrated circuit or a hybrid may need to be designed for conducting specific tasks. Detailed discussion on this topic is beyond the scope of this book.

Packaging/Enclosure Design

There are four topics that I categorize under packaging and enclosure design and analysis. These are electromagnetic, thermal, mechanical, and thermomechanical analyses. We do not cover electromagnetics here; however, its importance cannot be overstated. Unfortunately, much of the analysis for electromagnetic interference (EMI) or electromagnetic compatibility is done as an after event. Testing is done once the system is developed and often coupling and interactions are ignored. EMI is difficult to calculate exactly; however, back-of-the-envelope estimates may be developed to ensure higher end product compliance. Thermal analysis is concerned with calculating the component-critical temperatures. Mechanical analysis is concerned with the housing of the electronics (from component housing to PCB to enclosure and finally to the rack) as well as the ability of this housing to maintain its integrity under various loading conditions such as shock and vibration. In addition, the gradual wear and degradation of other moving or nonmoving components are covered under the mechanical analysis umbrella. Thermomechanical management is concerned with the impact of thermal loads on the mechanical behavior of the system.

In this book, we set the foundation for thermal and mechanical analyses of electronics packaging/enclosure design.

Reliability

Although in my view thermal, mechanical, thermomechanical, and EMI analyses are subsets of reliability analysis, most engineers consider reliability calculations to cover areas such as mean time to failure or mean time between failures. This information helps us develop a better understanding of maintenance and repair scheduling as well as warranty repairs and merchandise returns caused by failure.

This book provides an overview of the topic of reliability as well.

2

Basic Heat Transfer—Conduction, Convection, and Radiation

Basic Equations and Concepts

As electric current flows through electronic components, it generates heat. This heat generation is proportional to both the current level as well as the electrical resistance of the component.

Should heat be generated in a component and not escape, the component temperature begins to rise and it will continue to rise, until the component melts and the current is disconnected. To prevent this temperature rise, heat must be removed to a region of lesser temperature. There are three mechanisms for removing heat; namely, conduction, convection, and radiation.

Conduction takes place in opaque solids, where, using a simple analogy, heat is passed on from one molecule of the solid to the next. Mathematically, it is usually expressed as

$$Q = \frac{KA}{L}\left(T_{\text{hot}} - T_{\text{cold}}\right) \tag{2.1}$$

In this equation, Q is heat flow, T is temperature, K is thermal conductivity, A is cross-sectional area, and L is the length heat travels from the hot section to the cold.

Convection takes place in liquids and gases. The molecules in fluids are not as tightly spaced as solids; thus, heat packets move around as the fluid moves. Therefore, heat transfer is much easier than conduction. Mathematically, it is expressed as

$$Q = hA\left(T_{\text{hot}} - T_{\text{cold}}\right) \tag{2.2}$$

In this equation, Q and T are heat flow and temperature as in Equation 2.1. However, h is defined as the coefficient of heat transfer, and A is cross-sectional area between the solid generating heat and the fluid carrying it away.

Radiation takes place as direct heat transfer from one region to another. Similar to light, it does not require a medium to travel. It is expressed as

$$Q = \sigma\varepsilon A\left(T_{\text{hot}}^4 - T_{\text{cold}}^4\right) \tag{2.3}$$

In this equation, Q is heat flow and T is temperature as in the other two equations. A is the area of the radiating surface, ε is emissivity—a surface property discussed later, and σ is a universal number called Stefan–Boltzmann constant.

These equations will be discussed in some detail, and you will learn how these equations will enable you to either evaluate the thermal performance of an existing system or set design criteria for new systems to be developed. Bear in mind that, in general, these equations express physical concepts but do not produce "locally exact" solutions.

For now, let me draw your attention to a few important points. First, there has to be a temperature differential for heat to flow; next, heat rate depends on the cross-sectional area; finally, although the relationship between heat flow and temperature difference is linear for conduction and convection, for radiation, this relationship is extremely nonlinear.

General Equations

If you need to obtain a locally exact solution, you should employ a more general set of equations. These equations are based on conservation of mass, conservation of momentum, conservation of energy, and a constitutive relationship. This general form of equations for fluid flow and heat transfer is as follows:

$$\begin{aligned}
&\rho_{,t} + \left(\rho u_i\right)_{,j} = 0 \\
&\left(\rho u_j\right)_{,t} + \rho u_i u_{j,k} = -p_{,j} + (\lambda u_{k,k})_{,j} + \left(\mu\left(u_{i,j} + u_{j,i}\right)\right)_{,i} + \rho f \\
&\left(\rho C_p T\right)_{,t} + \rho C_p u_k T_{,k} = -p u_{k,k} + (K T_{,j})_{,j} + \lambda\left(u_{k,k}\right)^2 + \mu\left(u_{i,j} + u_{j,i}\right) u_{j,i} + \dot{Q} \\
&p = p(\rho, T)
\end{aligned} \tag{2.4}$$

In these equations, a comma denotes taking a derivative. Indices t, i, j, and k denote time and spatial directions x, y, and z, respectively.

Clearly, unlike the previous set of equations, these equations are not simple to solve nor can they be readily used to evaluate system performance or to set design criterion. Generally, it takes sophisticated computer hardware and software to solve these equations. The significance of these equations may be numerated as follows:

1. They produce exact solutions of any thermal/flow problems.
2. These equations could be reduced to the simpler forms introduced earlier.
3. They may be used to develop a set of parameters that enables us to evaluate system parameters and design criterion above and beyond the information given to us by the previous set of equations. These parameters are nondimensional and can be used as a means of comparing various variables among systems that have different configurations such as size or heat generation rate.

Nondimensional Groups

Most often, results of engineering research and works in fluid flow and heat transfer are expressed in terms of nondimensional numbers. It is important to develop a good understanding of these nondimensional numbers. The set of interest to us is as discussed in subsequent sections.

Nusselt Number

Nusselt number shows the relationship between a fluid's capacity to convect heat versus its capacity to conduct heat.

$$Nu = \frac{hL}{K}$$

Grashof Number

Grashof number provides a measure of buoyancy forces of a particular fluid.

$$Gr = \frac{L^3 \rho^2 g \beta \Delta T}{\mu^2}$$

Prandtl Number

Prandtl number shows the relationship between the capacity of a fluid to store heat versus its conductive capacity.

$$Pr = \frac{C_p \mu}{K}$$

Reynolds Number

Reynolds number gives a nondimensional measure for flow velocity.

$$Re = \frac{\rho U L}{\mu}$$

We will revisit these equations and their significance in electronics enclosure thermal evaluation later.

3

Conductive Cooling

Introduction

As mentioned in Chapter 2, conduction takes place in opaque solids, where, using a simple analogy, heat is passed on from one molecule of the solid to the next.

Let us look at an example:

Consider a layer of epoxy with a thermal conductivity of 0.15, a thickness of 0.01, and a cross-sectional area of 1. A heat source on the left-hand side generates a heat load of 100. The surface temperature on the right-hand side is 75 (note that all units have been purposely ignored). What is the surface temperature on the left-hand side? For the sake of brevity, ignore the units.

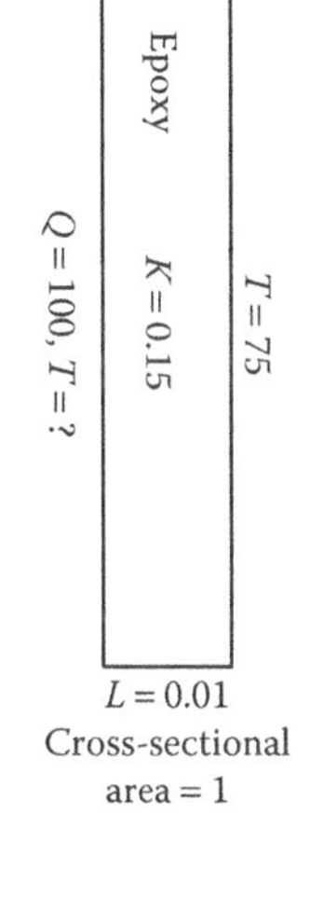

$L = 0.01$
Cross-sectional area = 1

$$Q = \frac{KA}{L}(T_{hot} - T_{cold}) \text{ or } Q = \frac{KA}{L}(\Delta T)$$

$$100 = \frac{(0.15)(1)}{0.01}(\Delta T) \Rightarrow \Delta T = 6.67$$

$$T = 75 + 6.67 = 81.67$$

Notice that this formula only gives the temperature at one point; namely, the left-hand side. However, the temperature distribution in the epoxy is not known. This distribution can only be calculated by using other mathematical formulae.

Thermal Resistance

Similar to electrical resistance to current flow, any given material also resists heat flow. This concept is very useful and can be developed to provide a systematic approach to solving heat flow problems.

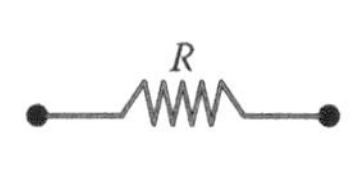

In electricity the relationship between the electric potential and resistance is defined as

$$\Delta E = IR$$

where I is the electrical current. A similar relationship may also be developed for temperature, thermal resistance, and heat flow:

$$\Delta T = QR \text{ if } R = \frac{L}{KA}$$

The previous example may now be solved using this approach. By using the concept of thermal resistance we obtain

$$R = \frac{L}{KA} \text{ or } R = \frac{0.01}{0.15 \times 1} = 0.0667$$

$$\Delta T = QR \Rightarrow \Delta T = 100 \times 0.0667 = 6.67$$

$$T = 75 + 6.67 = 81.67$$

Although a very simple problem was used to demonstrate the thermal resistance concept, this method can be applied to complicated problems with relative ease.

Sample Problem and Calculations

Consider this geometry of a typical chassis wall. Find the hot side temperature if the wall temperature on the cold side is maintained at 75°F; each opening is 5 × 1 in; sheet metal is 0.050 in thick aluminum (6061). The length is in inches, heat flow in BTU/hr, and temperature in degrees Fahrenheit.

Before tackling this problem, we need to know about thermal resistance networks so that this and similar problems may be modeled properly.

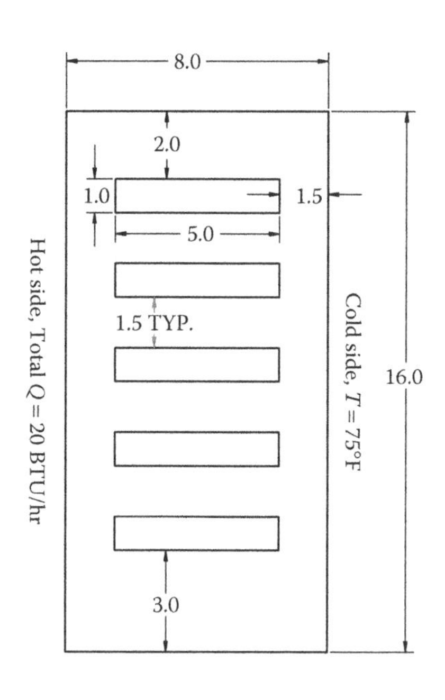

Resistance Network

Similar to the flow of electricity through a network of various components, each having a different electric resistance, heat, too, may flow through different paths in parallel and/or in series, each having different thermal resistance. Thermal networks developed in this fashion provide a powerful tool to find an equivalent resistance for the entire network, hence allowing us to evaluate a temperature difference.

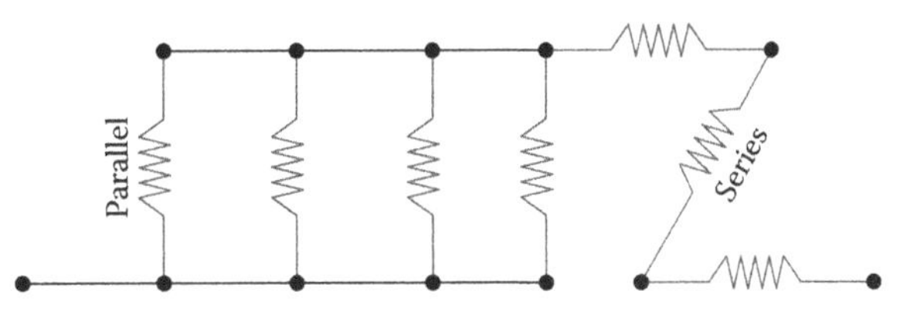

Network Rules

Since the elements of this network are either in series or in parallel, we first need to know how to find the equivalent resistance for each one.

Series Rules

When components are placed in series, the overall thermal resistance of such a network increases.

$$R_{\text{total}} = R_1 + R_2 + R_3 + \ldots$$

Parallel Rule

When components are placed in parallel, the overall thermal resistance of such a network decreases.

$$\frac{1}{R_{\text{total}}} = \frac{1}{R_1} + \frac{1}{R_2} + \frac{1}{R_3} + \ldots$$

Sample Problem and Calculations

Consider the chassis wall again. We need to find the hot side temperature if the wall on the right-hand side is maintained at 75°F temperature. The first step is to develop the representative network, then reduce it and finally find the equivalent resistance (R_{total}). This process is depicted in Figure 3.1.

In Table 3.1, the length, area, and resistance of each element is tabulated. Recall that $R = (L / KA)$ and thermal conductivity for aluminum is 7.5 (Btu/[hr in °F])

Now we need to find the equivalent resistance for the elements in parallel:

$$\frac{1}{R_9} = \frac{1}{R_2} + \frac{1}{R_3} + \frac{1}{R_4} + \frac{1}{R_5} + \frac{1}{R_6} + \frac{1}{R_7}$$

$$\frac{1}{R_9} = \frac{1}{6.67} + \frac{1}{8.89} + \frac{1}{8.89} + \frac{1}{8.89} + \frac{1}{8.89} + \frac{1}{4.44}$$

$$R_9 = 1.21$$

This enables us to replace the network representative with a simpler one in which the resistance elements are in series:

$$R_T = R_1 + R_9 + R_8$$

$$R_T = 0.25 + 1.21 + 0.25 \Rightarrow R_T = 1.71$$

$$\Delta T = QR \Rightarrow \Delta T = 20 \times 1.71 \Rightarrow \Delta T = 34.2°\text{F}$$

$$T_{\text{hot}} = 75 + 34.2 \Rightarrow T_{\text{hot}} = 109.2°\text{F}$$

The following two points must be noted here:

1. No temperature variation in the vertical direction has been taken into account.
2. We have only calculated the temperature at the high point. No other temperature information is known to us through this calculation. If critical components are placed inside, we need to find a way to ensure that we have not exceeded their operating temperature range.

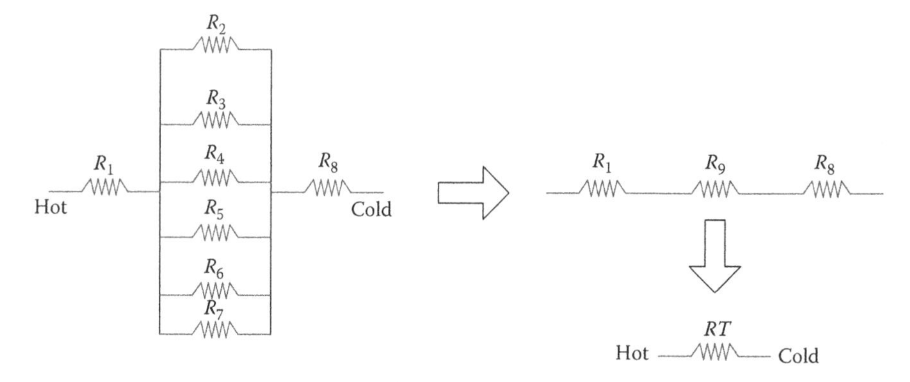

FIGURE 3.1
The thermal resistance network of the chassis example.

TABLE 3.1
The Information Pertinent to the Chassis Example

Length (in)	Area (in²)	Resistance
$L_1 = 1.5$	$A_1 = 16 \times 0.05$	$R_1 = 0.25$
$L_2 = 5.0$	$A_2 = 2 \times 0.05$	$R_2 = 6.67$
$L_3 = 5.0$	$A_3 = 1.5 \times 0.05$	$R_3 = 8.89$
$L_4 = 5.0$	$A_4 = 1.5 \times 0.05$	$R_4 = 8.89$
$L_5 = 5.0$	$A_5 = 1.5 \times 0.05$	$R_5 = 8.89$
$L_6 = 5.0$	$A_6 = 1.5 \times 0.05$	$R_6 = 8.89$
$L_7 = 5.0$	$A_7 = 3 \times 0.05$	$R_7 = 4.44$
$L_8 = 1.5$	$A_8 = 16 \times 0.05$	$R_8 = 0.25$

Before considering these two comments, we need to verify that we have a good solution here. Since this is a relatively simple problem, we can find a solution with a high degree of accuracy using finite element methods.

Comparison with Exact Results

Figure 3.2 may be considered as the exact results for this problem using finite element analysis (FEA). The maximum temperature from this analysis is also 109.2°F. However, one will notice that the temperature distribution along the left edge is not uniform. The resistance network method has predicted that the temperature all along the hot side is uniform.

FIGURE 3.2
The "exact" solution obtained from finite element methods.

Assumptions

The reason for this discrepancy is that in the previous technique, it is assumed that heat flow is uniform along the direction of the thermal resistance. Effectively, this means that the heat conduction problem is one dimensional. Clearly, this assumption does not hold true all the time as in the corners of this example problem; however, it has validity if used with caution.

Temperature at Intermediate Points

To calculate internal temperature distribution, one needs to bear in mind that the relationship $\Delta T = QR$ holds true not only for the entire network but for each element as well. Therefore, temperatures at interior points may be calculated. However, instead of the total resistance of the entire network, the proper resistance associated with the location of interest must be used. Furthermore, one should keep in mind that Q is constant throughout the system and flows in the direction from the hot to the cold spots.

For example, temperature on the right side of the chassis openings is

$$\Delta T = QR_8 \Rightarrow \Delta T = (2\times 3.41)\times 0.25 \Rightarrow \Delta T = 1.705°\text{F}$$

$$T_{\text{right side}} = 75 + 1.705 \Rightarrow T_{\text{right side}} = 76.705°\text{F}$$

Similarly, the temperature on the left-hand side of the same openings is

$$\Delta T = Q(R_8 + R_9) \Rightarrow \Delta T = (2\times 3.41)\times(0.25 + 1.21) \Rightarrow \Delta T = 9.96°\text{F}$$

$$T_{\text{left side}} = 75 + 9.96 \Rightarrow T_{\text{left side}} = 84.96°\text{F}$$

Exercise: IC Temperature Determination

One area of thermal modeling is the heat flow in between various layers of materials. An example of this configuration is heat flow from a chip through the printed circuit board (PCB) into a heat sink as shown in Figure 3.3.

In this application, all the heat is transferred via conduction. As a result, spreaders must be used to transfer heat efficiently. In the selection of spreaders, care must be exercised to choose materials that have compatible coefficients of thermal expansion (CTE). Neighboring materials with incompatible CTEs could potentially cause failures. We will consider this topic in Chapter 11, but for now, we will ignore any potential CTE mismatches and their impact. The following information is provided to us:

Integrated circuit (IC) generates 2 W of heat.

Top two adhesives are 0.008 in thick, with a thermal conductivity equal to 0.450 BTU/hr ft °F.

Lower adhesive is 0.003 in thick, with a thermal conductivity equal to 0.450 BTU/hr ft °F.

Silver spreaders are 0.05 in thick, with a thermal conductivity equal to 280 BTU/hr ft °F.

Electrical insulation is 0.005 in thick, with a thermal conductivity equal to 0.2 BTU/hr ft °F.

The cross-sectional area of the spreaders as well as adhesives and insulation is 1 in^2.

There are 50 via holes each having a 0.025 in diameter plated with 2 oz copper.

The thickness of the PCB is 0.032 in.

Thermal conductivity of copper is 220 BTU/hr ft °F.

The metallic core is maintained at 85°F.

There are several pitfalls requiring our attention. First, the set of units as shown in Figure 3.3 is not consistent. Next, the thickness of copper in via holes is given in terms of its weight. Finally, the proper conduction area for the vias must be calculated. With these in mind, the number crunching is straightforward. The thermal network is shown in Figure 3.4.

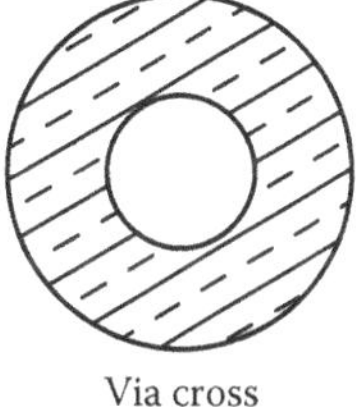

As pointed out, it is customary to specify the copper thickness in terms of its weight. Each ounce of copper denotes a thickness of 0.0014 in. Therefore, 2 oz copper provides a thickness of 0.0028 in. As for proper via area calculation, one only has to be mindful that typically the via has a hole in the middle and the area for conduction is the area of the donut shape.

The thickness of the copper is 0.0028 (2 oz copper) leading to a hole diameter of 0.0194 [$0.0194 = 0.025 - (2)(0.0028)$]. The conduction area, therefore,

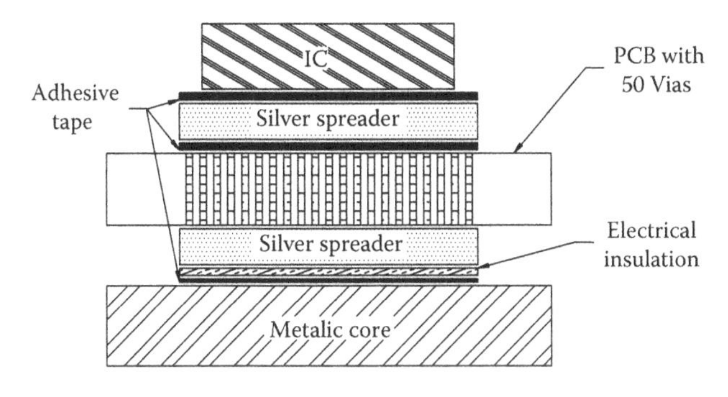

FIGURE 3.3
Heat flow from an IC through the PCB into the heat sink.

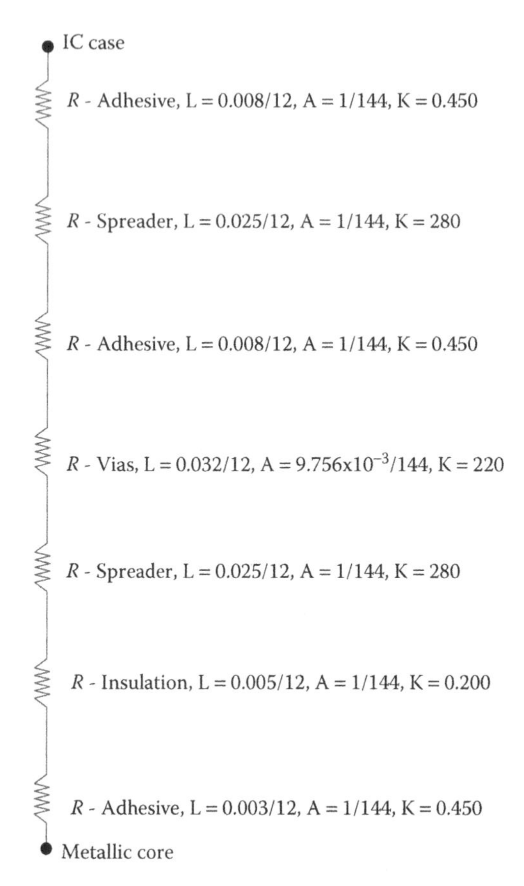

FIGURE 3.4
The heat flow network.

is the area of the via minus the area of the hole. This leads to a value of 1.95×10^{-4} in^2. There are 50 vias so the final area is 9.75×10^{-3} in^2. Bear in mind that all units must be consistent, so all lengths must be converted to feet (Table 3.2).

Now, the temperature of the IC may be calculated:

$$\Delta T = QR_T \Rightarrow \Delta T = \left(2\,\text{W} \times 3.41 \frac{\text{BTU / hr}}{\text{W}}\right) \times 0.988 \Rightarrow \Delta T = 6.736°\text{F}$$

$$T_{\text{IC}} = 85 + 6.736 \Rightarrow T_{\text{IC}} = 91.736°\text{F}$$

Notice that 2 W of heat generation must be converted into BTU/hr, hence it is multiplied by 3.41 conversion factor.

TABLE 3.2

Element Data for the Integrated Circuit Heat Flow Problem

Element	Length (ft)	Area (ft^2)	Conductivity	Resistance
Adhesive	0.008/12	1/144	0.450	0.213
Spreader	0.025/12	1/144	280	0.001
Adhesive	0.008/12	1/144	0.450	0.213
50 Vias	0.032/12	$9.76 \times 10^{-3}/144$	220	0.179
Spreader	0.025/12	1/144	280	0.001
Insulator	0.005/12	1/144	0.200	0.300
Adhesive	0.003/12	1/144	0.450	0.080
			Total resistance	0.988

A strong feature of this approach is its ease of modeling: making changes to the design and comparing results. For example, should we decide to add an adhesive to the bottom of the vias and the silver spreader, we could easily observe the impact on the IC temperature (Table 3.3).

$$\Delta T = QR_T \Rightarrow \Delta T = (2 \times 3.41) \times 1.201 \Rightarrow \Delta T = 8.191°\text{F}$$

$$T_{\text{IC}} = 85 + 8.191 \Rightarrow T_{\text{IC}} = 93.191°\text{F}$$

Another feature is that the elements with the greatest resistance may readily be identified and hence optimized, should the need be.

One may argue that while using the full area of the spreader under the 50 vias may be justifiable, using the full area (1 in^2) of adhesive may be erroneous because of its low conductivity. With the same logic, it may be said

TABLE 3.3

Element Data for the Modified Heat Flow Problem

Element	Length (ft)	Area (ft^2)	Conductivity	Resistance
Adhesive	0.008/12	1/144	0.450	0.213
Spreader	0.025/12	1/144	280	0.001
Adhesive	0.008/12	1/144	0.450	0.213
50 Vias	0.032/12	$9.76 \times 10^{-3}/144$	220	0.179
Adhesive	**0.008/12**	**1/144**	**0.450**	**0.213**
Spreader	0.025/12	1/144	280	0.001
Insulator	0.005/12	1/144	0.2	0.300
Adhesive	0.003/12	1/144	0.450	0.080
			Total resistance	1.201

that a better approximation to the area of the adhesive may be to use an area equivalent to that of the vias. On this basis, we would predict a chip temperature of nearly 240°F. In reality, however, heat spreads and for a better approximation we need to take heat spreading into account.

Heat Spreading

In the previous example, an underlying assumption was that the 2 W generated in the IC is uniformly distributed over the one-square-inch area. In real life, however, as heat flows away from its source, it spreads in a spherical fashion. In other words, it progressively covers a larger area. The formed cone angle depends on the thermal conductivity of the substrate material.

Thermal Modeling

The solution methodology is the same as explained previously; one needs to identify various thermal paths and calculate their thermal resistances. Since heat spreads, one has to consider and calculate a larger area. A good approximation is to use an average area of the top and bottom surfaces. For the one-dimensional problem depicted in Figure 3.5, heat spreading is assumed to be in the x–y plane alone and the depth is ℓ

$$x = y \tan\theta$$

$$\text{Average area} = \frac{\{d\ell + (d+2x)\ell\}}{2}$$

$$\text{Average area} = (d+x)\ell$$

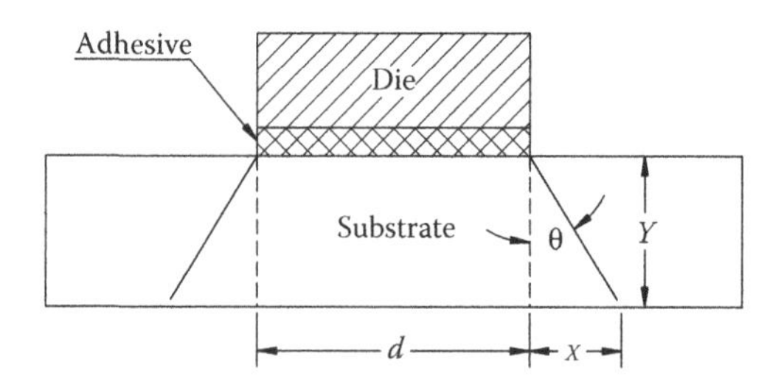

FIGURE 3.5
Heat spreading cone.

To calculate this area, the spread angle must be known. The following two empirical formulae developed based on the data provided by Leatherman (1996) are available:

$$\theta = 90 \ \tanh\left\{0.355\left(\frac{\pi K}{180}\right)^{0.60}\right\} \text{ for conductivity in W/(m°C)} \tag{3.1}$$

$$\theta = 90 \ \tanh\left\{0.510\left(\frac{\pi K}{180}\right)^{0.55}\right\} \text{ for conductivity in BTU/(hr ft °F)} \tag{3.2}$$

where K is substrate's thermal conductivity and θ is in degrees. Note that in this formulation, we have assumed that heat does not spread in the z direction. Should this be a suitable assumption, we would have the following relationship for the average area:

$$\text{Average area} = \frac{\{d\ell + (d+2x)(\ell+2x)\}}{2}$$

$$\text{Average area} = d\ell + (d+\ell)x + 2x^2$$

Krum (2000) provides the following relationship for calculating the spread angle:

$$\theta = \tan^{-1}\left(\frac{K_1}{K_2}\right)$$

where K_1 and K_2 are the thermal conductivities of the current and the underlying layers and θ is the spread angle in degrees. Note that if $K_1 = K_2$, then $\theta = 45$. This implies that the spread angle is always 45° regardless of the material used. This is not accurate.

Example

A silicon chip measuring 0.07 × 0.07 in and 0.025 in thick is mounted to an alumina case (0.2 × 0.2 × 0.025 in) with conductive epoxy 0.003 in thick. This assembly is potted such that the heat flow path is only through the substrate, which is mounted on a heat sink that is maintained at a 75°F temperature. Determine the chip temperature if the heat dissipation is 0.35 W. Thermal conductivity of Alumina is 17 BTU/(hr ft °F) and of thermal epoxy is 1.25 BTU/(hr ft °F).

The first step is to calculate the top and bottom area:

$$\text{Top area} = 0.07 \times 0.07 = 0.0049 \ \text{in}^2$$

$$\text{Bottom area} = (0.07 + 2x) \times 0.07 = 0.0049 \ \text{in}^2$$

where $x = y\tan(\theta)$ and θ is calculated from Equation 3.2.

$$\theta = 90\tanh\left\{0.51\left(\frac{\pi K}{180}\right)^{0.55}\right\} = 90\tanh\left\{0.51\left(\frac{3.14\times 17}{180}\right)^{0.55}\right\} \Rightarrow \theta = 23°$$

From here,

$$x = 0.025\tan(23)$$

$$x = 0.0106 \text{ in}$$

$$\text{Bottom area} = 0.0064 \text{ in}^2$$

$$\text{Avereage area} = (0.0064 + 0.0049)/2$$

$$\text{Average area} = 0.0056 \text{ in}^2$$

$$R_{\text{total}} = R_{\text{epoxy}} + R_{\text{alumina}}$$

$$R_{\text{total}} = \left(\frac{L}{KA}\right)_{\text{epoxy}} + \left(\frac{L}{KA}\right)_{\text{alumina}}$$

$$R_{\text{total}} = \left(\frac{0.003/12}{1.25\times(0.0049/144)}\right) + \left(\frac{0.025/12}{17\times(0.0056/144)}\right) = 9.005$$

$$\Delta T = QR_{\text{total}} \Rightarrow \Delta T = (0.35\times 3.41)\times 9.005 \Rightarrow \Delta T = 10.7\,°\text{F}$$

$$T_{\text{Chip}} = 75 + 10.7 \Rightarrow T_{\text{Chip}} = 85.7\,°\text{F}$$

Note that lengths must be converted to feet and heat generation to BTU/hr.

If we were to include heat spreading in two directions, the average area would be calculated as follows:

$$\text{Bottom area} = (0.07 + 2x)\times(0.07 + 2x) = 0.0083 \text{ in}^2$$

$$\text{Average area} = (0.0083 + 0.0049)/2 = 0.0066 \text{ in}^2$$

Temperature may now be calculated with the average areas:

$$R_{\text{total}} = \left(\frac{L}{KA}\right)_{\text{epoxy}} + \left(\frac{L}{KA}\right)_{\text{alumina}}$$

$$R_{\text{total}} = \left(\frac{0.003/12}{1.25 \times (0.0049/144)} \right) + \left(\frac{0.025/12}{17 \times (0.0066/144)} \right) = 8.547$$

$$\Delta T = QR_{\text{total}} \Rightarrow \Delta T = (0.35 \times 3.41) \times 8.547 \Rightarrow \Delta T = 10.2°\text{F}$$

$$T_{\text{Chip}} = 75 + 10.2 \Rightarrow T_{\text{Chip}} = 85.2°\text{F}$$

Notice that inclusion of the heat spreading in the *z* direction did not make a significant change in the temperature of the chip. The reader is encouraged to evaluate die (chip) temperature, ignoring the spreading effect, and compare the results with the above values.

Applications

Because the generated heat spreads over larger and larger areas, the temperature difference between the source and the sink tends to drop. This physical phenomenon is exploited extensively in the semiconductor packaging industry to provide cooler die temperatures. In the development of the PCB layout and design, heat spreading needs to be taken into account to calculate the number and the location of via holes.

Another application of heat spreading is in the selection of heat sinks and interface materials. The example shown in Figure 3.6 depicts a TO-220 package dissipating 1 W in a 70°F environment. Note that the interface material is modeled to be larger than the size of the component. This is to ensure that the spreading is captured within the interface material.

Junction-to-Case Resistance

Many chip manufacturers provide a set of data commonly referred to as junction-to-case resistance, which is defined as follows:

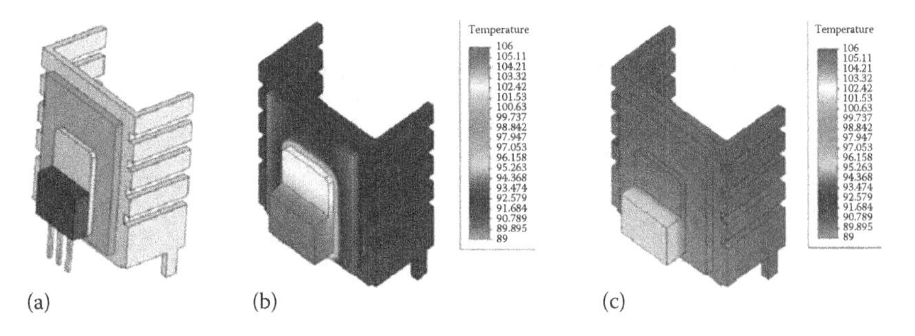

FIGURE 3.6
(a) Initial model, (b) maximum temperature rise is 36°F, and (c) maximum temperature rise is 25°F—a temperature drop of nearly 11°F due to heat spreading.

$$\theta_{j-c} = \frac{T_j - T_c}{q_j}$$

where j denotes the junction of the chip, c the outer casing, q_j the heat generated by the chip, and θ_{j-c} is the thermal resistance between the die and its outer casing. It may be used—with caution—to develop a system's resistance network and hence calculate the die temperature without the exact knowledge of the material(s) used in the chip encapsulation. Furthermore, this is a handy formula for evaluating the temperature of various chips in conjunction with choosing a heat sink or a fan–heat sink combination.

Contact Interface Resistance

In many applications, components must be attached together to develop the needed configuration. For example, PCB assemblies (PCBAs) must be placed on edge guides, or power supply units must be bolted to the chassis. This requires that two separate surfaces be joined at an interface allowing heat to flow across this interface. Because of surface irregularities, the actual contact area is much smaller than the apparent contact area. As we now know the magnitude of heat flow area has a direct impact on the temperature difference; the larger the area, the lower the temperature. Therefore, a smaller contact area at the interface leads to higher temperature rises than expected.

This problem is illustrated below. In Figure 3.7a, a close-up of such an interface is depicted. Where the surfaces meet, the temperatures are lower (Figure 3.7b) and, consequently, where there is a gap, the temperatures are higher. Similarly, in Figure 3.7c, the heat flux through the interface is shown. Clearly, the interface presents a thermal barrier that needs to be addressed in thermal design or analysis.

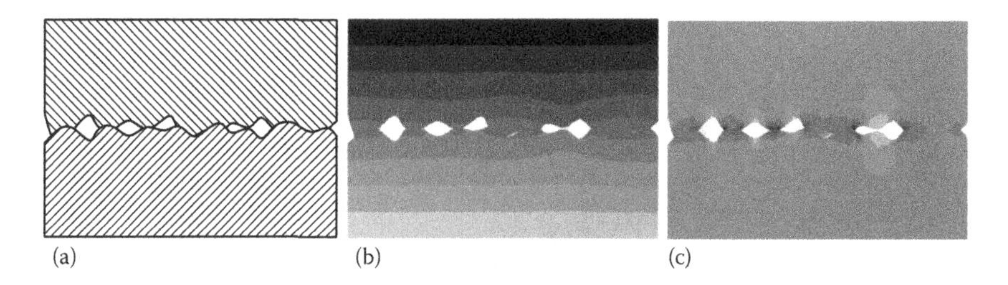

FIGURE 3.7
Heat flow conditions at the junction of two surfaces. (a) Depiction of the interface, (b) temperature variation, and (c) flux variation.

One remedy is to apply pressure. Another solution is to apply interface materials such as thermal grease. As pressure increases, the interface resistance decreases; however, this decrease is bound by an asymptotic value as shown in Figure 3.8.

Modeling the Interface

To model the presence of the interface in a network the following relationship may be used:

$$\Delta T = QR_{\text{int}} \text{ where } R_{\text{int}} = \frac{1}{h_{\text{in}}A} \tag{3.3}$$

In this relationship, A is the apparent contact surface area and h_{in} is the interface resistance. Note that h_{in} increases with both temperature and pressure, leading to an overall decrease in resistance. It is worth mentioning that while high altitude does have an impact on this resistance, it is not significant for most earth-bound applications. However, for space applications care must be exercised in devising appropriate interface pressure.

Calculating h_{in} is not a trivial task. Yovanovich et al. (1997) provide a relatively simple approach that takes into account the surface roughness and

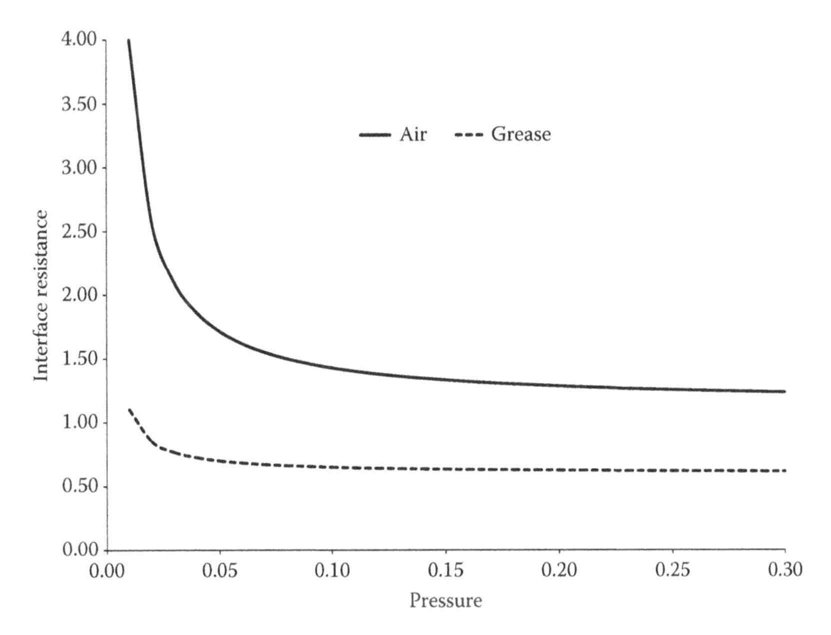

FIGURE 3.8
The impact of pressure on interface resistance. Added pressure reduces contact resistance.

microhardness as well as its mechanical and thermal properties to evaluate its resistance. Briefly, there are two factors contributing to h_{in}; first is conduction through the metal caused by physical contact of asperities, and second is the conduction through the gap. Needless to say, as more pressure is applied to the interface, surface asperities deform and provide a larger contact area, leading to a reduction of interface resistance. Interface materials such as grease provide a better conduction path through the gaps that help lower the resistance.

Among the factors that affect the interface resistance is the presence of oxide layers and/or surface treatments such as surface finishing or a coating. For instance, by electropolishing a metallic surface, not only does the surface become smoother, but oxide layers may be removed, leading to a better thermal conductivity at the surface. Now, if the same surface is coated with a material such as industrial diamonds, heat can spread to a wider area, leading again to lower interface resistance values. MIL-HDBK-251 (1978) provides a discussion of this topic as well as a relatively extensive set of figures providing the resistance between various surfaces. For instance, for Al 7075, h_{in} varies from about 500 [BTU/(hr ft^2 °F) at zero pressure and 25 W/in^2] to nearly 5000 [BTU/(hr ft^2 °F) at 400 psi pressure and 150 W/in^2] (MIL-HDBK-251 1978). For a study of the impact of surface finishes for electronics, see Feldstein and Dumas (2000). For a general discussion of thermal contact conductance, see Madhusudana (1996).

Exercise—Calculate the Component Temperature

Several heat-generating components dissipating a total of 12 W are placed on the long end of an L-shaped aluminum bracket as shown in Figure 3.9. This bracket is mounted on a chassis maintained at 100°F. Determine the component temperature provided that conduction is the only available heat path. The bolt exerts 25 psi of pressure. As a means of comparison, a finite element model was developed and analyzed under similar thermal conditions including the interface condition. The FEA temperature was 161.88°F. Appendix A provides details of the numerical analysis.

There are three different ways to model this problem depending on the thermal paths to be considered; namely, the heat flow through the bracket and the heat flow through the interface. These will be examined here. All approaches, however, assume a 1D conduction only heat flow. Furthermore, we will not treat the bolt any differently than the rest of the system.

First Approach

Bracket

Segment 1: From heat source to the chassis; heat path length is measured from the center of heat source to the chassis; heat path area is the cross-sectional area of the bracket.

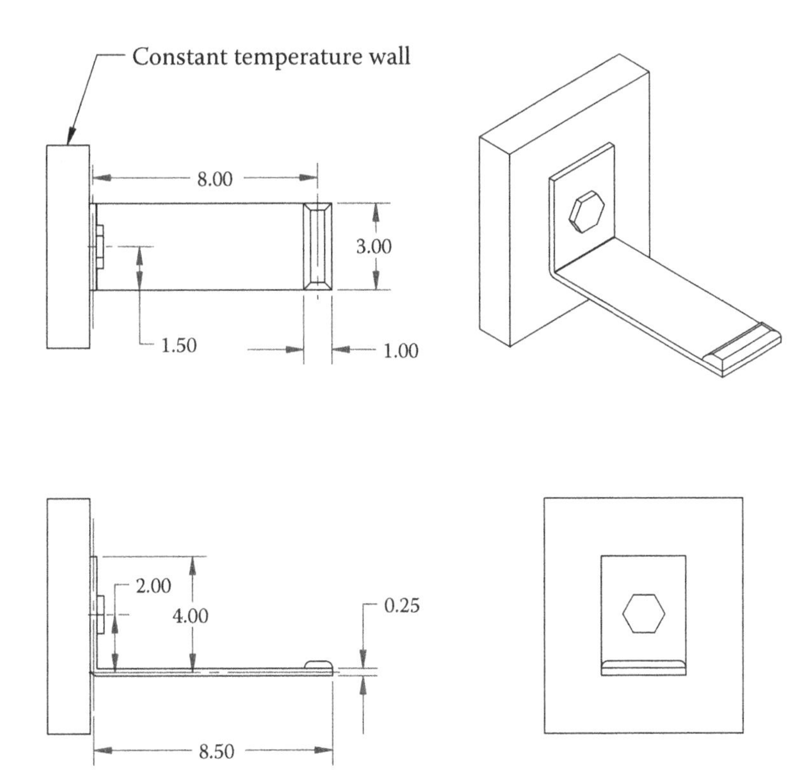

FIGURE 3.9
Bracket chassis and heat flow problem.

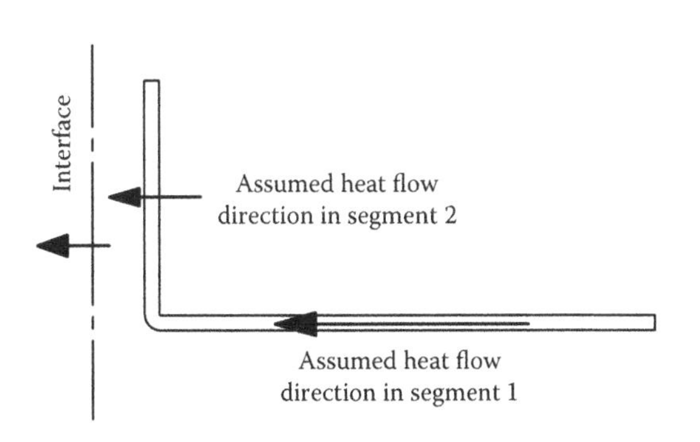

$$\text{Length} = L = 8 \text{ in}$$

$$\text{Area} = 3 \times 0.25 = 0.75 \text{ in}^2$$

$$K = 90 \text{ Btu} / (\text{hr ft°F})$$

$$R_{\text{segment 1}} = \frac{8/12}{90 \times (0.75/144)} = 1.422$$

Segment 2: The portion of the bracket in contact with the chassis; heat path length is the bracket's thickness; heat path area is assumed to cover only the area between the bolt and segment 1. This is justified on the basis that heat will not spread to the entire area under the bolt.

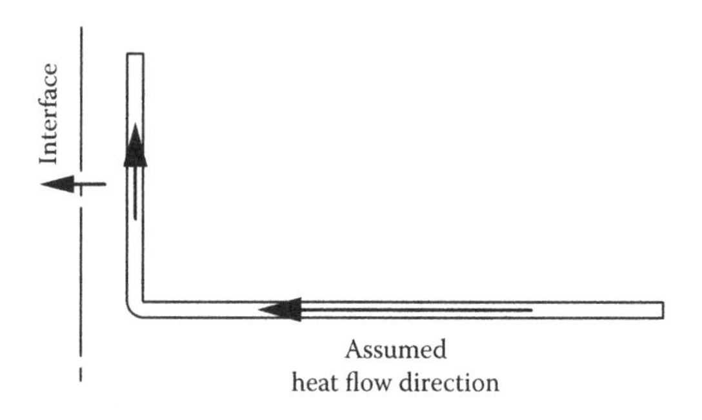

$$\text{Length} = L = 0.25 \text{ in}$$

$$\text{Area} = 2 \times 3 = 6 \text{ in}^2$$

$$K = 90 \text{ Btu}/(\text{hr ft°F})$$

$$R_{\text{segment 2}} = \frac{0.25/12}{90 \times (6/144)} = 0.0056$$

$$R_{\text{bracket}} = R_{\text{segment 1}} + R_{\text{segment 2}}$$

$$R_{\text{bracket}} = 1.422 + 0.0056 = 1.428$$

Interface

Interface resistance may be evaluated based on Equation 3.3.

$$\text{Area} = A = 2 \times 3 = 6 \text{ in}^2$$

For Al 7075, h_{int} varies from about 500 Btu/(hr ft^2 °F) at zero pressure and 25 W/in^2 to nearly 5000 BTU/(hr ft^2 °F) at 400 psi pressure and 150 W/in^2 (MIL-HDBK-251 1978). Based on a linear interpolation of data, h_{int} may be found.

$$h_{\text{int}} = 781.25 \text{ Btu}/(\text{hr ft}^2 \text{ °F}) \text{ at 25 psi pressure}$$

$$R_{\text{interface}} = \frac{1}{781.25 \times (6/144)} = 0.031$$

Combination

$$R_{\text{total}} = R_{\text{bracket}} + R_{\text{interface}}$$

$$R_{\text{total}} = 1.428 + 0.031 = 1.458$$

$$\Delta T = QR_{\text{total}} \Rightarrow \Delta T = \left(12\,\text{W} \times 3.41 \frac{\text{BTU/hr}}{\text{W}}\right) \times 1.458 \Rightarrow \Delta T = 59.7°\text{F}$$

$$T_{\text{Component}} = 100 + 59.7 \Rightarrow T_{\text{Component}} = 159.7°\text{F}$$

Second Approach

For the sake of argument, let us consider a slightly different thermal path and compare the results. In this approach, consider the length of the bracket to be 12 in (= 8 + 4) and the full area of the interface.

Bracket

$$\text{Length} = L = 8 + 4 = 12 \text{ in}$$

$$\text{Area} = A = 3 \times 0.25 = 0.75 \text{ in}^2$$

$$K = 90 \text{ BTU} / (\text{hr ft °F})$$

$$R_{\text{bracket}} = \frac{12/12}{90 \times (0.75/144)} = 2.13$$

Interface

$$\text{Area} = A = 4 \times 3 = 12 \text{ in}^2$$

Notice that a larger area is calculated here because we are assuming that the heat is transferred from the entire back plate area.

$$h_{\text{int}} = 781.25 \text{ BTU} / (\text{hr ft}^2 \text{ °F}) \text{ at 25 psi pressure}$$

These data were linearly interpolated from the information given earlier that at 0 psi, $h_{\text{int}} = 500$, and at 400 psi, $h_{\text{int}} = 5000$.

$$R_{\text{interface}} = \frac{1}{781.25 \times (12/144)} = 0.0154$$

Combination

$$R_{\text{total}} = R_{\text{bracket}} + R_{\text{interface}}$$

$$R_{\text{total}} = 2.13 + 0.015 = 2.149$$

$$\Delta T = QR_{\text{total}} \Rightarrow \Delta T = \left(12\,\text{W} \times 3.41 \frac{\text{BTU / hr}}{\text{W}}\right) \times 2.149 \Rightarrow \Delta T = 87.92°]$$

$$T_{\text{Component}} = 100 + 87.92 \Rightarrow T_{\text{Component}} = 187.92°\text{F}$$

Clearly, this is a very conservative solution.

Third Approach

This approach is similar to the first approach; however, we use the spread angle to estimate the effective length of the bend in the bracket. For a material with the thermal conductivity of 90 BTU/(hr ft °F), the spread angle is approximately 52° based on Equation 3.2 (Figure 3.10).

Bracket

Segment 1:

$$\text{Length} = L = 8 \text{ in}$$

$$\text{Area} = 3 \times 0.25 = 0.75 \text{ in}^2$$

$$K = 90 \text{ BTU / (hr ft°F)}$$

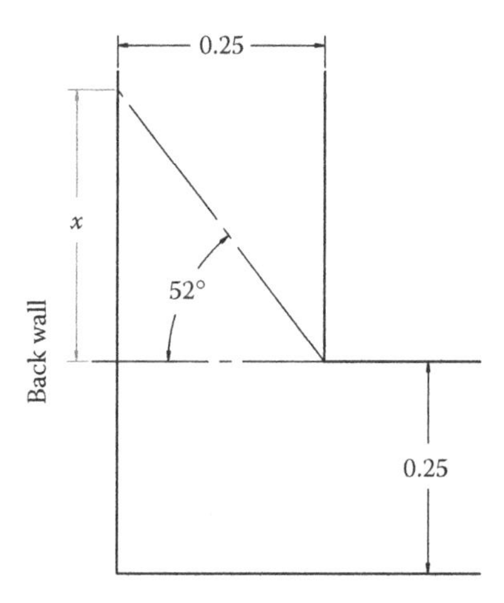

FIGURE 3.10
A close-up of the interface and heat spread angle.

$$R_{\text{segment 1}} = \frac{8/12}{90 \times (0.75/144)} = 1.422$$

Segment 2:

$$\text{Length} = L = 0.25 \text{ in}$$

$$x = 0.25 \tan(52) = 0.316$$

$$\text{Average cross-sectional area} = A = 1/2\{(x + 0.25)(3) + (0.25)(3)\} = 1.224 \text{ in}^2$$

$$K = 90 \text{ BTU}/(\text{hr ft}°\text{F})$$

$$R_{\text{segment 1}} = \frac{0.25/12}{90 \times 1.224/144} = 0.027$$

$$R_{\text{bracket}} = R_{\text{segment 1}} + R_{\text{segment 2}} = 1.449$$

Interface

The interface area is calculated by adding the spread distance x to the thickness and then multiplying the results by the thickness of the bracket.

$$A_{\text{interface}} = (x + 0.25) \times 3$$

$$A_{\text{interface}} = (0.316 + 0.25) \times 3 = 1.698 \text{ in}^2$$

$$h_{\text{int}} = 781.25 \text{ Btu}/(\text{hr ft}^2\ °\text{F}) \text{ at 25 psi pressure}$$

$$R_{\text{interface}} = \frac{1}{781.25 \times 1.698/144} = 0.1086$$

Combination

$$R_{\text{total}} = R_{\text{bracket}} + R_{\text{interface}}$$

$$R_{\text{total}} = 1.449 + 0.109 = 1.558$$

$$\Delta T = QR_{\text{total}} \Rightarrow \Delta T = \left(12\,\text{W} \times 3.41 \frac{\text{BTU/hr}}{\text{W}}\right) \times 1.558 \Rightarrow \Delta T = 63.73°\text{F}$$

$$T_{\text{Component}} = 100 + 63.73 \Rightarrow T_{\text{Component}} = 163.73°\text{F}$$

TABLE 3.4

A Comparison of Component Temperatures

Approach	Component Temperature	Exact Solution
1	159.68	161.88
2	187.92	161.88
3	163.73	161.88

In Section "Exercise—Calculate the Component Temperature", it was mentioned that the finite element solution is 161.88°F. The solution by this approach is both conservative and yet close to the exact solution. For the sake of comparison, calculated component temperatures are tabulated in Table 3.4.

It is obvious that the first approach, although it provided the smallest margin of error, was not a conservative approach. The second approach provided a very conservative solution, and the third approach provided a conservative and yet realistic solution.

A Word on Edge Guides

In a number of packaging designs, board edge guides can be used to conduct heat away from the PCB. To take advantage of this feature, the ground plane of the PCB must be placed as close to the surface as possible. Furthermore, the higher the contact pressure between the PCB and the guide, the lower the thermal resistance. Typical values of thermal resistance range from 20°F in/W for the U or G guides to 10°F in/W for the Z type guides to 4°F in/W to wedge clamps. As an added benefit, these components also provide an added measure of vibration rigidity.

2D or 3D Heat Conduction

It is possible to form a network of thermal resistance elements to cover an area (two dimensional [2D]) or a volume (three dimensional [3D]) and then write the equations to balance the heat in and out of the system. This is, in a way, the basis of finite difference and/or elements analyses; however, from a practical point of view, for truly 2D or 3D problems, a commercial FEA package must be used to minimize both time and possibility of errors.

Thermal Conductance

Thermal conductance is defined as the inverse of thermal resistance and is employed in a similar manner. Its application is both more powerful and at the same time more time consuming. To use the thermal conductance approach, the conduction equation is rewritten as follows:

$$C\Delta T = Q \text{ if } C = \frac{KA}{L}$$

Now, one may conceive a more general form of this equation:

$$[\mathbf{C}]\{\mathbf{T}\} = \{\mathbf{Q}\}$$

In fact, it is this form of this equation that makes it particularly suitable for the 2D and 3D thermal analyses.

To develop a model, the conductance network is first developed, which is very similar to a resistance network. All heat sources (and sinks) must be placed on this network. Then, the heat flow is balanced at each node. To maintain consistency, it is assumed that at each node, all surrounding nodes have higher temperatures. The following provides an illustration.

Consider the configuration as shown in Figure 3.11. A plate is heated in its center and has a constant temperature (T_S) applied to one boundary. The procedure to develop the conductance matrix is as follows.

First, we need to divide the area into smaller segments where points of interest lie. Here, we need to identify nodes at the heat source (node 9) as well as on the boundaries (nodes 1, 7, and 8). Then, we develop the conductance network that is very similar to a resistance network. Care must be exercised in identifying the correct values of conduction length as well as conduction cross-sectional area. For instance, in the example shown in Figure 3.11, the conduction length between nodes 6 and 7 is longer than between nodes 5 and 6 but the cross-sectional area is the same.

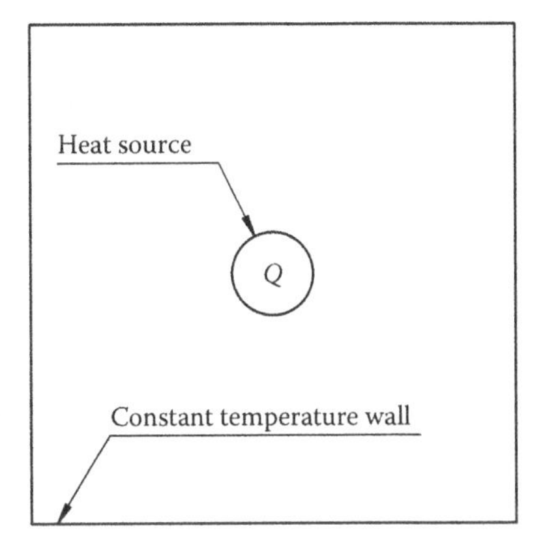

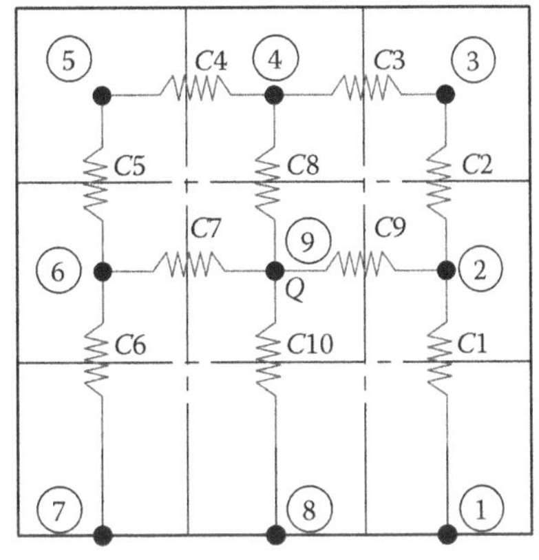

FIGURE 3.11
Heat source on the center of a plate and its associated thermal model.

To create the set of simultaneous equations to calculate temperatures, we start by writing the heat balance at node 2. Recall that we assume heat flows into each node under consideration.

At node ② $C_1(T_1 - T_2) + C_9(T_9 - T_2) + C_2(T_3 - T_2) = 0$
At node ③ $C_2(T_2 - T_3) + C_3(T_4 - T_3) = 0$
At node ④ $C_3(T_4 - T_3) + C_3(T_5 - T_4) + C_8(T_9 - T_4) = 0$
At node ⑤ $C_4(T_4 - T_5) + C_5(T_6 - T_5) = 0$
At node ⑥ $C_6(T_7 - T_6) + C_5(T_5 - T_6) + C_7(T_9 - T_6) = 0$
At node ⑨ $Q + C_9(T_2 - T_9) + C_8(T_4 - T_9) + C_7(T_6 - T_9) + C_{10}(T_8 - T_9) = 0$

By rearranging these equations, we can arrive at the following set of simultaneous equations:

$$\begin{bmatrix} \begin{pmatrix} C_1 + C_2 \\ +C_9 \end{pmatrix} & -C_3 & 0 & 0 & 0 & -C_9 \\ -C_2 & (C_2 + C_4) & -C_3 & 0 & 0 & 0 \\ 0 & -C_3 & \begin{pmatrix} C_3 + C_4 \\ +C_8 \end{pmatrix} & -C_4 & 0 & -C_8 \\ 0 & 0 & -C_4 & (C_4 + C_6) & -C_5 & 0 \\ 0 & 0 & 0 & -C_5 & \begin{pmatrix} C_5 + C_6 \\ +C_7 \end{pmatrix} & -C_7 \\ -C_9 & 0 & -C_8 & 0 & -C_7 & \begin{pmatrix} C_7 + C_8 \\ +C_9 + C_{10} \end{pmatrix} \end{bmatrix} \begin{Bmatrix} T_2 \\ T_3 \\ T_4 \\ T_5 \\ T_6 \\ T_9 \end{Bmatrix}$$

$$= \begin{Bmatrix} C_1 T_S \\ 0 \\ 0 \\ 0 \\ C_6 T_S \\ Q + C_{10} T_S \end{Bmatrix}$$

Recall that $T_1 = T_7 = T_8 = T_S$. To solve this set of equations, we now need to evaluate the C_i's:

$$C_i = \frac{KA_i}{L_i}$$

In this example, L_i refers to the length between any two nodes where conduction takes place and A_i refers to the cross-sectional area of the same path.

The strength of this approach is that once these equations have been set up in a spreadsheet, various scenarios may be examined to answer particular design questions.

Example

An electronic control box contains three PCBAs as shown in Figure 3.12. The smaller board is 8.5 × 6.5 in and the larger boards are 11.5 × 6.5 in. The gap between the inner walls and the PCBAs is 0.25″ on all sides. Each board is 0.063 in thick and the enclosure wall thickness is 0.10 in. The top surface of the enclosure (on the smaller PCBA side) as well as two side walls are mounted to surfaces that are virtually thermal isolators. The voltage and current supplied to this system is 24 V at 10 A and it operates at 80% efficiency. The smaller top PCBA dissipates one half of the heat generated by the other two boards. Furthermore, the top surface of the enclosure is insulated and the enclosure surface temperature is maintained at 75°F. We need to know the approximate average temperature of each board. This information is needed for reliability analysis, and to determine whether to use potting.

In developing this network, we are, in essence, reducing a 3D problem into a 1D problem. Two underlying assumptions are that not only are the boards uniformly packed, but each component dissipates a similar level of heat. In many realistic problems, this is not the case, but this

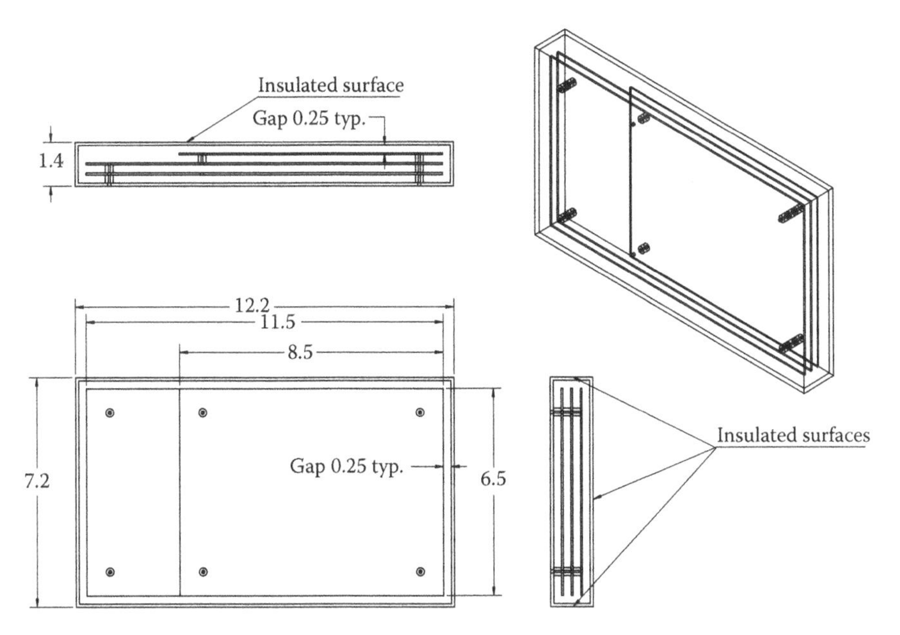

FIGURE 3.12
Three-dimensional view of the enclosure.

approach could very easily provide a ballpark estimate of component temperature.

In this situation, to calculate *S 1R* and *S 1L* resistances, the volume above PCBA 1 is divided into two regions; the right region and the left region (Figure 3.13). The heat flow distance (*L*) of each region is taken to be 6 in, and the cross-sectional area is assumed to be one half of the area of the strip covering the gap between PCBA 1 and the surface above it:

$$\left(\frac{2\times(12+7)\times1.5}{2}\right)$$

The same logic is also used for the gap between the boards. Because the boards are spaced closely together, the assumption is made that convection does not play a significant role in this problem and is thus ignored.

This resistance network may be simplified as shown in Figure 3.14. This resistance network is then converted to a conductance network (Figure 3.15). Recall that conductance is defined as the inverse of resistance.

Now, we need to balance the heat equation at each node. To ensure that a consistent heat flow direction is followed, assume that the temperature of each node is lower than its surrounding nodes.

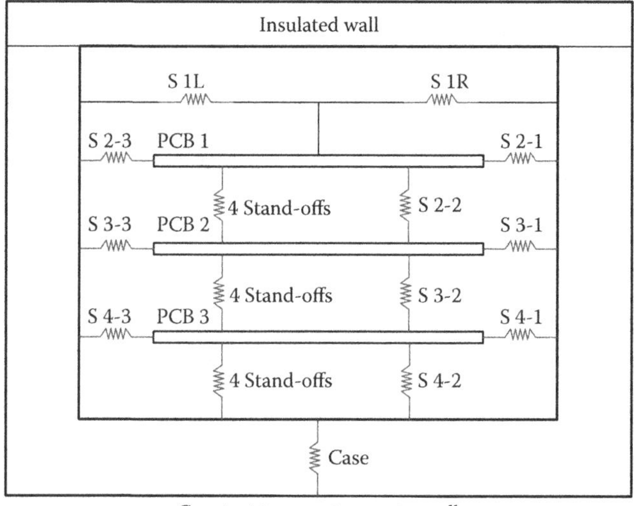

FIGURE 3.13
Initial resistance network.

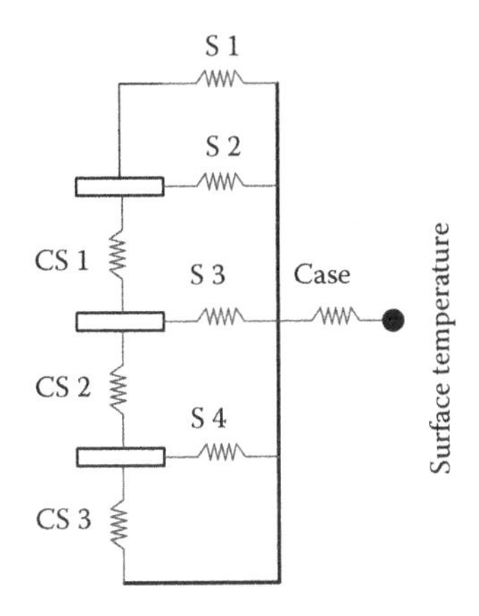

FIGURE 3.14
Simplified resistance network.

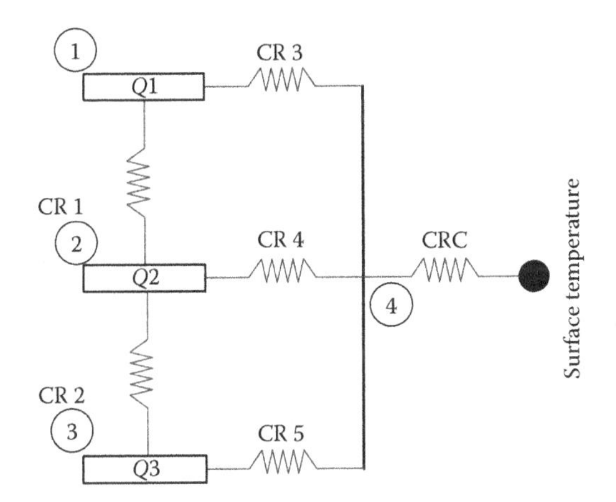

FIGURE 3.15
The conductance network.

$$Q_1 + CR_1(T_2 - T_1) + CR_3(T_4 - T_1) = 0$$

$$Q_2 + CR_1(T_1 - T_2) + CR_4(T_4 - T_2) + CR_2(T_3 - T_2) = 0$$

$$Q_3 + CR_2(T_2 - T_3) + CR_5(T_4 - T_3) = 0$$

$$CR_c(T_S - T_4) + CR_3(T_1 - T_4) + CR_4(T_2 - T_4) + CR_5(T_3 - T_4) = 0$$

This leads to the following matrix equation:

$$\begin{bmatrix} (CR_1 + CR_3) & -CR_1 & 0 & -CR_3 \\ -CR_1 & \begin{pmatrix} CR_1 + CR_2 \\ +CR_4 \end{pmatrix} & -CR_2 & -CR_4 \\ 0 & -CR_2 & (CR_2 + CR_5) & -CR_5 \\ -CR_3 & -CR_4 & -CR & \begin{pmatrix} CR_C + CR_3 \\ +CR_4 + CR_5 \end{pmatrix} \end{bmatrix} \begin{Bmatrix} T_1 \\ T_2 \\ T_3 \\ T_4 \end{Bmatrix} = \begin{Bmatrix} Q_1 \\ Q_2 \\ Q_3 \\ CR_C T_S \end{Bmatrix}$$

The solution to this equation will determine the temperature at each PCBA as well as the inside wall temperature (T_4). To develop a better feel for this approach, let us assign the dimensions and properties used in Table 3.5 and obtain the following set of linear equations:

$$\begin{bmatrix} 5.008 & -2.361 & 0 & -2.648 \\ -2.361 & 4.974 & -2.457 & -0.156 \\ 0 & -2.457 & 2.554 & -0.097 \\ -2.648 & -0.156 & -0.097 & 13769.567 \end{bmatrix} \begin{Bmatrix} T_1 \\ T_2 \\ T_3 \\ T_4 \end{Bmatrix} = \begin{Bmatrix} 32.74 \\ 65.47 \\ 65.47 \\ 103200 \end{Bmatrix}$$

Calculations of heat dissipation values need to be further explained. Earlier, it was specified that the voltage and current supplied to this unit was 24 V and 10 A, respectively. Furthermore, the unit was 80% efficient. Thus the total dissipated heat is $Q_{total} = 24 \times 10 \times (1-0.8) = 48$ W. As indicated the smaller board dissipated half as much as the other two. Thus $Q_1 = 9.2$ W and $Q_2 = Q_3 = 19.2$ W. It should be mentioned that the material and geometric properties in Table 3.5 are consistent with the American units. Thus to calculate the temperature values correctly, the dissipated heat needs to be converted to BTU/hr. With this note in mind, the solution to the above set of simultaneous equations is as follows:

$$\begin{Bmatrix} T_1 \\ T_2 \\ T_3 \\ T_4 \end{Bmatrix} = \begin{Bmatrix} 126.84 \\ 171.10 \\ 193.09 \\ 75.01 \end{Bmatrix}$$

To examine the impact of potting, we would replace the value of 0.0153 for thermal conductivity of air for an appropriate value for the thermal conductivity of potting, say, 0.12 (as shown in Table 3.6). Since the calculations were made using a spreadsheet, the new temperatures can be readily calculated.

TABLE 3.5

The Dimensions and Properties Used for Air

Station	Length (in)	Area (in²)	*K* BTU/(hr ft °F)	Resistance	Conductance
Standoff 1	0.25	0.05	34	1.765	0.567
Standoff 2	0.25	0.05	34	1.765	0.567
Standoff 3	0.25	0.05	34	1.765	0.567
S 1R	6	28.5	0.0153	165.119	6.06×10^{-3}
S 1L	6	28.5	0.0153	165.119	6.06×10^{-3}
S 1				12.822	7.80×10^{-2}
S 2-1	0.25	14.2	0.0153	13.808	7.24×10^{-2}
S 2-2	0.75	55.2	0.0153	10.656	9.38×10^{-2}
S 2-3	3.25	14.2	0.0153	179.508	5.57×10^{-3}
S 2				12.777	7.83×10^{-2}
S 3-1	0.25	9.5	0.0153	20.640	4.85×10^{-2}
S 3-2	0.5	74.7	0.0153	5.250	0.190
S 3-3	0.25	9.5	0.0153	20.640	4.85×10^{-2}
S 3				10.320	9.69×10^{-2}
S 4-1	0.25	4.75	0.0153	41.280	2.42×10^{-2}
S 4-2	0.25	74.7	0.0153	2.625	0.381
S 4-3	0.25	4.75	0.0153	41.280	2.42×10^{-2}
S 4				20.640	4.85×10^{-2}
Case	0.1	236	70	7.26×10^{-5}	1.38×10^{-4}
R1				0.424	2.36
R2				0.407	2.46
R3				0.378	2.65
RR1				6.400	0.156
RR2				10.320	9.69×10^{-2}
RR3				0.371	2.70
R-Case	0.1	236	70	7.26×10^{-5}	1.38×10^{-4}
CR1					2.36
CR2					2.46
CR3					2.65
CR4					0.156
CR5					9.69×10^{-2}
CRC					1.38×10^{4}

TABLE 3.6

The Dimensions and Properties Used for Potting Material

Station	Length (in)	Area (in^2)	*K* BTU/(hr ft °F)	Resistance	Conductance
Standoff 1	0.25	0.05	34	1.765	0.567
Standoff 2	0.25	0.05	34	1.765	0.567
Standoff 3	0.25	0.05	34	1.765	0.567
S 1 R	6	28.5	0.12	21.053	4.75×10^{-2}
S 1 L	6	28.5	0.12	21.053	4.75×10^{-2}
S 1				1.635	0.612
S 2-1	0.25	14.2	0.12	1.761	0.568
S 2-2	0.75	55.2	0.12	1.359	0.736
S 2-3	3.25	14.2	0.12	22.887	4.37×10^{-2}
S 2				12.777	7.83×10^{-2}
S 3-1	0.25	9.5	0.12	2.632	0.380
S 3-2	0.5	74.7	0.12	0.669	1.49
S 3-3	0.25	9.5	0.12	2.632	0.380
S 3				1.316	0.760
S 4-1	0.25	4.75	0.12	5.263	0.190
S 4-2	0.25	74.7	0.12	0.335	2.99
S 4-3	0.25	4.75	0.12	5.263	0.190
S 4				2.632	0.380
Case	0.1	236	70	7.26×10^{-5}	1.38×10^{4}
R1				0.333	3.00
R2				0.266	3.76
R3				0.190	5.25
RR1				1.449	0.690
RR2				1.316	0.760
RR3				0.177	5.63
R-Case	0.1	236	70	7.26×10^{-5}	1.38×10^{4}
CR1					3.00
CR2					3.76
CR3					5.25
CR4					0.690
CR5					0.760
CRC					1.38×10^{4}

$$\begin{bmatrix} 8.257 & -3.003 & 0 & -5.255 \\ -3.003 & 7.453 & -3.761 & -0.690 \\ 0 & -3.761 & 4.521 & -0.760 \\ -5.255 & -0.690 & -0.760 & 13773.371 \end{bmatrix} \begin{Bmatrix} T_1 \\ T_2 \\ T_3 \\ T_4 \end{Bmatrix} = \begin{Bmatrix} 32.74 \\ 65.47 \\ 65.47 \\ 103200 \end{Bmatrix}$$

$$\begin{Bmatrix} T_1 \\ T_2 \\ T_3 \\ T_4 \end{Bmatrix} = \begin{Bmatrix} 93.81 \\ 115.79 \\ 123.42 \\ 75.01 \end{Bmatrix}$$

Clearly, potting has a great impact on the temperature of the PCBAs. The reader is encouraged to develop the equations needed to calculate the thermal conductance values for this network and solve the simultaneous equations.

4

Radiation Cooling

Introduction

Radiation heat transfer takes place when two bodies "at a distance" exchange heat directly. It has an electromagnetic nature; hence, it does not need a medium. And as the temperature difference increases, radiation becomes more significant. In the past the impact of radiation was generally ignored when temperature levels were considered low because of the difficulty in handling the nonlinearity. However, this need not be the case with the advent of spreadsheets and their simple-to-use computing power.

There are two categories of radiation: wall-to-wall radiation and participating media. In wall-to-wall radiation, heat exchange takes place between two surfaces or between a surface and its external environment. The medium does not have any influence on the process. In participating media radiation, not only do the walls radiate to each other but also they exchange heat from the medium. Therefore, the medium influences the radiation process.

In a majority of cases in electronics packaging, one need not be concerned with the medium participating in the heat exchange process. However, should the environment where the electronics system is placed contain excess moisture, gases such CO_2, or particulate such as soot, this type of radiation must be considered and its impact evaluated. Herein, we are only concerned with the wall-to-wall radiation type.

Factors Influencing Radiation

Recall the general radiation equation:

$$Q = \sigma \varepsilon A\left(T_{\text{hot}}^4 - T_{\text{cold}}^4\right)$$

More specifically, the radiation heat exchange equation between two surfaces i and j is written as

$$Q = \sigma\left(\varepsilon_i A_i F_{ij} T_i^4 - \varepsilon_j A_j F_{ji} T_j^4\right) \tag{4.1}$$

σ is called the Stefan–Boltzmann constant (5.67×10^{-8} W/[m^2 °K^4] or 0.1713×10^{-8} Btu/[ft^{2}h °R^4]). A is the surface area of i or j. T is temperature in absolute scale (°R = °F + 460 or °K = °C + 273). F_{ij} is the view factor between surfaces i and j. In this formula, we have introduced two concepts:

1. Emissivity (ε), which is a function of surface properties
2. View factor, which is a function of surface geometry and configuration

Surface Properties

To develop a better feel for radiation, let us review some definitions.

In radiation, there is often a mention of blackbodies. A blackbody is a standard with which other surfaces are compared. The level of radiation emitted or absorbed by a blackbody depends only on temperature and is independent of wavelength.

Engineering materials including those used in electronics packaging, do not exhibit blackbody characteristics; hence, the emitted or absorbed radiation levels are less than an equivalent blackbody. The ratio of the emitted energy of a "real" surface to an equivalent blackbody is called emittance and is denoted by ε_λ (λ is wavelength). Similarly, the ratio of absorbed energies is called absorptance (α_λ). Generally, both emittance and absorptance are functions of wavelength and temperature as well as surface properties. For a blackbody, emittance and absorptance is equal to unity. This also means that what is absorbed by a blackbody is emitted.

A second, more practical idealization may be made. A gray body is defined as a body that emits and absorbs energy independent of wavelength. Thus, for a gray surface, emittance (now called emissivity) is equal to absorptance (now called absorptivity) and they are less than unity ($\alpha = \varepsilon < 1$). They may still be functions of temperature.

It is clear from this definition that a gray surface does not absorb all of the incident energy. The remainder has to be either transmitted through the body or be reflected away. Thus, by denoting ρ as reflectance, and τ as transmittance, one may easily conclude that $\alpha + \rho + \tau = 1$. Most solid materials are opaque, so $\tau = 0$. Therefore, $\alpha + \rho = 1$ or $\varepsilon + \rho = 1$. This relationship indicates that the higher the reflectivity of a gray surface, the lower its emissivity and absorptance become.

This finding holds true for many engineering materials used in electronics packaging. The heat generated within electronics enclosures is generally in a wavelength spectrum where gray body approximations are valid. It is interesting to note that the surface color will not have an impact in this spectrum, and black and white surfaces emit and absorb equally.

Solar radiation falls within a wavelength range where these assumptions do not hold and emissivity may not necessarily be equal to absorptivity. In this case the color of surface could have an impact.

Another factor impacting the value of emissivity is surface polish. Surface polish increases ρ and thus lowers ε. Nonmetallic surfaces tend to have larger emissivities than metallic surfaces. A polished metallic surface has low emissivity, but as the same surface oxidizes, its emissivity increases.

View Factor Calculations

Calculation of radiative energy exchange between any two surfaces requires determining the geometrical configuration factor, or view factor, between the two surfaces. This was demonstrated in Equation 4.1 as the factor F_{ij} or F_{ji}. For two blackbodies the view factor is defined as the fraction of the diffusely distributed radiant energy leaving one surface i that arrives at a second surface j:

$$F_{ij} = \frac{1}{A_i} \iint_{A_i A_j} \frac{\cos\beta_i \cos\beta_j \, dA_i \, dA_j}{\pi r^2} \tag{4.2}$$

where A_i and A_j are the areas of surfaces i and j, respectively, and β_i and β_j are the angles between the position-dependent normal vectors to surfaces i and j and a line of length r connecting the points of evaluation of the surface normal vectors (Figure 4.1).

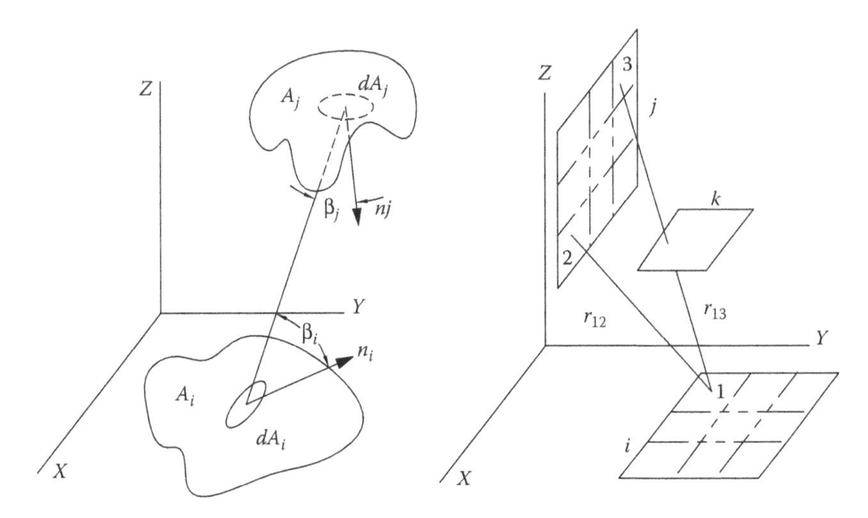

FIGURE 4.1
Surfaces view and shadow each other.

The derivation of Equation 4.2 can be found in Siegel and Howell (1981). The basic assumptions used in deriving Equation 4.2 are that the two surfaces are diffusely emitting and reflecting and the two surfaces are isothermal. As a result of these assumptions, the view factor depends only on the geometry of the system. It is also important to note that for each surface i

$$\sum_{j=1}^{N} F_{ij} = 1$$

where N is the number of surfaces.

Briefly, one may define F_{ij} as an indication of what percentage of one surface is seen by another surface. Therefore, unless one surface is completely enclosed by another, the view factor between them is less than one. This implies that the external surfaces of an electronics system will have a view factor of one in relationship to the environment; however, calculating F_{ij} for internal surfaces as well as fins may be difficult and time consuming.

The major complication with the view factor calculation is the possibility of partial blocking or "shadowing" between two surfaces by an intervening body. Three types of shadowing may exist between two surfaces: total self-shadowing, partial self-shadowing, and third surface shadowing. An in-depth discussion of this topic is beyond the scope of this book. In general, an accurate calculation of view factors is a time- and computation-intensive task.

In general, Equation 4.2 is not used directly; rather, different strategies are employed depending on whether the model is 3D, 2D, or axisymmetric. Details of these algorithms and a complete discussion of the entire view factor calculation can be found in Shapiro (1983).

The reader should not be dismayed by these complications. The view factors for known shapes and geometries are calculated and available in standard handbooks.

View Factor Calculations in Electronics Packages

It is often asked whether to include radiation exchange for an electronics enclosure. Generally speaking, it is relatively easy to include radiation exchange between an electronics package and its environment. In this case, the view factors are assumed to be one and the contribution of radiation heat transfer is included in the heat balance equation. By the same token, internal radiation may be ignored.

There are, however, very realistic problems where these assumptions do not hold any longer. An example of this type of problem is radiation heat transfer between the blades of a fin where precise view factors need to be calculated. Two more examples are illustrated below. For a more detailed discussion of radiation and view factor calculations, Krieth (1973) and Steinberg (1991) are recommended.

Examples and Illustrations

To illustrate an application of wall-to-wall radiation exchange in electronics packaging, the solutions to two example problems are presented in this section. The first problem consists of the flow of air past multiple heat-generating chips surrounded by walls of porous material. This problem involves conduction within the solid, porous, and fluid materials; convection within the fluid; as well as radiation boundary conditions. This example illustrates the impact of radiation with the presence of conduction and convection (a subject that will be discussed in Chapter 5). The second problem is flow over a heat-generating step in a vertical channel and illustrates again the same phenomenon; however, radiation is allowed to escape to the environment.

In these simulations, FIDAP, a computational fluid dynamics software, was used and four node linear quadrilateral finite elements were employed. For more details, see Engelman and Jamnia (1991) as well as the FIDAP *Theoretical Manual* (1989).

Electronics Packaging Problem

A two-cell vertical channel with heat-generating chips was modeled; the bottom and top portions as well as the sides of the channel were composed of porous materials. The left-hand channel had two heat-generating steps and the right-hand channel included only one chip. The two channels were connected by a heat-conducting material. The inner walls were assumed to be gray diffuse surfaces, and there was convection to the environment from the outer walls. The complete geometry is shown in Figure 4.2. Table 4.1 summarizes the various material properties. For this simulation, the full Navier–Stokes equations including buoyancy effects and the energy equation were solved.

For radiation boundary condition specification purposes, 49 radiating surfaces were defined; these surfaces are shown in Figure 4.3. Note that the 27 surfaces in the middle of the computational domain were defined as blocking surfaces for view factor calculation purposes. Figure 4.4 shows the resultant temperature distribution for an emissivity of 0.8,

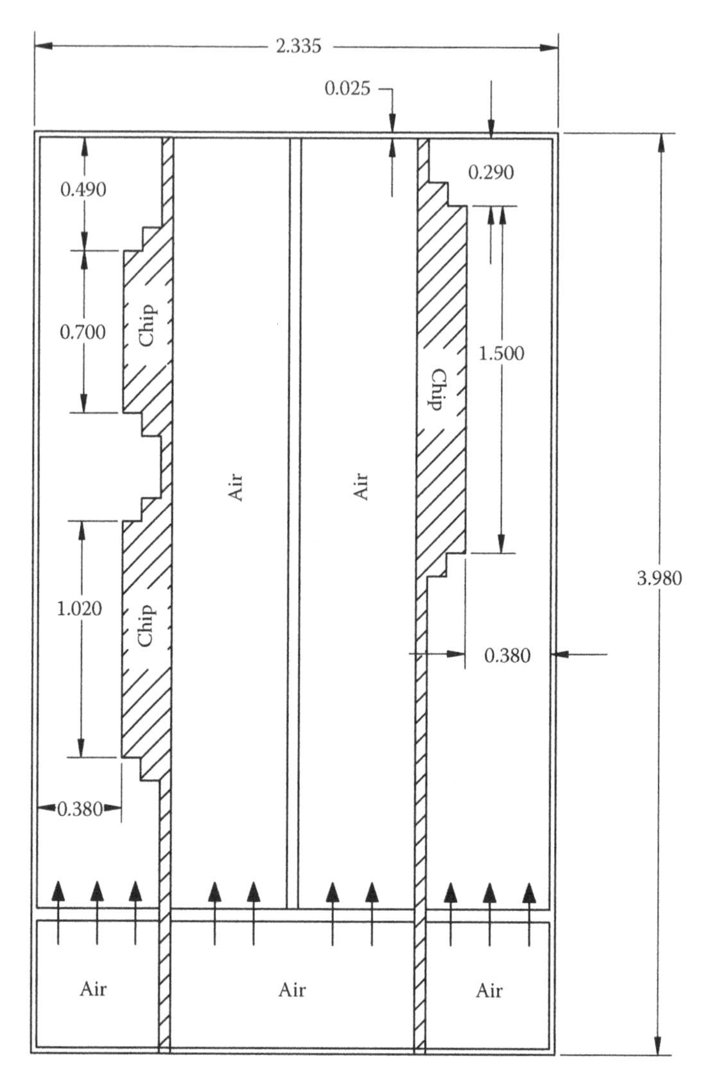

FIGURE 4.2
Electronics packaging problem: complex geometry and boundary conditions.

and Figure 4.5 shows the velocity field for the fluid as it passes through the system. To investigate the effects of radiation on the resultant temperature field, three different values of wall emissivity were employed ($\varepsilon = 0.0$, 0.1, 0.8). The chip emissivity remained constant ($\varepsilon = 0.88$) for all three cases. Figure 4.6 shows the temperature profile over the chips and the opposing wall in the left channel, and Figure 4.7 depicts the temperature profile over

TABLE 4.1

Material Properties of the Complex Problem

Properties of Air	Conductivity of the Solid and Porous Materials	Properties of the Radiative Surfaces	Permeability of the Different Porous Materials	Heat transfer Coefficient at the Outer Walls
Density = 1.929×10^{-5} Viscosity = 4.689×10^{-5} Specific heat = 1005 Conductivity = 6.655×10^{-4} Volume expansion = 0.0033 at a reference Temp = 300	Porous base = 0.94 Chip packages = 0.038 Package leads = 3.8 Central heat sink = 1.143 Caging = 1.70	Emissivity = 0.88 Stefan–Boltzmann constant = 3.6577×10^{-11}	Porosity = 0.48, Permeability = 10-5 in the direction normal to the plane of the material	$h = 6.45 \times 10^{-3}$ at reference temperature = 300

Note: All units are in watts, sec, inches.

the chip and the opposing wall in the other channel. Although the results indicate that for this particular system radiation does not have a significant effect on the resultant temperature distribution, the radiative interaction between the chip and the walls becomes more influential in reducing the temperature as the wall emissivity increases.

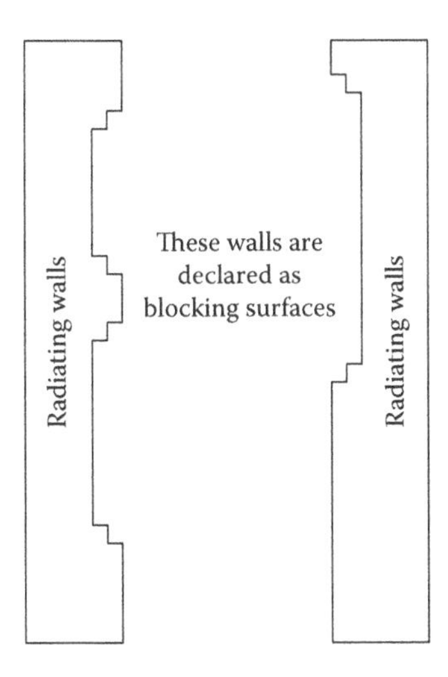

FIGURE 4.3
Electronics packaging problem: radiation surfaces.

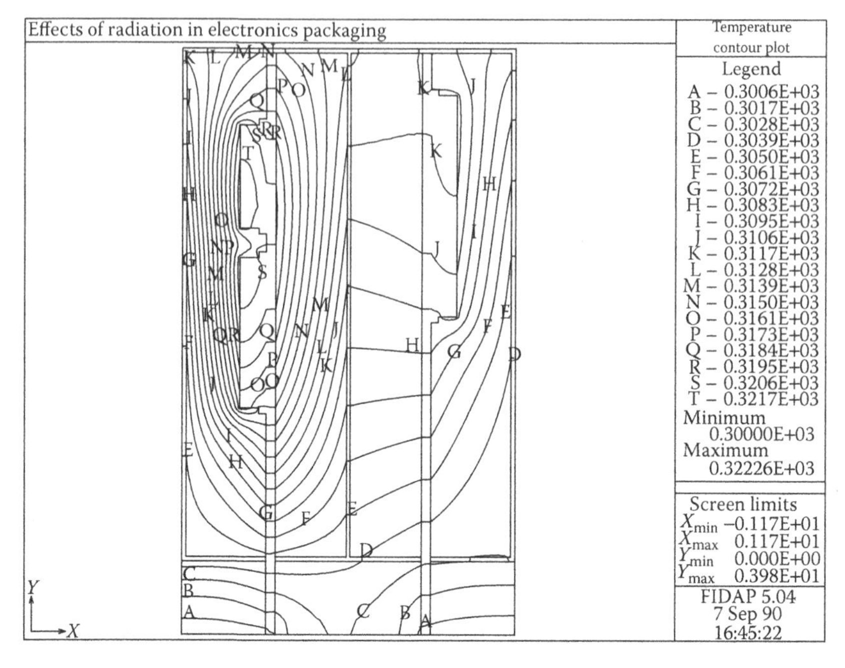

FIGURE 4.4
Electronics packaging problem: temperature distribution.

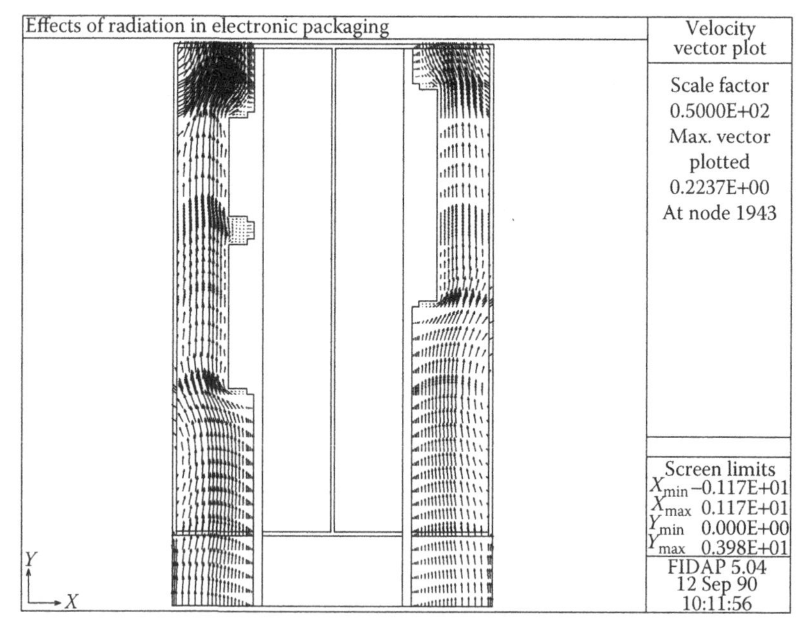

FIGURE 4.5
Electronics packaging problem: velocity field.

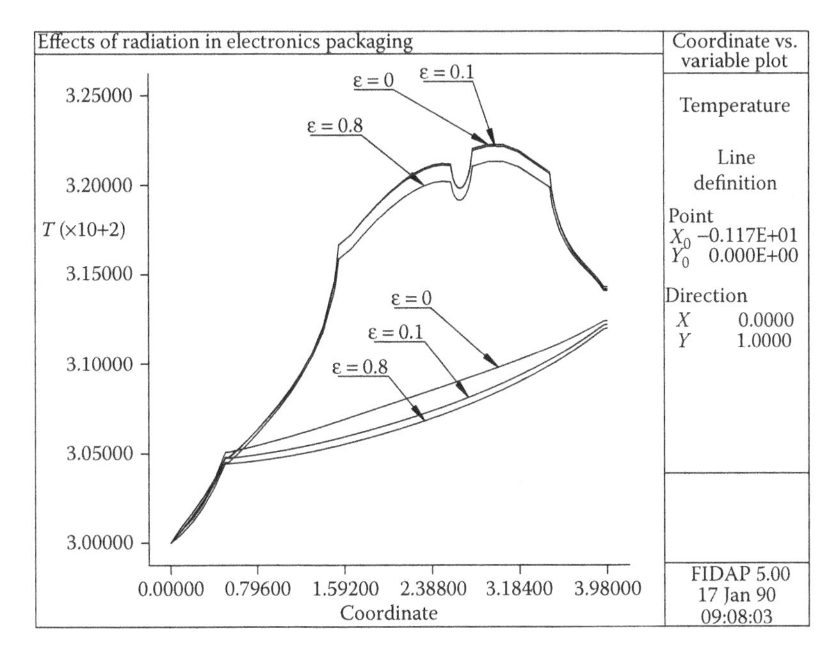

FIGURE 4.6
Electronics packaging problem: temperature along the two chips.

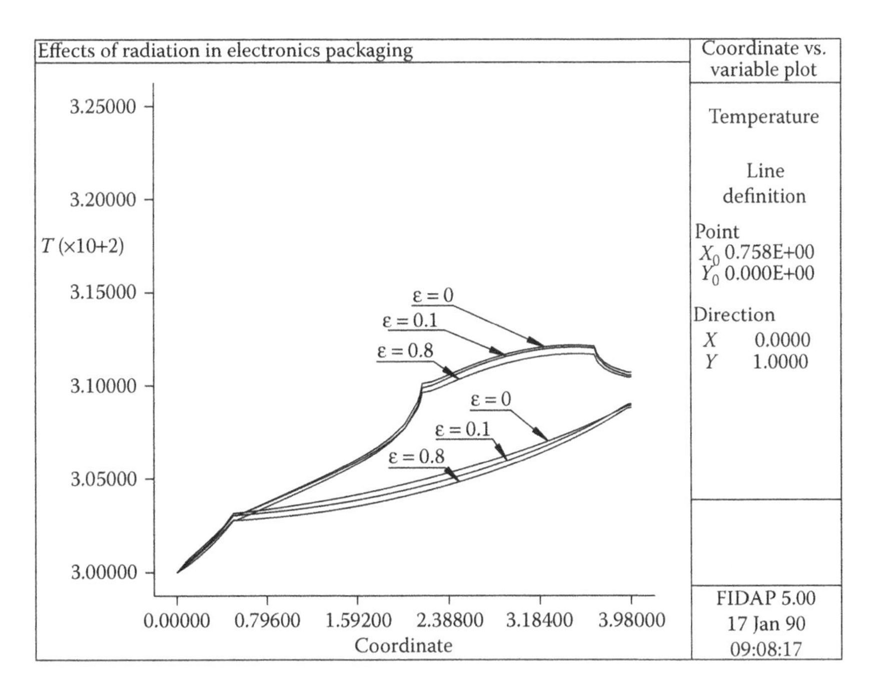

FIGURE 4.7
Electronics packaging problem: temperature along a line over the one chip.

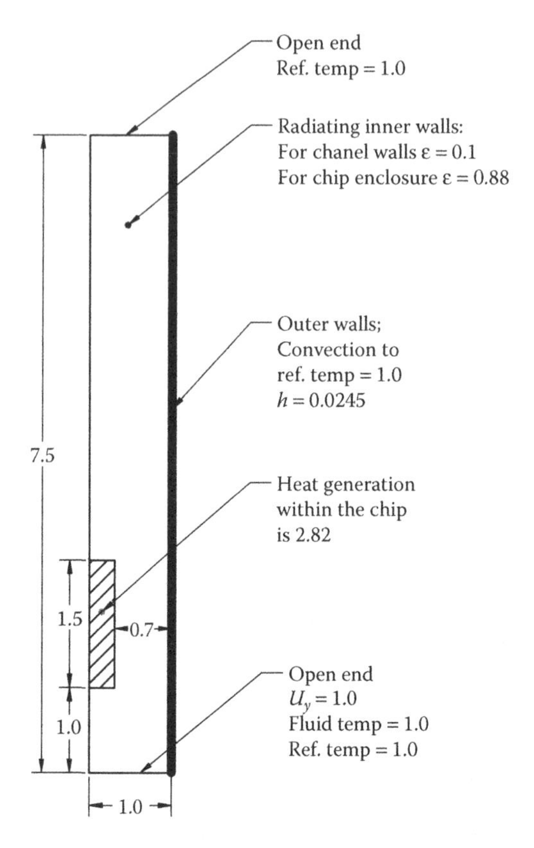

FIGURE 4.8
Geometry and boundary conditions.

Flow in a Vertical Open-Ended Channel

In this problem, the effect of different combinations of modes of heat transfer is studied. The geometry is a vertical 2D channel with a heat-generating step and convection cooling on the outer wall opposing the chip. The conditions are such that flow with a uniform temperature enters the channel and passes over the step and then leaves the system. Figure 4.8 shows the geometry as well as the boundary conditions and the input data. The solid was assumed to be 10 times as conductive as the fluid. A Prandtl number of 0.73 and a Reynolds number of 1.3698 were specified for the fluid. Four cases were studied: (1) conduction only, (2) conduction and radiation, (3) conduction and convection, and (4) all three modes combined. In this analysis, buoyancy effects were ignored and no slip wall (i.e., zero fluid velocity at the walls) conditions were assumed. Figure 4.9 shows the temperature distribution for each case. It is clear that radiation and convection both reduce the overall temperature. In this problem, radiation has a dominant effect because heat could escape to the environment via the radiative mechanism.

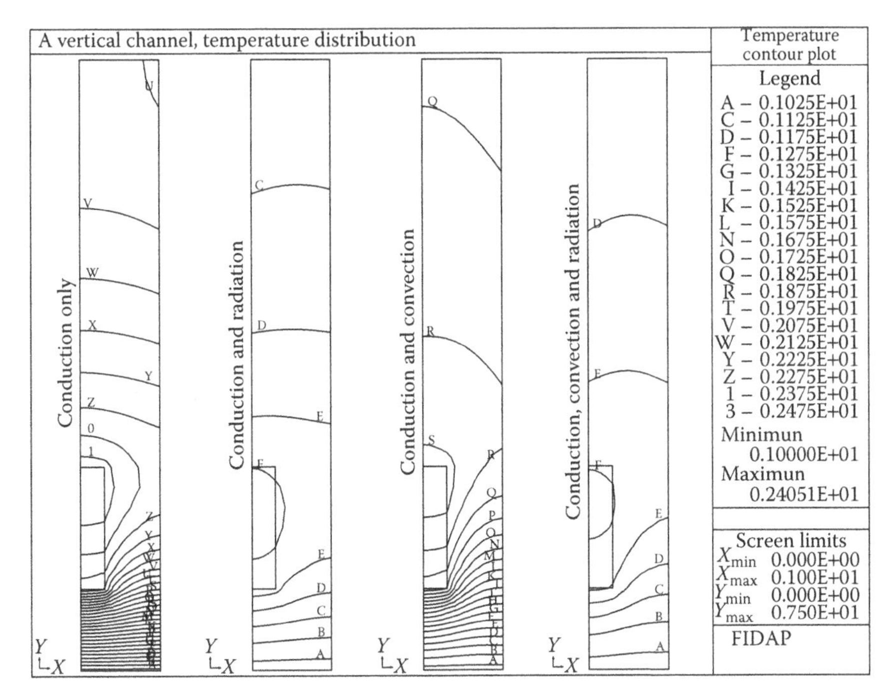

FIGURE 4.9
Open-ended channel: temperature distribution for the different modes of heat transfer.

Cabinet Surface Temperature

An electronics cabinet dissipates 800 W of energy and its physical dimensions are 12 × 18 × 26 in. The surface is painted white. Assuming an ambient temperature of 50°C and considering conduction and radiation alone (no solar energy), we need to estimate the interior temperature.

It is important to develop the ability to develop a "feel" for surface and interior temperatures of electronics enclosures with minimum information. The only information given about the cabinet is that it dissipates 800 W of energy and its physical dimensions are 12 × 18 × 26 in.

To make a back-of-the-envelope calculation, it is not necessary to know about how the components are mounted inside the enclosure. The fact is that the generated heat will escape, either to all of the surfaces or just a few, depending on the configuration used.

A good assumption is that the heat is uniformly distributed to all of the external surfaces. Another assumption is that surface and interior temperatures are equal. For the sake of comparison, this case is also solved assuming that the entire generated heat is only transferred to one of the largest surfaces.

First Approach

Eight hundred watts is distributed equally to all surfaces. Assume a sink temperature of 50°C (122°F) and no solar radiation. The generated heat should be set equal to the radiated heat (notice that no heat is conducted away from the cabinet):

$$\text{Generated heat in BTU/hr} = 800\ \text{W} \times 3.41 \frac{\text{BTU / hr}}{\text{W}} = 2728\ \text{BTU/hr}$$

Radiation heat transfer is calculated from the following formula:

$$Q = F\sigma\varepsilon A\left(T_{\text{surf}}^{4} - T_{\text{amb}}^{4}\right)$$

where F is the view factor, ε is the emissivity, A is the area, σ is the Stefan–Boltzmann constant (=1.713×10^{-9}), and T is temperature. The subscript surf stands for surface and amb is for ambient. The surface is painted white so emissivity is about 0.9, and since the cabinet is exchanging heat to the environment, the view factor F is one. One may note that the form of this equation is slightly different from Equation 4.1. The implicit assumption is that $F_{ij} = F_{ji} = F$. Also, since the heat exchange of the enclosure is of interest, only the surface area of the enclosure needs to be taken into account.

$$\text{Ambient absolute temperature} = 122 + 460 = 582°\text{R}$$

$$\text{Total area} = 2 \times (26 \times 12 + 18 \times 12 + 26 \times 18) / 144 = 13.83\ \text{ft}^2$$

$$\text{Heat balance} \Rightarrow 2728 = (1)(0.9)(13.83)\left(1.713 \times 10^{-9}\right)\left(T_{\text{surf}}^{4} - 582^{4}\right)$$

$$\Rightarrow T_{\text{surf}} = 702°\text{R} = 242°\text{F}$$

Using the SI units, we obtain

$$\text{Ambient absolute temperature} = 50 + 273 = 323°\text{K}$$

$$\sigma = 5.67 \times 10^{-8}\ \text{W} / (\text{m}^2 °\text{K}^4)$$

$$\text{Total area} = 2 \times (26 \times 12 + 18 \times 12 + 26 \times 18) \times (0.000625) = 1.245\ \text{m}^2$$

$$\text{Heat balance} \Rightarrow 2728 = (1)(0.9)(1.245)\left(5.67 \times 10^{-8}\right)\left(T_{\text{surf}}^{4} - 323^{4}\right)$$

$$\Rightarrow T_{\text{surf}} = 391°\text{K} = 118°\text{C}$$

One may notice that 242°F converted to degrees Celsius yields 116.6°C. Why, then, did we calculate 118°C instead? The answer to this question lies in the fact that computational errors have propagated because we have not carried enough precision in our calculations. For example, if we were to take 1 in as 0.025 m, then our total area would be 1.245 m^2. However, if we carry more precision and take 1 in as 0.0254 m, then our total area would be calculated as 1.285 m^2. In fact, should the reader carry the calculations with this new updated value of area, he or she would discover that the results are much closer.

Second Approach

Eight hundred watts is distributed to only one of the largest surfaces. All other assumptions remain the same.

$$\text{Total area} = 26 \times 18 / 144 = 3.25 \text{ ft}^2$$

$$\text{Heat balance} \Rightarrow 2728 = (1)(0.9)(3.25)(1.713 \times 10^{-9})(T_{\text{surf}}^4 - 582^4)$$

$$\Rightarrow T_{\text{surf}} = 901°\text{R} = 441°\text{F} = 227°\text{C}$$

Clearly, every effort must be made to increase the heat transfer surface area.

A Few Design Tips

When designing thermal management systems based on radiation heat transfer, a few points need to be kept in mind.

First, it is important to maximize the surface emissivity. This may be done by either painting or anodizing the surface if possible. Typical values of emissivity are reported as 0.04 to 0.3 for metals which are clean and recently processed; about 0.7 to 0.8 for metals that have been in service and have developed a relatively thick oxidation layer. Anodized metallic surfaces also have a typical emissivity of 0.7 to 0.8. Plastics exhibit values of 0.7 to 0.9; and, painted surfaces 0.85 to 0.95.

Second, as it was demonstrated with the view factor calculations (Equation 4.2), radiation heat transfer is directly related to surface orientation. As a result, a good design rule is to orient heat radiating surfaces such that they face either an opening to the outside or have an unencumbered view of cooler surfaces.

Finally, to the extent possible, minimize the thermal path resistance from the heat-generating element to the radiating surface in order to increase the temperature difference between the hot and cold surfaces for maximum radiative heat transfer.

5

Fundamentals of Convection Cooling

Introduction

With the exception of space applications, there are hardly any practical thermal cooling problems that do not include convection. Convection involves heat transfer between the surface of a solid and its surrounding fluid. The rate at which heat is transferred is evaluated by

$$Q = hA\left(T_{\text{wall}} - T_{\infty}\right)$$

This equation only looks simple! h—the heat transfer coefficient—varies depending on the flow regime, that is, laminar or turbulent, geometry, and fluid properties.

If we rewrite the convection equation slightly differently, we obtain

$$q = h\left(T_{\text{wall}} - T_{\infty}\right)$$

$q = (Q/A)$ is called the heat flux and is heat transfer rate per unit area. The units for heat transfer coefficient (h) could be

$$\frac{\text{W}}{\text{m}^2\,{}^{\circ}\text{C}} \text{ or } \frac{\text{BTU}}{\text{hr}\,\text{ft}^2\,{}^{\circ}\text{F}}$$

T_{∞} is called bulk temperature and is (theoretically) unaffected by the heat input from the solid. In electronics packaging problems, however, this assumption may not hold true at all times.

For a fluid to remove heat from a surface, heat must first be conducted into the fluid to be removed. Therefore, convection heat transfer depends not only on how fast the fluid flows, but also on how well it "conducts" heat near the surface. The ratio between the fluid's ability to conduct heat and then move it away is called the "Nusselt number." It ties both conduction and convection together.

$$Nu = \frac{h_c L}{K_f}$$

In this equation, h_c is the heat transfer coefficient, L is a characteristic length, and K_f is the fluid conductivity. Inherently, this relationship provides a localized value because at various locations, these properties change. However, it is also customary to have a similar relationship that reflects averaged values over an area,

$$\overline{Nu} = \frac{\overline{h_c} L}{K_f}$$

where a bar over h or Nu indicates its average value.

Regardless of local or average value, the Nusselt number enables us to compare various fluids under various flow regimes and conditions for their ability to remove heat from a surface.

Flow Regimes, Types, and Influences

In fluid flow, there are three regimes; namely, laminar where a coherent flow pattern exists, transition region, where this pattern begins to break down, and turbulent where no coherent flow pattern exists at all. These regimes are depicted in Figure 5.1.

The value of the heat transfer coefficient depends on the characteristics of the dominant flow pattern. For example, consider the impact of turbulence on the Nusselt number for long ducts carrying gasses or liquids (Krieth, 1973). For laminar flow ($Re < 2100$, $Pr > 0.7$):

FIGURE 5.1
Different regions in a fluid flow. (Adapted from Van Dyke, M., *An Album of Fluid Motion*, Stanford, CA: Parabolic Press, 1982. With permission.)

$$Nu = 1.86\left(Re\,Pr\frac{D}{L}\right)^{0.33}$$

For turbulent flow ($Re > 6000$, $Pr > 0.7$):

$$Nu = 0.023\ Re^{0.8}\ Pr^{0.33}$$

In these two relationships, *Re* is Reynolds number, *Pr* is Prandtl number, *D* is hydraulic diameter, and *L* is length. This relationship is graphically depicted in Figure 5.2. It is quite clear from this figure that turbulent flow has a higher Nusselt number and hence a greater ability to remove heat. Although this figure is extended for the entire Reynolds number range, it should be kept in mind that the region $2000 < Re < 6000$ is not governed by either of the two relationships and it remains, for this example, unclear. This region is referred to as the transition region and the flow maintains a certain level of pattern, while at the same time it is obvious that the pattern is breaking down. As a result, calculating any flow property in this regime is difficult and subject to error.

To identify the flow regime, one needs to calculate the Grashof number for natural convection or the Reynolds number for force convection. Generally speaking, in natural convection, laminar flows have a Grashof number less than 10^7, and transition regions between 10^8 and 10^{10}. For values greater than 10^{10}, the flow is considered to be turbulent. For a vertical heated wall, with lengths up to 18–20 in from the bottom of the wall, the flow remains laminar, and beyond 24 in, it is turbulent.

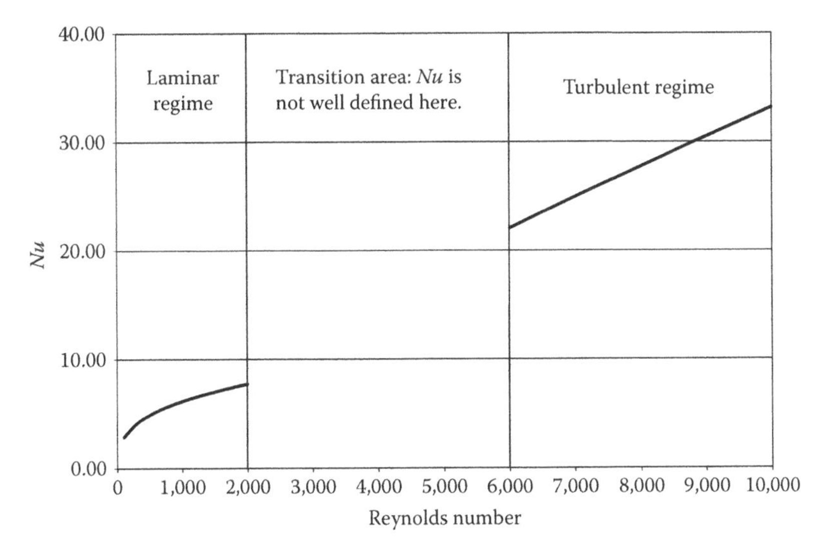

FIGURE 5.2
A graphical depiction of Nusselt numbers for laminar and turbulent flows. For this example, a length to diameter ratio of 20 and a Prandtl number of 0.75 was used.

For flows through ducts, laminar flows have a Reynolds number less than 2000 and turbulent flows have a Reynolds number greater than 6000, with the region in between being the transition region.

Fluid flow may also be divided into two types: natural (free) convection and forced convection. Similar to turbulent flow, forced convection has a higher capability to remove heat than natural convection.

One potential pitfall that the design engineer must be aware of is the impact of geographic altitude on the convection and its ability to remove heat. Commonly, equipment temperatures for many electronics packages are calculated for sea level conditions. An appreciable change in altitude will create a significant variation in both pressure and density. These two variables will impact temperature levels. Consequently, temperature values need to be recalculated when the same equipment is used in higher altitudes—either on airborne systems or in high-altitude places. We will discuss this later.

Free (or Natural) Convection

One technique used in cooling electronics is to take advantage of free (natural) convection. By taking advantage of heat rising, it is possible to design a system whereby cool air enters the enclosure from the bottom and warm air exits from the top. Note that the words "cool" and "warm" as opposed to "cold" and "hot" are used. When free convection is used to cool a system, temperature rises must be moderate.

In free convection, fluid flow is developed under the influence of buoyancy forces. Depending on the orientation of the heat source with respect to gravity, the flow field may be nonexistent, simple, or very complicated. It should also be noted that flow might start as laminar but develop into turbulent.

To study the relationship between the buoyancy and the ability of the fluid to remove heat, we need to reexamine the following nondimensional relationships:

$$Nu = \frac{hL}{K}$$

$$Nu = \frac{L^3 \rho^2 g \beta \Delta T}{\mu^2}$$

$$Pr = \frac{C_p \mu}{K}$$

where L is a characteristic length, ρ is the density of the fluid, g is gravity, β is the coefficient of fluid expansion, ΔT is the change in temperature, μ is fluid viscosity, C_p is fluid specific heat, h is the coefficient of heat transfer, and K is fluid conductivity. Experiments (Krieth 1973) have shown that the following relationship may be developed:

$$Nu = \text{Constant} \times Gr^m \times Pr^n$$

or

$$\frac{hL}{K} = C\left[\frac{L^3\rho^2 g\,\beta\,\Delta T}{\mu^2}\right]^m \left[\frac{C_p\,\mu}{K}\right]^n \tag{5.1}$$

This equation illustrates the relationship between the heat transfer coefficient, the ability of the fluid to absorb heat (C_p), its ability to conduct (K), as well as its inherent buoyancy forces. Constants C, m, and n are dictated by the flow regime and geometry. Most standard heat transfer textbooks provide these data for a variety of geometries. In electronics packaging, the vertical and horizontal walls (heated side facing up or down) are quite frequently used to model the system's enclosure. Table 5.1 provides the constants for these geometries.

Estimates of Heat Transfer Coefficient

The equation for the Nusselt number (Equation 5.1) may be reduced (or rearranged) to a specific equation for the heat transfer coefficient. This provides a simplified means for back-of-the-envelope engineering calculations. Rearrange the terms in Equation 5.1 to obtain the following:

$$h = \Psi C\left[\frac{L^3\,\Delta T}{L^{\frac{1}{m}}}\right]^m$$

where

$$\Psi = K\left[\frac{C_p\,\mu}{K}\right]^n \left[\frac{\rho^2\beta}{\mu^2}\right]^m$$

For example, consider properties of air at various temperatures as presented in Table 5.2.

TABLE 5.1

Empirical Data for Nusselt Number Constants

	Laminar Flow, $Gr < 10^8$			Turbulent Flow, $Gr > 10^{10}$		
Wall Orientation	C	m	n	C	m	n
Vertical	0.55–0.57	0.25	0.25	0.11–0.13	0.33	0.33
Horizontal heat facing up	0.54–0.71	0.25	0.25	0.130.16	0.33	0.33
Horizontal heat facing down	0.25–0.35	0.25	0.25	0.08–0.10	0.33	0.33

TABLE 5.2

Properties of Air at Various Temperatures

Temperature (°F)	0	32	66	100	200	300
ρ (lbm/ft^3)	0.86	0.081	0.076	0.071	0.060	0.052
C_p (BTU/lbm-F)	0.239	0.240	0.24	0.240	0.241	0.243
μ (lbm/ft-s)	1.110×10^{-5}	1.165×10^{-5}	1.225×10^{-5}	1.285×10^{-5}	1.440×10^{-5}	1.610×10^{-5}
K (Btu/h-ft-F)	0.0133	0.0140	0.0147	0.0154	0.0174	0.0193
β (1/F)	2.180×10^{-3}	2.030×10^{-3}	1.91×10^{-3}	1.790×10^{-3}	1.520×10^{-3}	1.320×10^{-3}

Now, evaluate Prandtl and Grashof numbers and heat transfer coefficient (h) based on an air temperature of 66°F.

$$Pr = \frac{(0.24\,\text{BTU/lbm F})(1.225 \times 10^{-5}\,\text{lbm/ft s})(3600\,\text{s/h})}{0.0147\,\text{BTU/h ft F}}$$

$$Pr = 0.72(\text{dimensionless})$$

$$Gr = \frac{L^3 (0.076\,\text{lbm/ft}^3)^2 (32.2\,\text{ft/s}^2)(1.91 \times 10^{-3}\,1/\text{F})\Delta T}{(1.225 \times 10^{-5}\,\text{lbm / ft s})^2}$$

$$Gr = 2.3673 \times 10^6 L^3 \Delta T$$

$$\Psi = 0.0147(0.72)^n (2.3673 \times 10^6)^m$$

From Table 5.1 for laminar flows and a vertical wall, we obtain $m = n = 0.25$. Thus, we calculate $\Psi = 0.53$ and the following relationship for the heat transfer coefficient h may be developed for air at 66°F temperature:

$$h = 0.53C\left(\frac{\Delta T}{L}\right)^{0.25} \tag{5.2}$$

For turbulent flows, a similar relationship may be developed based on $m = n = 0.33$.

$$h = 1.76C(\Delta T)^{0.33} \tag{5.3}$$

Notice that in this relationship, h is independent of length L. This is consistent with other reports that although the heat transfer coefficient in a laminar flow depends on the flow length, once the flow becomes turbulent, h loses its dependency on L (McAdams 1954). It is generally accepted that for laminar flows L must be less than 2 ft for laminar flow to exist (MIL-HDBK-251 1978, Steinberg 1991).

A similar calculation for a variety of temperatures between 0°F and 300°F has been conducted and the results are presented in Table 5.3. A close examination of the values of Ψ reveals that it does not change greatly and an average value may be used to develop a generalized relationship for h. Therefore, the heat transfer coefficient for air (0°F–300°F) in free convection may be calculated with some confidence from the following relationships. For laminar flows, the equation becomes

$$h = 0.51C\left(\frac{\Delta T}{L}\right)^{0.25} \quad (5.4)$$

and for turbulent flows, we would have

$$h = 1.76C(\Delta T)^{0.33} \quad (5.5)$$

Notice that ΔT is not generally known a priori, and an iterative solution must be used. The value of C varies from 0.25 for horizontal plates with the heated side facing downward to 1.45 for small components such as resistors and wires. Its value is nearly 0.5 for all other configurations of plates. Steinberg (1991) as well as MIL-HDBK-251 (1978) provide values of C for a variety of components.

It is extremely important to remember that these are empirical formulae. Therefore, the following dimensions must be used:

h_c is in BTU/h ft^2 °F
ΔT is in degrees F
L is in feet, a characteristic length

The value of L is generally determined by the flow path over the heated surfaces. For example, for a flat vertical plate, L is the height of the plate; for a flat horizontal plate

$$\frac{1}{L} = \frac{1}{\text{length}} + \frac{1}{\text{width}}$$

Keep in mind that to maintain a laminar flow, L should be less than 2 ft.

TABLE 5.3

Impact of Temperatures on Ψ

Temperature (°F)	0	32	66	100	200	300	Ave.
Ψ (Laminar)	0.55	0.54	0.53	0.52	0.49	0.46	0.51
Ψ (Turbulent)	1.92	1.84	1.76	1.67	1.48	1.32	1.66

Solution Procedure

Our ultimate goal in doing any thermal analysis is to ensure that our critical component temperature remains below a safe threshold. So far, as we have discussed the heat transfer coefficient value, we noticed that it depends on the surface temperature, which is what we are trying to find in the first place. Therefore, it becomes obvious that we need to employ an iterative solution technique to find both the heat transfer coefficient and the surface temperature. The very first step in accomplishing this task is to write the heat equation and balance the outgoing heat versus the incoming heat. Hereon, one of two approaches may be taken.

Approach 1

- Assume a temperature difference (ΔT).
- Calculate h.
- Calculate the temperature difference based on the generated heat, h, and the thermal resistance.
- Compare the assumed and calculated temperature difference (ΔT).
- Iterate if needed.

Approach 2

- Assume a h.
- Calculate the temperature difference (ΔT) based on the generated heat, h, and the thermal resistance.
- Calculate h based on calculated temperature difference (ΔT).
- Compare the assumed and calculated h.
- Iterate if needed.

There are three separate issues related to the critical component temperature calculation that need to be contemplated here. First, both approaches produce identical results; however, the second one may be easier to use and hence, more practical. The reason is that the range of variations for h is by far narrower than that of temperature. Furthermore, we may have empirical data on the heat transfer coefficient applicable to our problem. For instance, for a moderate-size (about 2 ft in length) enclosure, h is in the neighborhood of 1 BTU/hr ft^2 °F. As the size decreases, h typically increases and for relatively small enclosure (on the order of a few inches), it may be as high as 3 to 4 BTU/hr ft^2 °F.

Second, we assume that we "know" the level of the dissipated heat. It is generally understood that the power is equal to voltage times the current. This power, however, is the dissipated heat when and only when there are no outputs from the system. For a system with input as well as output, the dissipated heat is the difference between the power entering and the power leaving. A good rule of the thumb is that most systems are 75%–85% efficient, meaning that the dissipated heat is between 15% and 25% of the total input power.

Third, heat flow may take different paths through the heat flow network. Accordingly, depending on each path and the heat flux through it, we obtain different temperatures for each branch of the network. Thus, to calculate the temperature of each branch, we need to know not only the heat flux but the resistance of the branch as well. For instance, the generated heat inside an enclosure must find various thermal paths to the surface of the enclosure so that it may be dissipated to the outside world. These paths are conduction (through component leads to ground planes to chassis and finally to the outer surface), possibly radiation to the outer surface (if a component is much hotter than the surrounding components), and convection through the air inside of the enclosure. Based on the foregoing argument, it is clear that the only factor impacting the air temperature inside of an enclosure is the portion of the heat that contributes to the internal convection. Therefore, if we are interested in the interior air temperature, then we must consider convection path alone. The more efficient the conduction paths in the system, the smaller the portion of dissipated energy that would be used in the internal convection. Again, a good rule of thumb is that only 15%–25% of total heat passes through this path.

High Altitudes

As we ascend to higher altitudes, the density of the air drops. This drop could have a detrimental impact on air-cooled electronics packages. Recall that the heat transfer coefficient was derived from the Nusselt number, which is the product of the Grashof and Prandtl numbers. Also recall that the Grashof number is a function of density. Therefore, as density drops, the heat transfer coefficient and, along with it, the system's ability to remove the heat is lowered. So,

$$\frac{h_{\text{sea level}}L}{K} = C\left[\frac{L^3\rho^2_{\text{sea level}}\,g_{\text{sea level}}\,\beta\,\Delta T}{\mu^2}\right]^m\left[\frac{C_p\,\mu}{K}\right]^n$$

and

$$\frac{h_{\text{altitude}}L}{K} = C\left[\frac{L^3\rho^2_{\text{altitude}}\,g_{\text{altitude}}\,\beta\,\Delta T}{\mu^2}\right]^m\left[\frac{C_p\,\mu}{K}\right]^n$$

By dividing these two equations, we obtain the following relationship:

$$\frac{h_{\text{altitude}}}{h_{\text{sea level}}} = \left(\frac{\rho^2_{\text{altitude}}\,g_{\text{altitude}}}{\rho^2_{\text{sea level}}\,g_{\text{sea level}}}\right)^m$$

Through the gas equation of state, we realize that density is a direct function of pressure ($P = \rho RT$; R is a universal gas constant) and $m = 0.25$ for a large class of electronics systems. Furthermore, the change in gravitational constant is not very significant for the same class of problem. Thus, it can be shown that

$$h_{\text{altitude}} = h_{\text{sea level}} \sqrt{\frac{\rho_{\text{altitude}}}{\rho_{\text{sea level}}}} = h_{\text{sea level}} \sqrt{\frac{P_{\text{altitude}} T_{\text{sea level}}}{P_{\text{sea level}} T_{\text{altitude}}}} \tag{5.6}$$

Pressure and temperature may be obtained from atmospheric data. Appendix B provides a simple formulae for calculating the atmospheric data.

Board Spacing and Inlet–Outlet Openings

It has been suggested that for a system that is cooled based on natural convection, the board spacing must be at least 0.75 in (Steinberg 1991). Bejan et al. (1996) provide the means of calculating an optimum gap spacing for natural flow and the associated maximum heat removal:

$$D_{\text{opt}} = 2.3L\,Ra^{-1/4} \tag{5.7}$$

$$Q_{\max} \cong 0.45K\Delta T \frac{HW}{L} Ra^{1/2} \tag{5.8}$$

where:

$$Ra_{\text{L}} = \left[\frac{g\beta(T_{\text{outlet}} - T_{\text{inlet}})L^3}{\alpha\nu}\right]$$

H is the transverse length of the entire package
L is the length of the boards along the flow
W is the width of the stack
g is gravitational constant
α is thermal diffusivity
β is the coefficient of volumetric thermal expansion
μ is viscosity
ν is kinematic viscosity ($= \mu/\rho$, where ρ is density)

Hill and Lind (1990) provide the following relationships to calculate the inlet and outlet openings:

$$A_{\text{inlet}} = 0.3 \frac{Q}{\rho_a + \rho_i} \sqrt{\frac{T_i + 460}{l\Delta T^3}}$$

$$A_{\text{outlet}} = A_{\text{inlet}} \frac{T_i + 460}{T_a + 460}$$

where:

A is area (in^2)
Q is heat dissipation inside the enclosure (W)
ρ is air density (lb/ft^3)
l is height difference between inlet and outlet (ft)
ΔT is the required air temperature rise (°F)
T is temperature (°F)
subscripts i and a stand for inside the enclosure and ambient

Design Tips

In designing enclosures using free convection one has to take the following points into account:

1. The level of heat to be dissipated.
    a. Free convection is most effective in cooling relatively low component densities—thus, moderate temperature rises. An approximate heat flux limit for free convection is 1 W/in^2 for each board.
2. The enclosure is sealed.
    a. Consider the printed circuit board assembly (PCBA) orientation with regard to gravity. Properly oriented PCBAs allow for internal free convection to circulate the medium and lower the overall temperatures.
    b. Also consider using fins. Again, the orientation of the fins (parallel to gravity) is crucial to developing proper flow patterns.
3. The enclosure is not sealed and there are airflow intakes and outlets.
    a. When possible, ensure that the vertical dimension is at least twice as long as the horizontal dimensions. This allows for a tower effect and creates an efficient flow environment.
    b. The intake openings must be placed on the bottom of the enclosure and exhaust vents must be placed near the top for this configuration.
    c. In placing the components, locate the ones with the highest heat dissipation on the bottom of the enclosure near intake vents.
4. Ensure an efficient conduction path. One way of doing this is by placing the highest power-dissipating boards near the walls of the enclosure to shorten the conduction paths. If there are openings in the enclosure, radiation may play a role in heat removal as well. Also, place the boards at a minimum pitch of 0.75 in.
5. The efficiency of this cooling technique drops considerably by increasing altitude.

Cabinet Interior and Surface Temperature

Recall the electronics cabinet used previously that was dissipating 800 W of energy. Its physical dimensions are 12 × 18 × 26 in. The surface is painted white. Assuming an ambient temperature of 50°C and considering conduction, convection, and radiation as well as solar energy, we need to estimate the interior temperature. The system will be used in Orlando, Florida (nearly sea level conditions) as well as in Denver at 7000 ft. Do not take wind effects into account (Figure 5.3).

This method makes use of simple free convection and radiation assumptions to evaluate the interior and surface temperatures of the electronics cabinet considered previously. Briefly, the input heat energy and the output heat energy must be balanced to provide the overall surface temperature. It is assumed that no wind is present. The approach is first to calculate the surface temperature, and next to evaluate the interior temperature based on the calculated surface temperature.

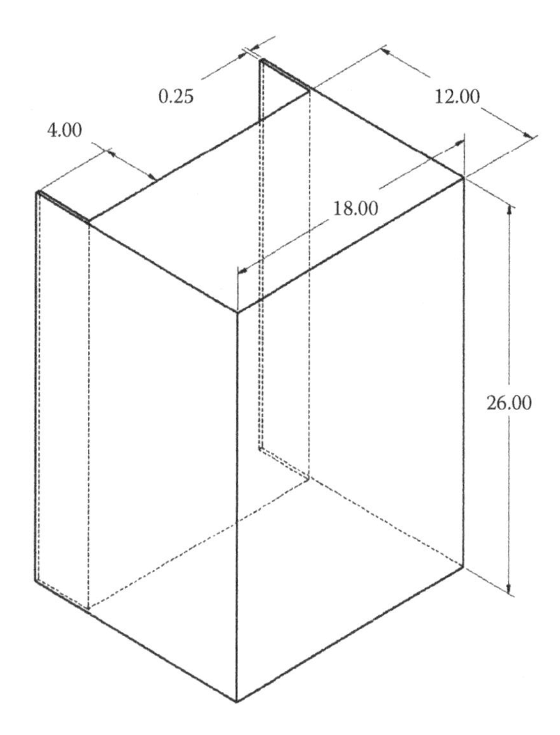

FIGURE 5.3
Electronics cabinet example—overall dimensions and sizes.

Calculating Surface Temperature

We start by developing an understanding of the sources of input heat into the system and the paths it may take to leave the system:

$$\text{Input Heat} = Q_{\text{electronics}} + Q_{\text{solar}}$$

$$\text{Output Heat} = Q_{\text{conduction}} + Q_{\text{convection}} + Q_{\text{radiation}}$$

where:

a. $Q_{\text{electronics}} = 800\,\text{W}\,(\text{given})$
b. $Q_{\text{solar}} = \text{Solar Absorptivity} \times \text{Solar Radiation} \times \text{Projected Area}$
c. $Q_{\text{conduction}} = (K\,A_{\text{conduction}}\;(T_{\text{surface}} - T_{\text{ambient}}))/L$
d. $Q_{\text{convection}} = h\,A_{\text{convection}}\left(T_{\text{surface}} - T_{\text{ambient}}\right)$
e. $Q_{\text{radiation}} = F\varepsilon A\sigma\left(T_{\text{surface}}{}^{4} - T_{\text{ambient}}{}^{4}\right)$

Heat Balance

The next step is to calculate the contribution of each factor as indicated in the previous segment and set the input values equal to the output values.

a. $Q_{\text{electronics}} = 800\,\text{W} \times 3.41\,\dfrac{\text{BTU}}{\text{hW}} = 2728\,\dfrac{\text{BTU}}{\text{h}}$
b. To calculate the solar radiation, assume that the Sun is shining on the largest surface alone:

$$\text{Projected Area} = A = \frac{26 \times 18}{144} = 3.25\,\text{ft}^2$$

$$Q_{\text{solar}} = 0.2 \times 355 \times 3.25 = 230.75\,\frac{\text{BTU}}{\text{h}}$$

Note that white paint has a solar absorptivity of about 0.2. Furthermore, only one of the largest surfaces (area = 3.25 ft^2) is assumed to receive solar radiation, which at noon is assumed to be a maximum of 355 BTU/hr ft^2.

Clearly, for a more realistic problem, the actual number of exposed surfaces must be used to calculate the projected area. Also, solar radiation varies depending on the time of day and the geographical location of the unit.

$$Q_{\text{Total Input}} = 2728 + 230.75 = 2958.75\,\frac{\text{BTU}}{\text{h}}$$

c. $Q_{\text{conduction}} = \dfrac{K A_{\text{conduction}} \left(T_{\text{surface}} - T_{\text{ambient}}\right)}{L}$

The enclosure is made from aluminum 6061 T6 thus $K = 90$ BTU/(h ft °F).

$$A_{\text{conduction}} = 2\frac{(0.25 \times 26)}{144} = 0.09 \text{ ft}^2$$

There are two flanges on each side.

$$L = \frac{4}{12} = 0.33 \text{ ft}$$

Note that T_{ambient} is 582°R. We use an absolute scale for temperature because we need to include the impact of radiation in our calculations.

$$Q_{\text{conduction}} = \frac{90 \times 0.09\left(T_{\text{surface}} - 582\right)}{0.33}$$

$$Q_{\text{conduction}} = 24.5\left(T_{\text{surface}} - 582\right)$$

d. $Q_{\text{convection}} = h A_{\text{convection}} \left(T_{\text{surface}} - T_{\text{ambient}}\right)$

$$A_{\text{convection}} = 2\frac{26 \times 12 + 26 \times 18 + 12 \times 18}{144} \quad \Rightarrow \quad A_{\text{convection}} = 13.83 \text{ ft}^2$$

The heat transfer coefficient is not known at this time, so it is treated as an unknown. Recall that T_{ambient} is 582°R.

$$Q_{\text{convection}} = 13.83 h\left(T_{\text{surface}} - 582\right)$$

e. $Q_{\text{radiation}} = F \varepsilon A \sigma \left(T_{\text{surface}}^{4} - T_{\text{ambient}}^{4}\right)$

Surface emissivity is 0.9, thus the radiation equation may be written as follows assuming that the box radiates on all six sides.

$$Q_{\text{radiation}} = 1 \times 0.9 \times 13.83 \times 1.713 \times 10^{-9}\left(T_{\text{surface}}^{4} - 582^{4}\right)$$

$$Q_{\text{radiation}} = 21.32 \times 10^{-9}\left(T_{\text{surface}}^{4} - 582^{4}\right)$$

By summing the output heat and equating the algebraic sum to the total input heat, a nonlinear equation for solving T_{surface} may be written:

$$\begin{aligned} &\text{Input Heat} = \text{Output Heat} \\ &2958.75 = Q_{\text{conduction}} + Q_{\text{convection}} + Q_{\text{radiation}} \\ &2958.75 = 24.5(T_{\text{surface}} - 582) + 13.83h(T_{\text{surface}} - 582) \\ &\qquad + 21.32 \times 10^{-9}\left(T_{\text{surface}}^{4} - 582^{4}\right) \end{aligned} \tag{5.9}$$

Notice that this equation has two unknowns, namely, the surface temperature T and heat transfer coefficient h. Depending on the flow regime, either Equation 5.4 or 5.5 could serve as a second relationship between temperature and coefficient of heat transfer. Considering that turbulent flow generally does not develop for lengths of 24 in and below, we can safely assume that our flow regime is laminar; thus, in this case equation (5.4) applies. Let us examine this equation once again.

$$h = 0.51C\left(\frac{\Delta T}{L}\right)^{.25} \tag{5.10}$$

To obtain T_{surface}, h is first estimated, then Equation 5.9 is solved. Next, this calculated temperature is used in Equation 5.10 to get a better estimate for h. This procedure is repeated until estimated and calculated values of h are very close. This approach is very simple once a spreadsheet is employed. It is clear, however, that we need to assign a value or values for C before we can calculate h. Considering that there are both vertical and horizontal plates, it may be confusing as to which value of C to choose. One approach would be to calculate the contribution of each side separately and then integrate each contribution into Equation 5.9. Another approach is to find an average value of C that would describe the entire enclosure. In this example, all four vertical walls as well as the top and bottom surface participate in the convection mechanism; thus, a weighted average of C would be a good approximation. Here in, a value of 0.55 for C is used. Equation 5.10 becomes

$$h = 0.28\left(\frac{\Delta T}{L}\right)^{0.25}$$

A good starting estimate for h is 1.

For $h = 1$, Equation 5.9 becomes

$$\begin{aligned} 2958.75 &= 24.5(T_{\text{surface}} - 582) + 13.83(1)(T_{\text{surface}} - 582) \\ &\quad + 21.32 \times 10^{-9}\left(T_{\text{surface}}^{4} - 582^{4}\right) \end{aligned}$$

A description of a means of solving a nonlinear equation is beyond the scope of this book. It will suffice to say that the solution is as follows

$$T_{\text{surface}} = 633.5°\text{R} = 173.5°\text{F}$$

$$\Delta T = 633.5 - 582 = 51.5$$

$$h_{\text{calculated}} = 0.28\left(\frac{\Delta T}{L}\right)^{0.25} = 0.28\left(\frac{51.5}{L}\right)^{0.25}$$

L is a characteristic length, and may be evaluated based on the following relationship

$$L = \frac{3 \times \text{Height} \times \text{Width} \times \text{Depth}}{\text{Height} \times \text{Width} + \text{Width} \times \text{Depth} + \text{Height} \times \text{Depth}}$$

$$L = \frac{3 \times 26 \times 18 \times 12}{26 \times 18 + 18 \times 12 + 26 \times 12} = 16.9\,\text{in}$$

L must be in feet, thus:

$$L = \frac{19.9}{12} = 1.4\,\text{ft}$$

$$h_{\text{calculated}} = 0.28\left(\frac{51.4}{1.4}\right)^{0.25}$$

$$h_{\text{calculated}} = 0.71$$

For $h = 0.71$, Equation 5.9 becomes

$$2958.75 = 24.5(T_{\text{surface}} - 582) + 13.83(0.71)(T_{\text{surface}} - 582)$$
$$+ 21.32 \times 10^{-9}\left(T_{\text{surface}}{}^4 - 582^4\right)$$

$$T_{\text{surface}} = 637.1°\text{R} = 177.1°\text{F}$$

$$\Delta T = 637.5 - 582 = 55.1$$

$$h_{\text{calculated}} = 0.28\left(\frac{55.1}{1.4}\right)^{0.25}$$

$$h_{\text{calculated}} = 0.73$$

Similarly, for $h = 0.73$, we get

$$T_{\text{surface}} = 636.9°\text{R} = 176.9°\text{F}$$

$$\Delta T = 636.9 - 582 = 54.9$$

$$h_{\text{calculated}} = 0.28\left(\frac{54.9}{1.4}\right)^{0.25}$$

$$h_{\text{calculated}} = 0.73$$

This solution converges to the surface temperature of 176.9°F and $h = 0.73$. The choice of a characteristic length may be puzzling in some cases. In general, the temperature is not super sensitive to this variable. Here, had we chosen a value of 0.5 for L, the results would not have been significantly different (174.3°F with $h = 0.928$).

Calculation of Internal Temperature

Earlier, it was demonstrated (in a general sense) that only a fraction of total dissipated energy heats the air inside the enclosure. Typical values are between 15% and 25% of the generated heat that is distributed inside of the cabin. The more efficient the thermal paths, the lower this value. Furthermore, one may assume that for a metallic enclosure, there are no thermal gradients through the thickness of the enclosure. In other words, the surface temperature of the enclosure is the same immediately on the inside and outside of the enclosure.

To solve the internal air temperature, we use the same logic as before:

1. Input

$$Q_{\substack{\text{electronics}\\\text{convection}}} = 0.25 \times 800 \times 3.41 = 682\frac{\text{BTU}}{\text{h}}$$

2. Output

$$Q_{\text{output}} = hA\left(T_{\text{internal}} - T_{\text{surface}}\right)$$

3. Heat balance

$$682 = h(13.83)\left(T_{\text{internal}} - 176.9\right)$$

Similar to the previous case, to solve this equation, first h is estimated and this equation is solved for T_{internal}. Note that the temperature scale is not in an absolute scale because internal radiation, if any, has been ignored. Once this temperature is calculated it may be used to get a better estimate for h. A good starting value for h is 0.75 for internal flows.

For $h = 0.75$, we have

$$682 = 0.75(13.83)(T_{\text{internal}} - 176.9)$$

$$T_{\text{internal}} = 242.7°\text{F}$$

$$\Delta T = 242.7 - 176.9 = 65.8$$

$$h_{\text{calculated}} = 0.28\left(\frac{65.8}{1.4}\right)^{0.25}$$

$$h_{\text{calculated}} = 0.76$$

For $h = 0.76$, we have

$$682 = 0.76(13.83)(T_{\text{internal}} - 176.9)$$

$$T_{\text{internal}} = 241.9°\text{F}$$

$$\Delta T = 241.9 - 176.9 = 65.0$$

$$h_{\text{calculated}} = 0.28\left(\frac{65.0}{1.4}\right)^{0.25}$$

$$h_{\text{calculated}} = 0.76$$

This solution has converged and the interior air temperature is about 242°F.

Calculation of Component Temperature

In order to calculate the component or board temperature, we can use the same approach again with a starting value for $h = 0.76$ and $T_{\text{Bulk}} = T_{\text{internal}} = 242°\text{F}$. Suppose that in this example one board of interest dissipates 25 W (85 BTU/h). The board temperature is, therefore, calculated as follows, assuming that the board is 8 × 12 in and is standing vertically along its longest dimension:

$$Q_{\text{output}} = hA(T_{\text{board}} - T_{\text{internal}})$$

$$h = 0.29\left(\frac{\Delta T}{L}\right)^{0.25}$$

Notice that the coefficient in the h equation is now 0.29 because the PCBA is standing vertically.

$$L = \frac{12}{12} = 1\,\text{ft}$$

$$L = \frac{8 \times 12}{144} = 0.67\,\text{ft}^2$$

By manipulating these two equations, we obtain:

$$\left(T_{\text{board}} - 242\right) = \left(\frac{85}{0.29}\right)^{0.8}$$

$$T_{\text{board}} = 336°\text{F}$$

I agree with the reader that these temperature values are incredibly high. However, we should bear in mind that on the one hand, we started with a very harsh starting point, that is, a 122°F ambient condition and a huge heat dissipation level. On the other hand, we need to assess temperature levels before we can devise appropriate thermal solutions. Therefore, this is just the starting point in the design and not the final stage.

Fin Design

Fins greatly improve the efficiency of convection cooling. However, there are a variety of fin configurations and the designer still faces the issue of selecting proper fins particularly for the external surfaces of the enclosure. In this segment, we discuss a technique for a simple, first-order calculation for the most basic shape of the fin, that is, plate fins.

Basic Procedure

In this procedure, one must first evaluate the temperature distribution of the system without the presence of fins based on the techniques described previously, and calculate an overall heat transfer coefficient. The next step is to increase the heat transfer coefficient to lower the critical/design temperature to the values set by the design criterion. Then, based on the following formula, calculate the fin geometry.

$$\text{Original area} \times \text{newly calculated } h = \text{fin surface area} \times \text{original } h$$

This formula is not an exact relationship but is useful for a quick back-of-the-envelop calculation. For a better explanation of this approach, let us solve the previous example again and select a fin.

RF Cabinet Free Convection Cooling

The initial thermal characteristics of an RF cabinet were investigated as a preliminary stage in the design cycle. Now, we need to select a heat sink to maintain a maximum interior temperature of 82°C (179.6°F), assuming an ambient temperature of 50°C (122°F). Recall that the RF cabinet dissipates 800 W of energy and its physical dimensions are 12 × 18 × 26 in. The surface of the enclosure is to be painted white.

Analytical Approach

For the sake of simplicity, the solution procedure will not be repeated here. Only Equation 5.9 is rewritten here:

$$2958.75 = 24.5\left(T_{\text{surface}} - 582\right) + 13.83\ h_{\text{effective}}\left(T_{\text{surface}} - 582\right) + 21.32 \times 10^{-9}\left(T_{\text{surface}}^{\ 4} - 582^4\right)$$

The only difference is that h has been replaced with $h_{\text{effective}}$. The solution procedure is as before; however, there is no need to balance $h_{\text{effective}}$ against $h_{\text{calculated}}$. The results for a variety of $h_{\text{effective}}$ are presented in Table 5.4. Since, the design criterion is based on maintaining the interior temperature to 82°C (179.6°F), let us consider this portion of analysis.

Calculation of Internal Temperature

It will be assumed that compared to the previous study, we have increased the heat conduction efficiency of our system and now 85% of the generated heat escapes to the surface through conduction. We have achieved this by mounting the heat-generating elements directly onto the wall and through the use of copper straps and better thermal interface materials. Thus, only 15% of the generated heat is distributed inside of the cabin. Following the same heat balance principle, we have

TABLE 5.4

$h_{\text{effective}}$, Internal and Surface Temperatures for a 122°F Ambient Temperature

$h_{\text{effective}}$	Surface Temperature	Interior Temperature	Interior h
1	173.45	216.72	0.68
2	163.75	207.02	0.68
3	157.06	200.33	0.68
4	152.20	195.47	0.68
5	148.50	191.77	0.68
6	145.61	188.88	0.68
7	143.28	186.55	0.68

1. Input

$$Q_{\substack{\text{electronics}\\ \text{convection}}} = 0.15 \times 800 \times 3.41 = 409.2\,\frac{\text{BTU}}{\text{h}}$$

2. Output

$$Q_{\text{output}} = h_{\text{internal}}\,A\left(T_{\text{internal}} - T_{\text{surface}}\right)$$

3. Heat balance

$$409.2 = h_{\text{internal}}\left(13.83\right)\left(T_{\text{internal}} - T_{\text{surface}}\right)$$

The solution procedure is exactly the same as before. Table 5.4 gives T_{internal} and T_{surface} for a variety of $h_{\text{effective}}$ values.

Note that the value of interior h has remained constant through the range of surface temperatures. Furthermore, increasing the exterior heat transfer coefficient has a diminishing return. And, even at the large value of $h_{\text{effective}} = 7$, the interior temperatures are above the design criterion. What is a possible solution? Let us assume that we could increase the interior heat transfer coefficient by employing a fan, per se. How would this impact the design criterion? Table 5.5 presents the same results for Interior $h = 1$ and 2. Clearly, the design criterion could be achieved for interior $h = 2$, and $h_{\text{effective}} = 3$.

It is important to be able to determine what fin size and spacing would produce the effective heat transfer coefficient calculated above. Basically, for the same ΔT value:

$$h\,A_{\text{fin}} = h_{\text{effective}}\,A_{\text{sides}}$$

Suppose that the fin height is ℓ_1, the thickness is t, and the spacing is ℓ_2. η is the fin efficiency, n is the number of fins, and L is the total length to be covered by the heat sinks. The following relationship may be deducted:

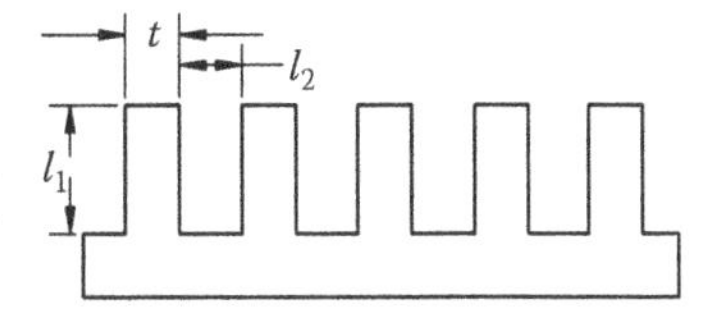

TABLE 5.5

$h_{\text{effective}}$, Internal and Surface Temperatures for a 122°F Ambient Temperature

	Interior $h = 1$		**Interior $h = 2$**	
$h_{\text{effective}}$	**Surface Temp.**	**Interior Temp.**	**Surface Temp.**	**Interior Temp.**
1	173.45	203.04	173.45	188.24
2	163.75	193.34	163.75	178.54
3	157.06	186.65	157.06	171.85
4	152.20	181.79	152.20	166.99
5	148.50	178.09	148.50	163.29
6	145.61	175.20	145.61	160.40
7	143.28	172.87	143.28	158.07

$$n = INT\left\{\frac{h_{\text{effective}}L}{h(2\ell_1 + t + \ell_2)}\right\}$$

Since, the fin geometry (i.e., ℓ_1, t, and ℓ_2) is specified by the design engineer, these parameters must be subjected to the following geometric constraint.

$$L - \ell_2 \le n(t + \ell_2) \le L$$

For this particular example, note that the natural convection coefficient of heat transfer, h, is 0.73 and the perimeter of the cabinet is 42 in. For the above values of $h_{\text{effective}}$ and a fin efficiency of 0.9, the fin size and spacing shown in Table 5.6 may be selected.

Design Recommendations for This Example

It is quite possible for a 12 × 18 × 26 in. box to dissipate 800 W of energy in an ambient temperature of 50°C and maintain an 82°C internal temperature. However, one needs to exercise caution and careful analysis so that localized temperature rises do not exceed the maximum limit. This requires that all high-powered heat sources be attached directly to the internal surface of the external heat sink. All high-powered heat sources must be mounted with copper straps and proper thermal pads. An internal low velocity fan must be incorporated so that localized hot spots do not develop.

Fin Design Considerations

A total of 37 fins with the following characteristics need to be used.

The fin must be 2 in tall.

The fin must be 0.225 in thick.

The fin spacing must be 0.9 in.

TABLE 5.6

Proper Fin Spacing and Corresponding Thickness

$h_{\text{effective}}$	L (in)	ℓ_1 (in)	t (in)	ℓ_1 (in)	n
3	42	2	0.225	0.9	37
4	42	2.5	0.2	0.8	42
5	42	2.75	0.125	0.7	50

A More Exact Procedure

A more exact treatment of fin heat transfer is beyond the scope of this book. The primary reason is that, on the one hand, the developed flow may become quite complex depending on the geometry, and on the other hand, radiation may play a strong role in the heat transfer. The reader is encouraged to read Culham et al. (2000).

Forced Convection

There are a variety of forced-convection cooling techniques, which may be classified into three types: direct flow, cold plates and heat pipes. In direct flow, forcing air or other coolants directly over components cools the PCBA and its components. Cold plates cool PCBAs indirectly by either air or other fluids. Heat pipes take advantage of phase change properties of certain materials to transfer heat from one place to another.

In direct flow cooling—as shown in Figure 5.4, the designer must be concerned with the following issues before choosing a fan or a pump: pressure drop/losses; flow paths; component spacing and distribution; and in the case of air, effects of altitude.

In cold plate design—as depicted in Figure 5.5—heat is transferred to coolant through a heat sink; therefore, the flow path may be designed into the chassis. A cold plate design requires flow rate calculations to prevent either choking or lack of pressure as well as measures to prevent coolant leakage.

Heat pipes, as shown in Figure 5.6, are designed to transfer heat from a hot spot to a location where it can easily be removed and works best when heat is concentrated. They work on the basis of evaporation/condensation. Their performance (e.g., operating temperature) is based on wick as well as fluid media. We will not be concerned with heat pipe design or analysis issues in this book.

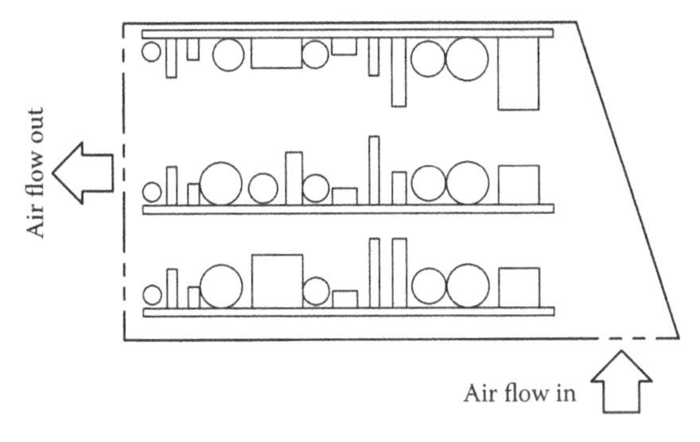

FIGURE 5.4
In direct flow, the components are cooled directly.

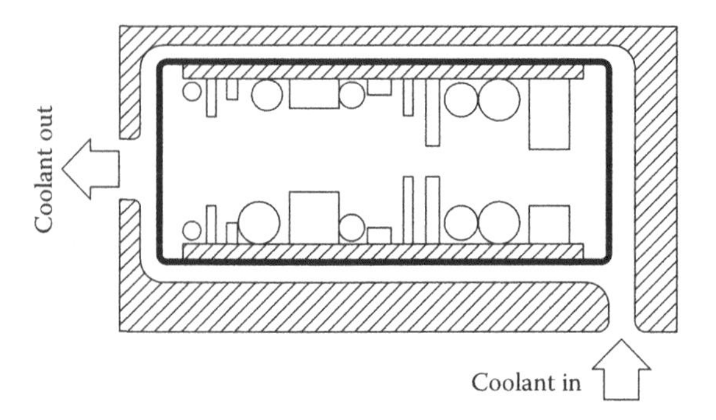

FIGURE 5.5
In cold plates, flow passes behind the PCBAs.

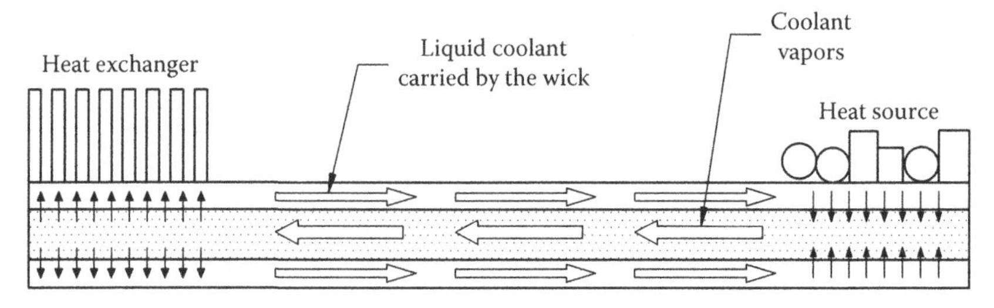

FIGURE 5.6
Heat pipes work best when heat source is concentrated in one area.

Direct Flow System Design

To design an enclosure based on the direct flow cooling principle is relatively simple. The steps needed to accomplish this are as follows:

1. Determine the total heat dissipation.
2. Establish the maximum allowable coolant temperature rise at the exit.
3. Calculate the flow rate that would maintain such a temperature rise for the given heat dissipation.
4. Determine the pressure drop curve throughout the system. This is also referred to as the system's "impedance curve".
5. Select a fan or pump based on the needed flow rate and the impedance curve.
6. Determine the distributed flow rate for each board.
7. Calculate the temperature of critical components on each board.

Generally speaking, as the design engineers, we have the total dissipation. Also, the maximum allowable coolant temperature is often dictated to us in various industry standards. Thus, the first step in our calculation is to determine the required flow to maintain the maximum temperature rise.

Required Flow Rate

Often, it is up to the design engineer to maintain a certain upper limit for the temperature rise. To do so, one has to evaluate the required flow rate. A simple one-dimensional relationship derived from the heat equation is as follows:

$$\dot{m} = \frac{Q}{C_p \Delta T}$$

where $\dot{m}$ is the mass flow rate, Q is the dissipated heat, C_p is the specific heat of the fluid, and ΔT is the temperature rise of the fluid from inlet to exit.

We need to keep in mind that the end result of this procedure is to select the proper fan. Many fan catalogs provide fan characteristics based on volume flow rate, such as cubic feet per minute (CFM). Furthermore, air is the coolant of choice for a large class of electronics enclosures, so it is only logical to rewrite this equation in these terms:

$$\rho \dot{V} = \frac{Q}{C_p \Delta T}$$
$$\dot{V} = \frac{Q}{\rho C_p \Delta T} \quad (5.11)$$

Now, consider the equation of state for air (as a perfect gas)

$$P_\circ = \rho R T_\circ$$

where:

$P_\circ$ is expressed in lb/ft^2

$T_\circ$ in degrees Rankine

R (1718 $\frac{\text{ft lb}}{\text{Sec °R}}$ for air) is specific gas constant

Equation 5.11 may be further reduced to

$$\dot{V} = \frac{QRT_\circ}{P_\circ C_p \Delta T}$$

By substituting 0.24 for air's specific heat and 1718 for gas constant—as well as making sure that the dimensions are consistent—the following relationship may be developed:

$$\dot{V} = \frac{3.7QT_\circ}{P_\circ \Delta T} \text{CFM (cubic feet per minute)} \quad (5.12)$$

Note that in this relationship, the following units "must be" maintained:

- Q is the power to be dissipated in BTU/h
- T_o is the inlet temperature in degrees Rankine (°F + 460)
- P_o is the inlet barometric pressure in lb/ft^2
- ΔT is the temperature rise across the equipment in degrees F

Sometimes barometric pressure is given in terms of inches of mercury ("Hg). In this case the flow rate equation may be rewritten as

$$\dot{V} = \frac{0.0524\ QT_o}{P_o \Delta T}\ \text{CFM} \tag{5.13}$$

Now the dimensions, which must be maintained, are as follows.

- Q is the power to be dissipated in BTU/h
- T_o is the inlet temperature in degrees Rankine (°F+460)
- P_o is the inlet barometric pressure in "Hg
- ΔT is the temperature rise across the equipment in degrees F

Some industry standards such as ARINC 600 require that the flow rate be calculated per kilowatt of dissipated heat. For this reason, the same equation may be manipulated as follows:

$$\dot{V} = \frac{178.8\ QT_o}{P_o \Delta T}\ \text{CFM} \tag{5.14}$$

Again, the dimensions, which must be maintained, are as follows.

- Q is the power to be dissipated in kW
- T_o is the inlet temperature in degrees Rankine (°F+460)
- P_o is the inlet barometric pressure in "Hg
- ΔT is the temperature rise across the equipment in degrees F

Exercise

An electronic box has the following components and maximum potential heat dissipation distribution:

1. PCBA 1—uP board, 26 W
2. PCBA 2—Memory board, 26 W
3. PCBA 3—Network board, 19.5 W
4. PCBA 4—Network board, 19.5 W
5. PCBA 5—Power Supply, 54 W
6. PCBA 6—E1 Network board, 19.5 W, quantity of two

What is the needed airflow (in CFM) to have an inlet to exit temperature rise of 10°C at sea-level condition? At sea level, standard temperature is 59°F and standard pressure is 29.9 "Hg (inches of mercury).

$$T = 460 + 59 = 519°\text{R}$$

$$Q = 26 + 26 + 19.5 + 19.5 + 54 + 2 \times 19.5 = 184\,\text{W}$$

$$Q = 184(\text{W}) \times 3.41\left(\frac{\text{BTU}}{\text{h}}\right) / (\text{W}) = 627.44\left(\frac{\text{BTU}}{\text{h}}\right)$$

$$\Delta T = 10°\text{C} = 18°\text{F}$$

$$\dot{V} = \frac{0.0524\ QT_{\circ}}{P_{\circ}\Delta T}$$

$$\dot{V} = \frac{0.0524 \times 627.44 \times 519}{29.9 \times 18}$$

$$\dot{V} = 31.7\ \text{CFM}$$

So, the required flow rate to maintain a 10°C inlet to exit temperature rise is 31.7 CFM.

Board Spacing and Configurations

The next step after calculating the required airflow is to specify board spacing. For instance, what should the board spacing for the previous system be? To answer this question, we need to develop a better understanding of fluid flow and its relationship to geometry as well as changes in pressure or pressure losses.

Flow Resistance

Fluid flows from one point to another only if a pressure difference exists between these two points—much the same as heat flow between two points requires a temperature difference. And similarly, the path in between these points resists this flow. In the case of fluid flow, this resistance is caused by friction and drag. The general governing equations are the Navier-Stokes equations, mentioned in Chapter 2, and an important subset of these is pipe flow equations. Piping systems and their associated flow equations closely approximate fluid flow through electronics enclosures.

The relationship between the pressure drop and flow rate across a pipe of length L and an arbitrary cross section is

$$\Delta P = N\left(\frac{\rho V^2}{2g}\right) \tag{5.15}$$

where the resistance factor (or number of heads lost) is represented by

$$N = f\frac{L}{D_h} \tag{5.16}$$

where:

f is the Darcy friction factor
L is pipe length
D_h is hydraulic diameter

Darcy friction factor, a well-studied subject for a variety of fluids and pipes, is directly proportional to surface roughness and is generally inversely proportional to the Reynolds number. For PCBAs, a typical value of f is about 0.02. For more details, see Baumeister et al. (1979).

Equation 5.15 may be rewritten in terms of flow rate ($\dot{V}$) as follows:

$$\Delta P = N\left(\frac{\rho\dot{V}^2}{2gA^2}\right) \tag{5.17}$$

In this work, we are primarily concerned with air as the fluid, pressure units in inches of water ("H_2O), and area expressed in terms of squared inches. To reflect these concerns, Equation 5.17 may be reduced to

$$\sigma\Delta P = \mathfrak{R}\left(\rho\dot{V}\right)^2 \text{ inches of water}\left(\text{"H}_2\text{O}\right) \tag{5.18}$$

where loss coefficient (or factor) is defined as follows:

$$\mathfrak{R} = 0.226\frac{N}{A^2}\frac{\text{"H}_2\text{O}}{(\text{lb}/\text{min})^2} \tag{5.19}$$

Note that $\dot{V}$ is the flow rate in CFM, ρ is air density at inlet (lb/ft^3), σ is the ratio of average density to standard density (0.0756 lb/ft^3), and A is the minimum flow area (in^2).

Number of Heads Lost (N)

As flow moves through various network elements, it may expand, contract, turn at a variety of angles, and so on. It is intuitively obvious that energy is required to move a packet of fluid from point A to B. This energy is shown in the form of a pressure drop. It is also obvious that if any obstacles appear on the flow path, more energy will be required to move the fluid. In fluid flow, expansions, contraction, turns, and the like, are considered obstacles and are referred to as head losses.

Obviously, evaluation of the final pressure drop depends on knowing the number of heads lost. Good sources for this are Steinberg (1991), Baumeister et al. (1979), and Sloan (1985). Some typical values are shown in Table 5.7.

TABLE 5.7

Velocity Head Losses in Electronics Enclosures

Configuration	Inlet	90°Turn	PCBAs	Outlet
N	1.0	1–1.5	0.5–2.5	1.0

The number of heads lost for an electronics enclosure may be best evaluated based on Equation 5.16 if proper experimental data are not available.

Earlier, it was mentioned that f depends both on surface roughness as well as the Reynolds number. For very smooth surfaces the following relationships hold (MIL-HDBK-251 1978):

1. Hagen–Poiseulle equation

$$f = \frac{64}{\mathrm{Re}}$$

2. Blasius equation

$$f = \frac{0.316}{\mathrm{Re}^{0.25}}$$

3. Prantle equation

$$f = \frac{0.184}{\mathrm{Re}^{0.2}}$$

Flow Networks

It was indicated earlier that flow through piping systems might be adapted for flow through electronics packages. The elements of such a network are either in series or in parallel or both.

Elements in Series

When the elements of a flow network are set in series configuration, the total loss coefficient is the sum of the individual element losses, and the total pressure loss is the sum of pressure losses in each segment. This may be expressed as follows:

$$\mathfrak{R}_{\text{total}} = \sum_i \mathfrak{R}_i = \mathfrak{R}_1 + \mathfrak{R}_2 + \mathfrak{R}_3 + \ldots$$

$$\Delta P_{\text{total}} = \sum_i \Delta P_i = \Delta P_1 + \Delta P_2 + \Delta P_3 + \ldots$$

Elements in Parallel

When the elements of a flow network are set in a parallel configuration, the physical laws of fluid flow dictate that the pressure losses be equal in all the branches. Based on this requirement, the total loss coefficient may be developed as follows:

$$\frac{1}{\sqrt{\Re_{\text{total}}}} = \sum_i \frac{1}{\sqrt{\Re_i}} = \frac{1}{\sqrt{\Re_1}} + \frac{1}{\sqrt{\Re_2}} + \frac{1}{\sqrt{\Re_3}} + \ldots$$

$$\Delta P_{\text{total}} = \Delta P_1 = \Delta P_2 = \Delta P_3 = \ldots$$

Flow Rate Distribution between Parallel Plates

Let us combine Equations 5.19 and 5.18 and review its implications:

$$\sigma \Delta P = 0.226 \frac{N}{A^2} \left(\rho \dot{V} \right)^2 \tag{5.20}$$

Here is what we can learn from this equation:

1. Pressure drop is directly related to the square of flow rate ($\dot{V}$).
2. Pressure drop is related to the square of density and by implication to altitude and geographical location of the equipment.
3. Pressure drop is directly related to head loss coefficients.
4. Pressure drop is inversely proportional to flow cross-sectional area and thus to geometry and board spacing.

Now that we understand that pressure drop is directly related to flow rate and loss coefficient and indirectly to geometry, we can begin to develop a means of calculating board spacing based on available flow rate and required temperature rise.

Consider that the flow (at a rate of $\dot{V}$) is first divided into two parallel channels and then joined again. It is clear that the sum of the flow rates in each channel is equal to the total flow rate:

$$\dot{V} = \dot{V}_1 + \dot{V}_2 \tag{5.21}$$

Also, since the channels are parallel, the pressure drops in each channel are equal.

$$\Delta P_1 = \Delta P_2$$

$$\Re_1 \left(\rho \dot{V}_1 \right)^2 = \Re_2 \left(\rho \dot{V}_2 \right)^2 \tag{5.22}$$

$$\Re_1 \left(\dot{V}_1 \right)^2 = \Re_2 \left(\dot{V}_2 \right)^2$$

Equation 5.21 along with Equation 5.22 form a set of two equations for two unknowns. The solution to this set of equations is as follows:

$$\dot{V}_1 = \left[1 - \frac{1}{\sqrt{\frac{\Re_2}{\Re_1}}}\right]\dot{V}$$

and

$$\dot{V}_2 = \left[\frac{1}{1 + \sqrt{\frac{\Re_2}{\Re_1}}}\right]\dot{V}$$

Earlier, we raised the question of calculating the proper board spacing. Theoretically, one may calculate the loss coefficient based on a specified board pitch and the total required flow, and then determine the flow rate in each channel. Should this channel flow rate prove to be insufficient, board pitch may be altered. The reader may develop the relationships for three or more channels.

Optimum Board Spacing and Heat Flow

Similar to the natural convection case, Bejan et al. (1996) provided the means of calculating an optimum gap spacing for forced flow and the associated maximum heat removal:

$$D_{opt} = 2.7 L \Pi_L^{-(1/4)}$$

$$Q_{max} = 0.6 K \Delta T \frac{HW}{L} \Pi_L^{1/2}$$

where:

$$\Pi_L = \left(\frac{\Delta P L^2}{\mu\alpha}\right)$$

H is the transverse length of the entire package
L is the length of the boards along the flow
W is the width of the stack
α is thermal diffusivity
μ is viscosity

ν is kinematic viscosity ($= \mu/\rho$, where ρ is density)
ΔP is pressure drop across the flow

System's Impedance Curve

The last step before specifying a fan or a pump is to develop the impedance curve of the system. This is really a fancy way of saying to plot the pressure drop as a function of flow rate. In fact, we already know from Equation 5.18 that the pressure drop is a quadratic function of flow rate as follows.

$$\sigma \Delta P = \mathfrak{R}_{\text{total}} \left(\rho \dot{V} \right)^2 \text{"} H_2O$$

So, in a way, to develop the impedance curve of a system is to find $\mathfrak{R}_{\text{total}}$. Following are the steps required for ΔP calculations:

1. Draw the flow network through the enclosure.
2. Assign numbers to each section.
3. Evaluate each segment's number of heads lost (N).
4. Evaluate the minimum flow area (A).
5. Calculate loss factor ($\mathfrak{R}$) for each segment.
6. Calculate the total enclosure loss factor ($\mathfrak{R}_{\text{total}}$).
7. Develop $\sigma \Delta P$ table versus $\dot{V}$ (flow rate).

Sample Problem

Consider the system in the previous exercise shown in Figure 5.7. The system dissipates a total of 184 W. We need to develop the system's impedance in order to select a proper fan. The operating conditions are between sea level and 7000 ft. Recall that the maximum potential heat dissipation distribution is as follows.

1. PCBA 1—uP board, 26 W
2. PCBA 2—Memory board, 26 W
3. PCBA 3—Network board, 19.5 W
4. PCBA 4—Network board, 19.5 W
5. PCBA 5—Power Supply, 54 W
6. PCBA 6—E1 Network board, 19.5 W, quantity of two

Each board is 6.5 × 10 in and minimum clearance for flow between boards is 0.25 in. To develop the flow diagram, we need the geometry information for the following stations. For details refer to Figure 5.7 and Table 5.8.

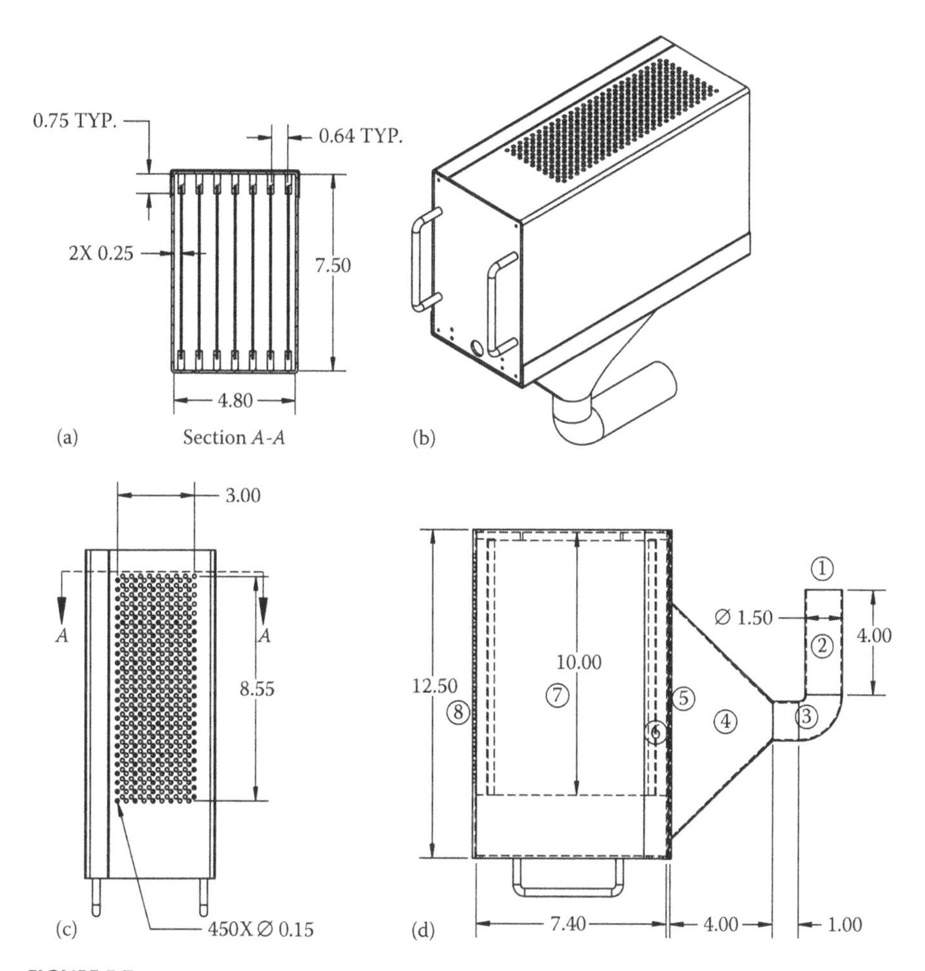

FIGURE 5.7
The system configuration.

Station 1: inlet contraction, area is

$$\pi r^2 = 3.14 \times \left(\frac{1.5}{2}\right)^2$$

Station 2: short pipe, area is

$$\pi r^2 = 3.14 \times \left(\frac{1.5}{2}\right)^2$$

Station 3: 90-degree turn, area is

$$\pi r^2 = 3.14 \times \left(\frac{1.5}{2}\right)^2$$

TABLE 5.8

Each Station's Loss Factor

Station	Area (in²)	N	$\Re$		$\Re_{ForStation}$
1	1.77	1	7.24×10^{-2}		0.0724
2	1.77	0.02	1.45×10^{-3}		0.0014
3	1.77	0.7	1.09×10^{-1}		0.0507
4	14.01	0.5	1.73×10^{-3}		0.0006
5	0.018	2	1.05×10^{-3}	$\frac{1}{\sqrt{\Re_{Station}}} = \sum_{450} \frac{1}{\sqrt{\Re_5}}$	0.0071
6	48	0.5	9.81×10^{-5}		0.0000
7	2.5	1	3.62×10^{-2}	$\frac{1}{\sqrt{\Re_{Station}}} = \sum_{6} \frac{1}{\sqrt{\Re_6}}$	0.0010
8	0.018	1	$1.05 \times 10^{+3}$	$\frac{1}{\sqrt{\Re_{Station}}} = \sum_{450} \frac{1}{\sqrt{\Re_8}}$	0.0034
System's total loss factor					0.1367

Station 4: expansion transition pipe, area is the average of the cross-sectional areas:

$$A_{\text{Ave.}} = \frac{1.77 + 26.25}{2}$$

Station 5: Perforated surface: area of each hole is

$$\pi r^2 = 3.14 \times \left(\frac{0.15}{2}\right)^2$$

Station 6: expansion into enclosure, notice that air expands into

$$A = 10 \times 4.8$$

Station 7: flow through PCBAs; minimum flow passage area for each PCBA is

$$A = 10 \times .25$$

Station 8: outlet perforated surface: area of each hole is,

$$\pi r^2 = 3.14 \times \left(\frac{0.15}{2}\right)^2$$

Now we can develop the impedance curve using a density of 0.0765 lb/ft³ leading to $\sigma = 1$. The impedance curve for this system at sea level is governed by the following relationship:

$$\Delta P = 0.1367\left(0.0765\dot{V}\right)^2 \text{ "H}_2\text{O}$$

$$\Delta P = 0.80\times 10^{-3}\left(\dot{V}\right)^2 \text{"H}_2\text{O}$$

At 7000 ft, the standard atmospheric conditions are as follows:

$$T = 493.73°\text{R}$$

$$P = 1.6331\times 10^3 \text{lb/ft}^2$$

$$\rho = 0.0620 \text{ lb/ft}^3$$

Now we can develop the impedance curve using a density of 0.062 lb/ft³. This leads to $\sigma = 0.81046$:

$$0.81046\Delta P = 0.1367\left(0.062\dot{V}\right)^2 \text{ "H}_2\text{O}$$

$$\Delta P = 0.648\times 10^{-3}\left(\dot{V}\right)^2 \text{ "H}_2\text{O}$$

Figure 5.8 depicts the impedance curves of this electronics system at sea level and 7000 ft.

It is noteworthy to consider that the number of heads lost for the PCBA section was set to 1. This resulted in a loss coefficient of 0.1367 and a maximum

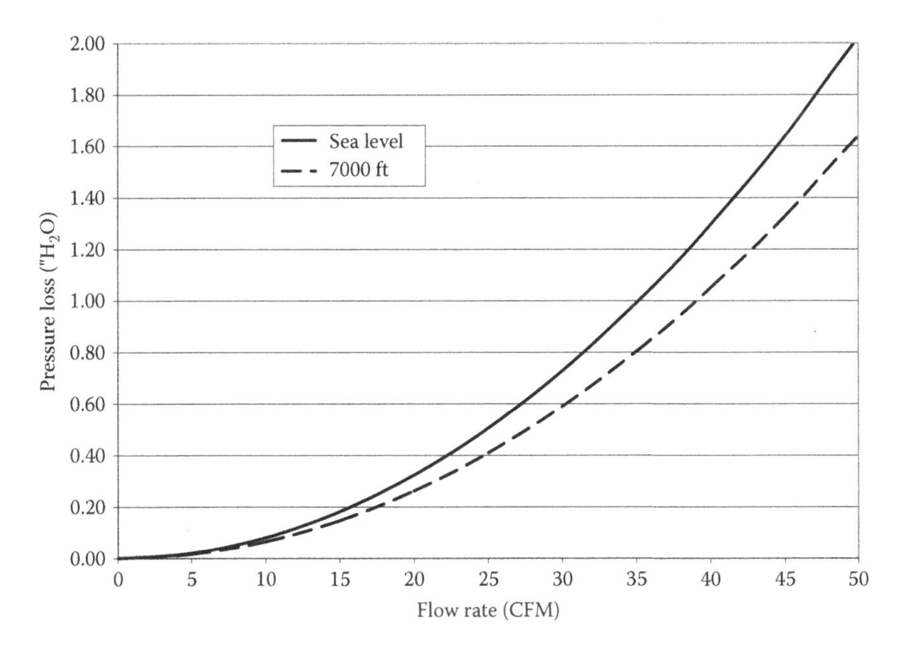

FIGURE 5.8
The system's impedance curves at sea level and 7000 ft.

sea level pressure equal to 2.00 in of water. Had we set that value to 2.5, the loss coefficient would have been 0.1382 and the maximum pressure of 2.02 in of water. It should be mentioned that the choice of head loss number for a PCBA depends on whether the components on the board are small and smooth, so as not to present a resistance to flow, or are large, bulky, and create a great deal of localized turbulence and resistance to flow. Clearly, the more resistance to flow, the higher the head loss number (Table 5.8).

Fan Selection and Fan Laws

From the point of view of this book, issues involved in fan selection center on finding the right fan that will deliver proper flow rate so that the temperature rise may be maintained below a critical value. However, there are other selection issues that need to be considered. These are

1. Noise, mechanical, and/or aerodynamic
2. Reliability
3. Power
4. Size and shape

A given fan can only deliver one flow rate for a given system impedance. Earlier, we developed a technique for evaluating the required flow rate for a given temperature rise (ΔT) as well as a technique for developing the impedance curve for a system. The task of fan selection is to find one with a performance curve such that it intersects with the impedance curve at the needed flow rate. Let us explain this by looking at the performance curve in more detail.

Fan Performance Curve

Previously, we developed a systematic approach for quantifying a system's resistance to airflow and hence heat removal. In the same vein, we can imagine that different fans behave differently when confronted with flow restrictions. Two extremes of flow restrictions are (1) when flow is completely blocked and (2) when there are no restrictions at all. If we were to measure the static pressure in front of the fan, in the first case we would measure a maximum value and in the second case we would measure near zero. If we were to devise a fixture that would allow a controlled change of flow, we would be able to measure and plot pressure as a function of flow rate. The resultant curve is called a "fan performance curve". Ideally, pressure is at maximum value when zero flow conditions exists, decreases monotonically as the flow increases, and goes to zero at the maximum flow. Figure 5.9 shows typical fan performance curves for propeller-style fans (vaneaxial) and backward inclined centrifugal fans

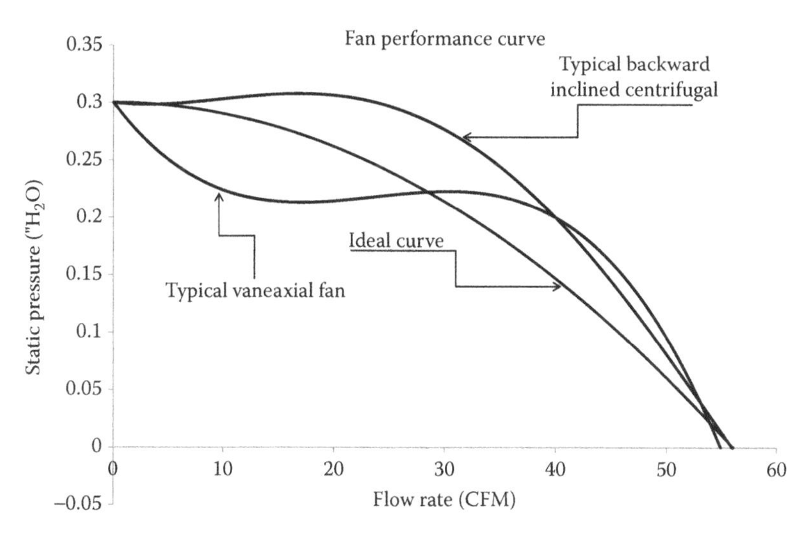

FIGURE 5.9
Typical and ideal fan performance curves.

(blowers). Notice that there is a sharp departure between an ideal curve and an actual one. The reasons for this departure are beyond the scope of this book.

So far, we have learned to calculate the required flow rate to keep the temperature rise at a prescribed value. We have also learned how to calculate a system's resistance to flow, and now we know about a fan's behavior. If we consider a complete system with the fan included, then the flow rate that passes through is at the cross point of the fan performance and impedance curves. This is shown in Figure 5.10 for three different fan curves. Notice that for any one fan-enclosure combination, there is only one operating condition and a corresponding flow rate. Therefore, we need to make sure that this operating flow rate is equal to the needed flow.

Example

Consider the system shown in Figure 5.7. Earlier, we calculated the required flow rate for a 10°C (18°F) temperature rise to be 31.7 CFM. Three fans' performance curves, as shown in Figure 5.10, intersect the system's impedance curve at sea level conditions (Figure 5.8). Fan A delivers 18 CFM, fan B delivers 26 CFM, and fan C delivers 32 CFM. Notice that if we were to use fan A, the temperature rise would be calculated as follows:

From Equation 5.13

$$\dot{V} = \frac{0.0524 Q T_\circ}{P_\circ \Delta T}$$

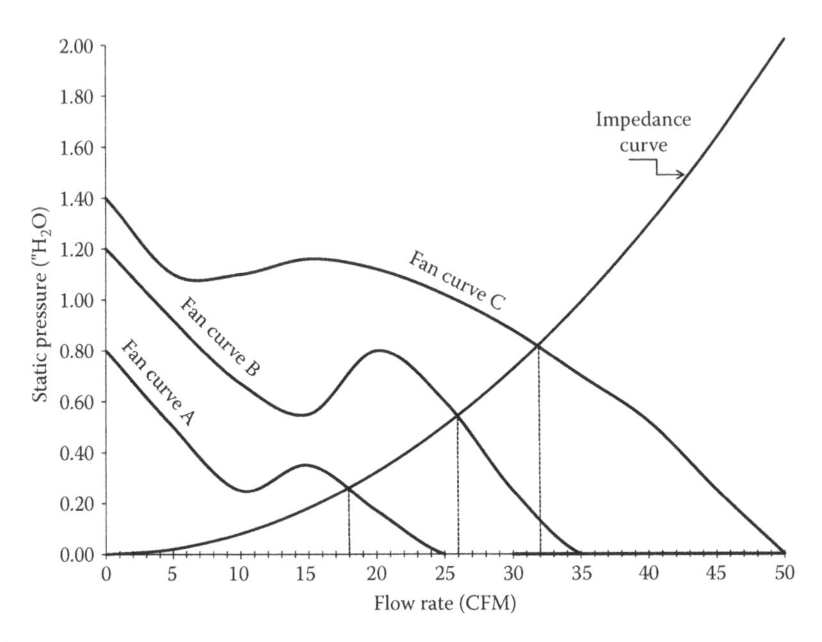

FIGURE 5.10
Intersection of system impedance curve and fan performance curves.

we obtain

$$\Delta T = \frac{0.0524 Q T_{\circ}}{\dot{V} P_{\circ}}$$

Now substitute the values and obtain

$$\Delta T = \frac{0.0524 \times 627.44 \times 519}{18 \times 29.9}$$

$$\Delta T = 31.7°\text{F or } 17.6°\text{C} \left(\text{for fan A}\right)$$

If we were to use fan B, our temperature rise would be 12.2°C (21.9°F). Clearly, neither of the two fans is acceptable. Notice that curve C delivers a 32 CFM flow rate to our system which is slightly better than the required flow rate.

Although in this scenario we assumed that performance curve C represented a third fan, it may be possible to use several fans either in parallel or series and thus develop a "suitable" fan curve. Although we just demonstrated that either A or B fans could not have been selected individually, they may be used in conjunction with one another to provide the needed flow rate. The concept of combining fans is similar to developing a fan network which is discussed next.

Fan Networks

Many systems require fans to be placed either in parallel or serial configurations. The impact of these configurations on the flow needs to be understood. In a parallel configuration, the flow rate increases but the maximum pressure drop remains the same (Figure 5.11). However, in a serial configuration, the pressure drop increases but the maximum flow rate remains the same (Figure 5.12). In practice, fan trays, that is, two or three fans in parallel, are used to increase the overall delivered flow rate to a system without increasing the size of a single fan. There is an added benefit as well. By having several variable speed fans in parallel, should one fail, others may be designed to increase speed to compensate for the failed unit. Fans in series are generally used to increase pressure, such as in a compressor.

More on Fan Performance Curves

Matching system impedance to a fan performance curve is only one aspect of fan selection. To choose an optimum fan, we need to ensure that the system operating point is within the fan's optimum range. To identify this optimum range, we need to first learn about the block-off and free flow points. In Figure 5.13, these points are identified as A and B, respectively.

Block-off is the condition when the fan is completely blocked, and while it is operating, no flow takes place (point A). Free flow is the opposite condition

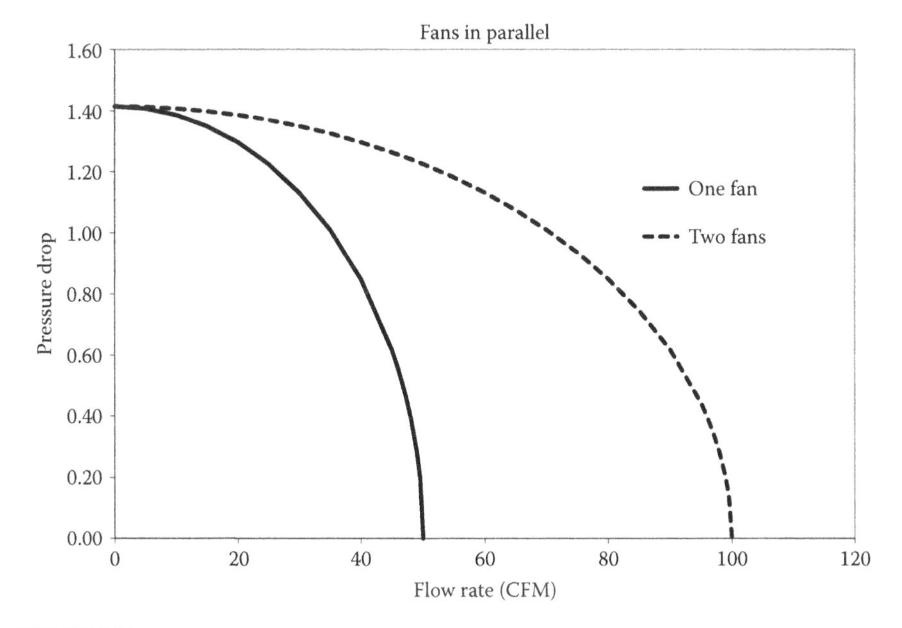

FIGURE 5.11
Two identical parallel fans.

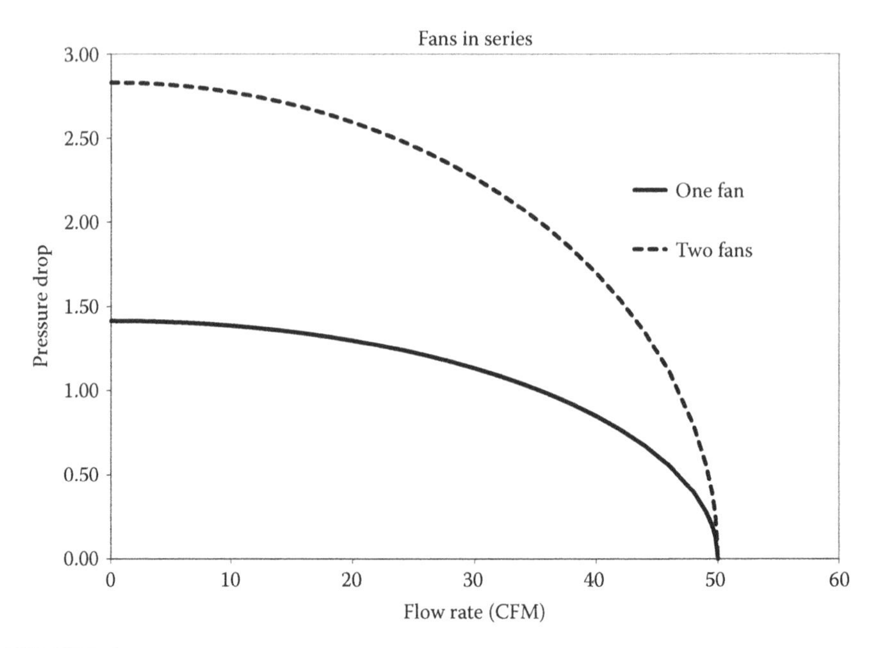

FIGURE 5.12
Two identical fans in series.

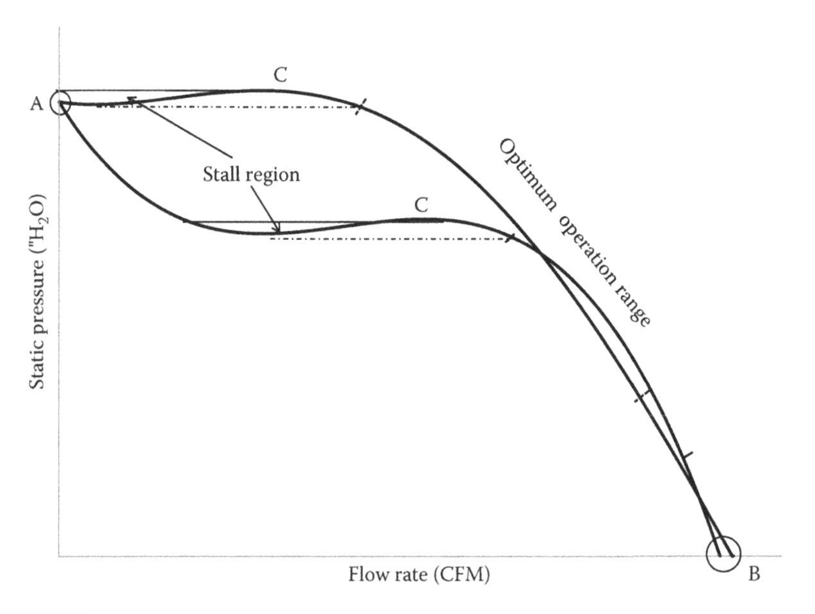

FIGURE 5.13
Various locations on a fan curve.

in which the fan is not blocked at all and air flows freely and, as a result, static pressure is zero (point B). While an ideal fan curve would depict a monotonic increase in pressure from a free-flow condition to block-off point, in reality, as the flow to the fan is restricted from a free-flow condition, static pressure would begin to rise to a maximum point C before the block-off point. If the flow is further restricted, the fan would stall and the pressure would begin to decrease with increasing restriction. Depending on the fan and its construction, this decrease would either continue to the block-off point, or a local minimum pressure may be reached, beyond which the pressure would begin to increase to a maximum at the block-off point.

If we were to divide the fan curve to two regions on either side of point C, the stall region falls on the left side of C and the optimum range is on the right side of C. It should be pointed out that unlike an airfoil where no lift is generated when it stalls, a fan still creates airflow in this region but its movement is erratic and a great deal of noise and vibration is generated.

Another consideration pointed out by Sloan (1985) is that the condition of flow prior to reaching the fan and immediately leaving it would impact the fan curve. If the flow profile is fully developed as in a duct, the performance is greatly enhanced. Thus, it is recommended that a duct be designed upstream of the fan, allowing the velocity profile to be fully developed before reaching the fan; however, placement of a duct downstream of the fan would reduce the performance because it would tend to create instability and irregular flow.

Fan Laws

Sometimes a fan evaluated for one set of conditions has to be used under a different set of conditions. Fan laws allow us to extrapolate current data to the new operating environment.

$$\dot{V}_A = \dot{V}_A \left(\frac{\mathrm{Size}_A}{\mathrm{Size}_B} \right)^3 \left(\frac{\mathrm{RPM}_A}{\mathrm{RPM}_B} \right) \tag{5.23}$$

$$\mathrm{Pressure}_A = \mathrm{Pressure}_A \left(\frac{\mathrm{Size}_A}{\mathrm{Size}_B} \right)^2 \left(\frac{\mathrm{RPM}_A}{\mathrm{RPM}_B} \right)^2 \left(\frac{\rho_A}{\rho_B} \right) \tag{5.24}$$

$$\mathrm{Power}_A = \mathrm{Power}_A \left(\frac{\mathrm{Size}_A}{\mathrm{Size}_B} \right)^5 \left(\frac{\mathrm{RPM}_A}{\mathrm{RPM}_B} \right)^3 \left(\frac{\rho_A}{\rho_B} \right) \tag{5.25}$$

In these equations, *Size* is fan diameter, *RPM* is the revolutions per minute, and *Power* is the required power to operate the fan. As an example, these equations indicate that all factors remaining the same, should we double the size of a fan, its flow rate would increase by a factor of 8, its pressure drop would increase by a factor of 4, and power consumption would increase by a factor of 32.

Component Hot Spot

The sole purpose of undertaking all the previous assumptions and calculations is to determine the temperature of the hottest spot in the system and the associated component. It will be shown later that the reliability of a component is degraded as its temperature increases. The two major contributors to this hot spot temperature are:

1. Heat accumulated by the coolant
2. Resistance of the film coefficient to carry heat away from the component

Based on these two factors, the hot spot temperature is:

$$T_{\text{spot}} = T_{\text{inlet}} + \Delta T_{\text{coolant}} + \Delta T_{\text{film}}$$

The design engineer usually specifies $\Delta T_{\text{coolant}}$, the maximum temperature gain of the coolant. At times, a particular standard or code prescribes this value. However, ΔT_{film} must be calculated. The basic convection heat transfer equation gives us the following relationship:

$$Q = h_c A \Delta T_{\text{film}}$$

In practice, this equation is elusive and at times difficult to solve because h_c varies depending on geometry, coolant type, and flow regime. Thus, its exact value may not be readily determined. It is in fact the subject of much discussion and research. However, for design purposes and in engineering practice, compromises and educated assumptions must be made.

Calculating h_c Using Colburn Factor

A stack of plug-in PCBAs may be assumed to behave very similarly to ducts. Therefore, the following formula may be applied (Steinberg 1991, K&K Associates 1999–2000):

$$h_c = JC_pG\left(\frac{C_p\mu}{K}\right)^{-(2/3)} \tag{5.26}$$

where:
$G = 60\left(\rho\dot{V}/A_{\text{total}}\right)$ is weight flow rate per unit area
$\dot{V}$ is flow rate (CFM)
J is Colburn factor
C_p is specific heat of fluid
μ is fluid viscosity
K is fluid thermal conductivity
ρ is fluid density
A_{total} is total flow cross-sectional area (ft^2)

Depending on the physical configuration and flow regime, the Colburn factor may be calculated as follows (Steinberg 1991, K&K Associates 1999–2000):

$$200 \le Re \le 2000 \text{ (Laminar Flow) Aspect Ratio} > 8,\ J = \frac{6}{Re^{0.98}},$$

$$200 \le Re \le 2000 \text{ (Laminar Flow) Aspect Ratio} \sim 1,\ J = \frac{2.7}{Re^{0.95}},$$

$$2000 \le Re \le 10{,}000 \text{ (Transition Flow)},\quad J = C_1 \left(\frac{\mu_b}{\mu_s}\right)^{0.14}$$

where:

μ_b is the viscosity (lb/ft h) evaluated at bulk fluid temperature
μ_s is viscosity evaluated at surface temperature
C_1 is evaluated empirically

K&K Associates (1999–2000) provides a graph for evaluating C_1 for a variety of tube and duct configurations. Appendix C provides a formula that has been used to fit a surface to these data using LABFit (Silva and Silva 2007).

$$10{,}000 \le Re \le 100{,}000 \text{ (Turbulent Flow)},\ J = \frac{0.023}{Re^{0.2}} \quad \text{where,}$$

$$Re = \frac{GD_h}{\mu} \text{ and,}$$

$$D_h = \frac{4 \times \text{Area}}{\text{Perimeter}} \text{ is the hydraulic diameter of the cross-sectional area}$$

Calculating h_c Using Flow Rate

The film temperature may be alternatively calculated from the following relationship (Sloan 1985):

$$h_c = \frac{3.867K}{b}\left[(1+a)^{1/3} - \frac{a}{3}\ln\left(\frac{a}{1+a}\right)\right]\frac{\text{W}}{\text{in}^2\,{}^\circ\text{F}} \tag{5.27}$$

where:

$a = 182.9\left(\frac{\rho\dot{V}}{YL}\right)D$
K is air conductivity W/in °F
ρ is air density (lb/ft^3)
$\dot{V}$ is flow rate (ft^3/min)
D is the air passage distance between the boards (in)
Y is the width of the board across the flow (in)
L is the length of the board along the flow (in)

Exercise

Recall the cabin telecommunication unit (CTU) in the previous exercise. It potentially has the following heat dissipation distribution:

1. PCBA 1—uP board, 26 W
2. PCBA 2—Memory board, 26 W
3. PCBA 3—Network board, 19.5 W
4. PCBA 4—Network board, 19.5 W
5. PCBA 5—Power Supply, 54 W
6. PCBA 6—E1 Network board, 19.5 W, quantity of two

Earlier, it was determined that the required flow rate to maintain a 10°C (18°F) inlet to exit temperature rise is 31.7 CFM. Furthermore, we learned that each board is 6.5 × 10 in and minimum clearance for flow between boards is 0.25 in. In this example, there are seven parallel boards but six ducts for air passage.

Now, we should determine the hot spot temperature. Assume the following conditions for air at 150°F:

$$C_p = 0.24 \frac{\text{BTU}}{\text{lb°F}}$$

$$\mu = 0.05 \frac{\text{lb}}{\text{ft h}}$$

$$K = 0.0164 \frac{\text{BTU}}{\text{h ft°F}}$$

and

$$\rho = 0.0765 \frac{\text{lb}}{\text{ft}^3}$$

Recall that

$$Q = h_c A \Delta T_{\text{film}}$$

Heat flow (Q) and surface area (A) are known. To calculate ΔT_{film}, we need to find h from Equation 5.26.

$$h_c = JC_pG\left(\frac{C_p\mu}{K}\right)^{-\frac{2}{3}}$$

For this equation, weight flow rate (G) and Colburn factor (J) must be calculated:

$$G = \frac{\rho\dot{V}}{\text{Total flow cross-sectional area}}$$

$$G = \frac{0.0765 \times 31.7}{\frac{0.25 \times 10}{144} \times 6}$$

$$G = 23.28 \frac{\text{lb}}{\text{min ft}^2}$$

The timescale must be changed to hours:

$$G = 23.28 \frac{\text{lb}}{\text{min ft}^2} \times \frac{60\ \text{min}}{\text{h}}$$

$$G = 1396.83 \frac{\text{lb}}{\text{h ft}^2}$$

The choice of the equation for the Colburn factor (J) depends on the value of the Reynolds number:

$$Re = \frac{GD_h}{\mu}$$

$$D_h = \frac{4 \times \text{Area}}{\text{Perimeter}}$$

$$D_h = \frac{4 \times \left(\frac{0.25 \times 10}{144}\right)}{2 \times \left(\frac{10 + 0.25}{12}\right)} \Rightarrow D_h = 0.0406\,\text{ft}$$

$$Re = \frac{1396.83 \times 0.0406}{0.05} \Rightarrow Re = 1134.20$$

The aspect ratio = 10/0.25 = 40; thus based on the values of Re and aspect ratio, the following relationship for the Colburn factor holds:

$$J = \frac{6}{Re^{0.98}}$$

$$J = \frac{6}{(1134.20)^{0.98}} = 0.00609$$

Now the heat transfer coefficient may be calculated:

$$h = 0.00609 \times 0.24 \times 1396.83 \times \left(\frac{0.24 \times 0.05}{0.0164}\right)^{-(2/3)}$$

$$h = 2.51 \frac{\text{BTU}}{\text{hr ft}^2\,{}^\circ\text{F}}$$

The highest heat dissipation of a board (PCBA 5) is 54 W (= 54 × 3.41 = 184.14 BTU/h). It may be assumed that heat is dissipated on both side of this PCBA. Thus we have

$$\Delta T_{\substack{\text{film}\\ \text{PCBA5}}} = \frac{184.14}{2\times\dfrac{6.5\times 10}{144}\times 2.51} = 81.26°\text{F}$$

$$T_{\text{component}} = 59 + 18 + 81.26 = 158.26\,°\text{F or } (70.1°\text{C})$$

The two E1 network PCBAs (19.5 W) have been placed on either side of the enclosure, so heat dissipation is done via one surface only:

$$\Delta T_{\substack{\text{film}\\ \text{PCBA6}}} = \frac{66.495}{\dfrac{6.5\times 10}{144}\times 2.51} = 58.69°\text{F}$$

$$T_{\text{component}} = 59 + 18 + 58.69 = 135.69\,°\text{F or } (57.6°\text{C})$$

In this exercise, we assumed the gap to be 0.25 in wide. Furthermore, we used the heat transfer coefficient model based on the Colburn factor. The implicit assumption here was that the channel is wide enough so that the heat from the walls does not impact the fluid bulk temperature. How would the solution (and hence our decision to design) be affected if we used a different model? Appendix E entertains these questions and provides a comparison of these different models and their implications.

Indirect Flow System Design

In the previous segment, steps needed to analyze and design a cooling system for electronics equipment were outlined and explained. The basis of that approach is that the cooling medium, namely air, is in direct contact with the components. Hence, the term "direct" flow system was used. Electronics equipment also takes advantage of heat exchangers and cold plates. Since these devices separate the electronics components from the coolant, they may be termed as "indirect" flow systems. Although from a design point of view, direct and indirect flow systems with their unique design issues are essentially different, from a physical scientific point of view, they are very similar. They both involve understanding pressure losses, fan/pump specifications, and mainly maintaining the critical component temperature below a design value. Depending on the heat dissipation levels, the indirect flow systems may employ fluids such as water, oils, and in the case of some aircrafts, even fuel. Therefore, design considerations should include not only choking but also temperature levels below the flash point of the fluids used.

The steps of designing a cold plate are very similar to the method outlined earlier and the reader is encouraged to consult other works (such as Sloan 1985). The relationship for flow rate remains the same as direct flow; however, one must make sure to use appropriate values of density and specific heat:

$$\dot{V} = \frac{Q}{\rho C_p \Delta T}$$

As before, Q is the dissipated heat, C_p is the specific heat of the fluid, and ΔT is the temperature rise of the fluid from inlet to exit. Similarly, the relationship between the pressure drop and flow rate across a pipe of length L and an arbitrary cross section is given by Equation 5.15. For more details on pipe flow, see Baumeister et al. (1979).

Resistance Network Representation

In the foregoing discussion, only pure convection was considered. In many practical problems, conduction and radiation may exist and their contribution must be considered. Earlier, we studied how to develop thermal resistance networks for conduction problems. The same concept can be extended for convection as well. As before:

$$\Delta T = QR, \text{ where } R = \frac{1}{hA}$$

Note that the difference with conduction thermal resistance is that K/A—the ratio of thermal conductivity to thermal path length—is replaced with h—heat transfer coefficient. In Chapter 6, we discuss this concept in more detail.

6

Combined Modes, Transient Heat Transfer, and Advanced Materials

Introduction

With the exception of the Electronic Cabinet example that we solved in Chapter 5, we have not considered how to combine the three modes of heat transfer to solve realistic problems. Furthermore, we have not discussed how time plays a role or the means by which we can calculate its impact.

In order to address this and other issues that might arise in the course of a thermal analysis, we will attempt to solve the problem depicted in Figure 6.1 and discuss various relevant points.

The system shown in Figure 6.1 is an automotive control module and is composed of a ceramic printed circuit board (PCB), a thin thermal pad, and the aluminum heat sink. The module is mounted in an airflow channel which is 0.25 in wide. The majority of the heat generated in the system is concentrated in the two components shown in detail B, which are capable of producing up to 2.5 W each. We need to identify the airflow rate needed to maintain the component junction temperature below its rated value. Consider the ambient temperature to be 75°F.

If we were to delve into solving this problem, we would begin by developing its resistance network as shown in Figure 6.2 for each rectifier. Properties needed to develop the resistance values are given in Table 6.1.

Notice that in Table 6.1 a component width and length is given for the PCB but not for the thermal pad nor for the heat sink. The reason is that we know the footprint of the component on the PCB, but as the heat spreads, we do not readily know the spread area on each layer. These areas will be calculated in the next step. Using Equation 3.2 and the values given in Table 6.1, we can form Table 6.2. It should be noted that the final area on one layer is assumed to be the initial area of the adjacent layer.

By using the average areas shown in Table 6.2, and the thermal conductivities and thermal lengths shown in Table 6.1, we can calculate the conduction as shown in Table 6.3 and reduce the overall resistance network as shown in Figure 6.3.

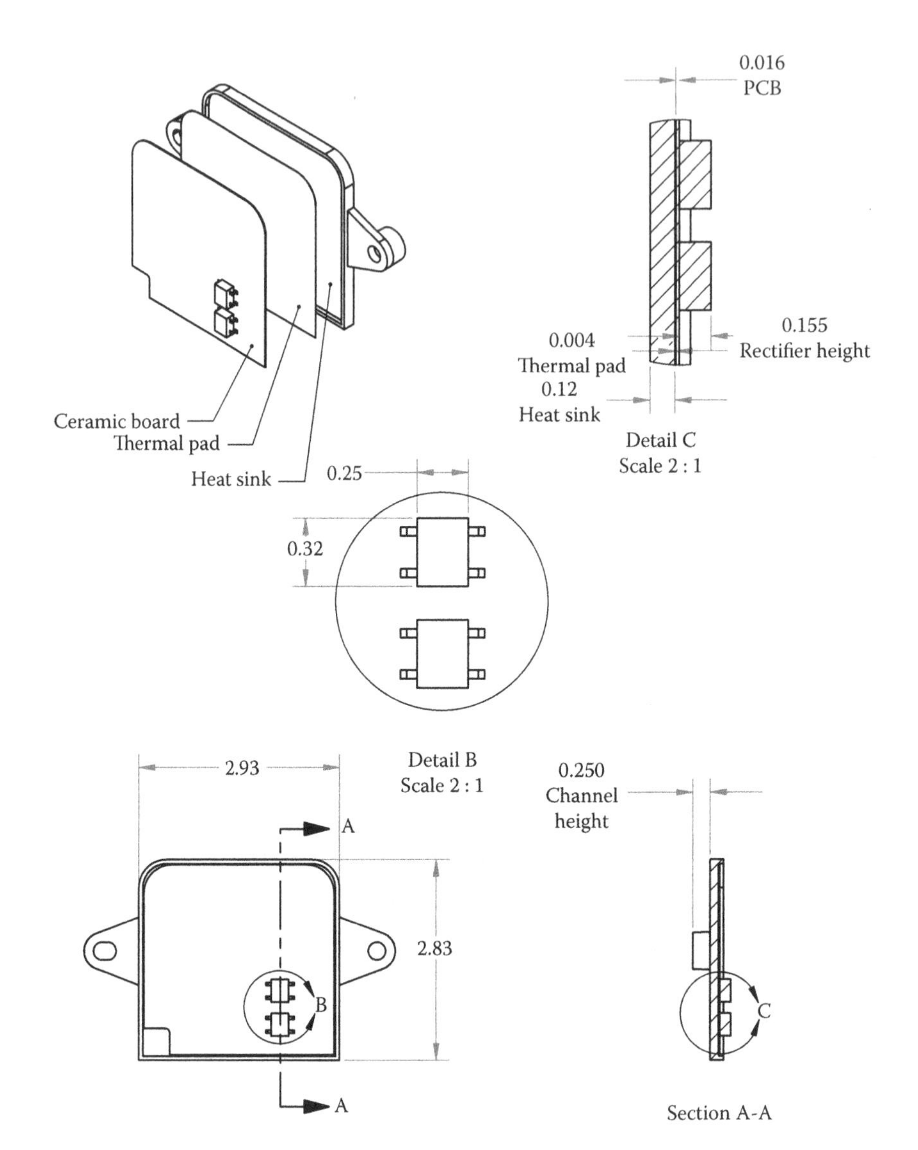

FIGURE 6.1
An engineering drawing of the electronic module.

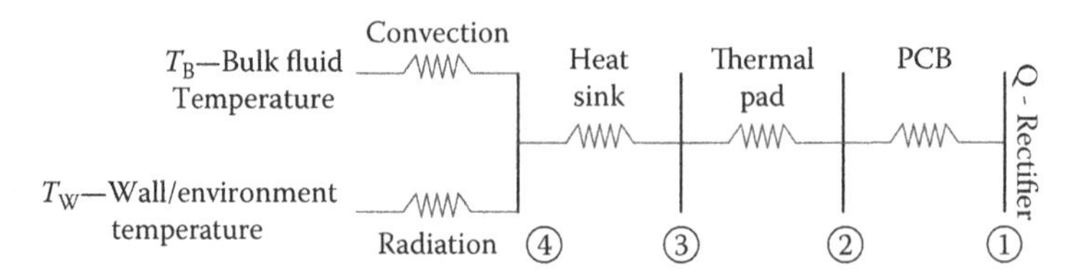

FIGURE 6.2
Thermal resistance network for the control module.

TABLE 6.1

Properties Needed to Develop Conduction Resistance Network

Component	Thermal Conductivity (BTU/hr ft °F)	Thickness (ft)	Component Width (ft)	Component Length (ft)
PCB (alumina)	15.0	1.33×10^{-3}	2.08×10^{-2}	2.67×10^{-2}
Thermal pad	7.1	3.33×10^{-4}	—	—
Heat sink (AL)	92.0	1.00×10^{-2}	—	—

TABLE 6.2

Calculated Properties for Conduction Resistance Network

Component	Component Width (ft)	Component Length (ft)	Initial Area (ft²)	Spread Angle	Spread Length (ft)	Final Area (ft²)	Average Area (ft²)
PCB (alumina)	2.08×10^{-2}	2.67×10^{-2}	5.56×10^{-4}	21.55	5.27×10^{-4}	6.07×10^{-4}	5.81×10^{-4}
Thermal pad (graphite)	2.19×10^{-2}	2.77×10^{-2}	6.07×10^{-4}	14.41	8.56×10^{-5}	6.15×10^{-4}	6.11×10^{-4}
Heat sink (aluminum)	2.21×10^{-2}	2.79×10^{-2}	6.15×10^{-4}	52.16	1.29×10^{-2}	2.56×10^{-3}	1.59×10^{-3}

Note: While only two decimal points are shown here, calculations have been carried out without truncating the results. Computations conducted with only two decimal points would lead to gross inaccuracies.

TABLE 6.3

Calculated Conduction Resistance Values

Component	Resistance
PCB (alumina)	0.1527
Thermal pad (graphite)	0.0771
Heat sink (aluminum)	0.0684
Total	0.2982

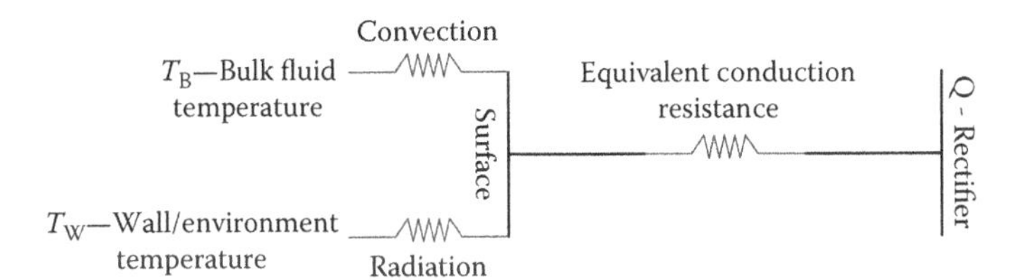

FIGURE 6.3
Reduced thermal resistance network.

TABLE 6.4

Calculated Properties for Conduction Resistance Network Based on Modified Heat Sink Area

Component	Component Width (ft)	Component Length (ft)	Initial Area (ft²)	Spread Angle	Spread Length (ft)	Final Area (ft²)	Average Area (ft²)
PCB (alumina)	2.08×10^{-2}	2.67×10^{-2}	5.56×10^{-4}	21.55	5.27×10^{-4}	6.07×10^{-4}	5.81×10^{-4}
Thermal pad (graphite)	2.19×10^{-2}	2.77×10^{-2}	6.07×10^{-4}	14.41	8.56×10^{-5}	6.15×10^{-4}	6.11×10^{-4}
Heat sink (aluminum)	2.21×10^{-2}	2.79×10^{-2}	6.15×10^{-4}	—	—	5.76×10^{-2}	2.91×10^{-2}

TABLE 6.5

Calculated Conduction Resistance Values Based on Modified Heat Sink Area

Component	Resistance
PCB (alumina)	0.1527
Thermal pad (graphite)	0.0771
Heat sink (aluminum)	0.0037
Total conduction resistance	0.2335

Table 6.2 provides the final heat spread area for the heat sink which is 2.56×10^{-3} ft². However, the discussion in Appendix D points to the fact that in the presence of convection, the spread area may be much larger. In fact, it may be a good assumption to use the entire surface of the heat sink. Thus Tables 6.2 and 6.3 are modified as shown in Tables 6.4 and 6.5.

Now we need to consider the convection and radiation links. Previously, we identified the governing equations for forced convection as well as radiation. These are

$$Q_{\text{Convection}} = hA\left(T_{\text{Surface}} - T_{\text{Bulk}}\right)$$

and

$$Q_{\text{Radiation}} = F\sigma\varepsilon A\left(T_{\text{Surface}}^{\ 4} - T_{\text{Ambient}}^{\ 4}\right)$$

Having developed the expression for convection and radiation modes of heat transfer, we can now write the heat balance equation:

$$Q_{\text{Rectifier 1}} + Q_{\text{Rectifier 2}} = Q_{\text{Convection}} + Q_{\text{Radiation}}$$

Recall that each rectifier dissipates up to 2.5 watts of heat

$$(2.5\times 3.41+2.5\times 3.41)\frac{\text{BTU}}{\text{hr}}=Q_{\text{Convection}}+Q_{\text{Radiation}}$$

Before we go any further, let us consider the impact of radiation. Let us anticipate that the surface temperature will not exceed 125°F. We will verify this assumption later. Furthermore, surface emissivity would be about 0.1 if the heat sink has a dull surface. Thus, the quantity of heat transferred via radiation is

$$Q_{\text{Radiation}}=1\times 1.713\times 10^{-9}\times 0.1\times 0.0576\left(585^4-535^4\right)$$

$$Q_{\text{Radiation}}=0.347\frac{\text{BTU}}{\text{hr}}$$

The total dissipated heat is 102.3 BTU/hr, but the contribution of radiation is nearly 0.3%. As a result, for this problem, we can focus on convection alone.

Now we can continue with the heat balance equation:

$$Q_{\text{Convection}}=hA\left(T_{\text{Surface}}-T_{\text{Bulk}}\right)=102.3\frac{\text{BTU}}{\text{hr}}$$

To calculate h, we would employ the Colburn approach and evaluate the data based on an air temperature of 100°F. The solution was obtained from developing a spreadsheet and the results are provided in Table 6.6. For a sample calculation refer to the exercise in Chapter 5.

In this example the value of the Reynolds number (= 8334.82) indicates that the fluid has surpassed the laminar regime and is in transition range going into turbulent flow but is not quite turbulent yet. As a result, to calculate the Colburn factor, the empirical formula in Appendix C was used. Based on the data provided in Table 6.6, the heat transfer coefficient is

TABLE 6.6

Air Properties at 100°F and Other Parameters Used in Calculations

ρ Lb/ft³	C_p BTU/lb F	K_f BTU/hr ft F	μ lb/(ft s)
0.071	0.24	0.0154	4.63E-02

Channel (Gap) Height (ft)	Channel Depth (ft)	Channel Width (ft)	CS Area Flow (ft²)	Flow (CFM)
0.0210	0.2442	0.2358	5.13×10^{-3}	12

Weigh Flow Rate	Hydraulic Diameter (ft)	Reynolds No.	Aspect Ratio	Colburn Factor	C1
9969.77	3.87×10^{-2}	8334.82	11.63	0.00448	0.0045

$$h = 13.35 \frac{\text{BTU}}{\text{hr ft}^2\ ^\circ\text{F}}$$

and

$$T_{\text{Surface}} = 97.18^\circ\text{F}$$

Using the surface temperature and the total conduction resistance, we can calculate each rectifier's temperature:

$$T = T_{\text{Surface}} + QR_{\text{Equiv.}}$$

$$T = 97.18 + (2.5 \times 3.41) \times .2335$$

$$T = 99.17^\circ\text{F}$$

Notice that to calculate the temperature at each rectifier, we need to use only the heat generated by that chip alone. Also, had we taken radiation into account, calculated surface temperature would be 97.00°F and component temperature would be 98.99°F.

Total System Resistance

So far, we have calculated the total conduction resistance of this system—as provided in Table 6.5. Chapter 5 suggested that thermal resistance caused by convection is easily calculated from the following formula:

$$R_{\text{Convection}} = \frac{1}{hA}$$

Furthermore, the equivalent thermal resistance between the rectifier and the environment is (Figure 6.4)

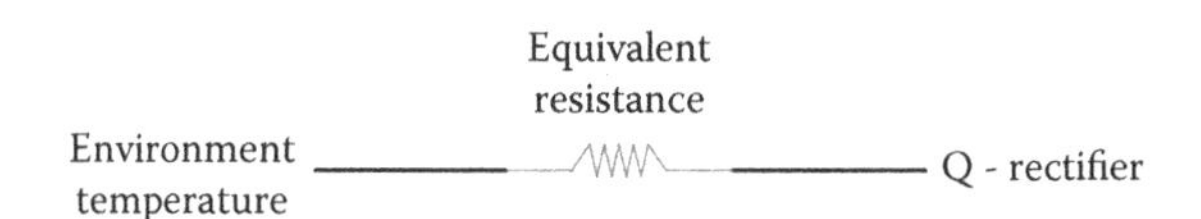

FIGURE 6.4
Total thermal resistance between rectifier and the environment.

$$R_{\text{Equivalent}} = R_{\substack{\text{Total}\\ \text{Conduction}}} + R_{\text{Convection}}$$

$$R_{\text{Equivalent}} = 0.2335 + \frac{1}{13.35 \times 0.0575}$$

$$R_{\text{Equivalent}} = 1.5340\left[\frac{\text{BTU}}{\text{hr°F}}\right]^{-1}$$

Time-Dependent Temperature Variation

To develop a better understanding of the nature of time-dependent problems, let us look at the general heat equation one more time. This is a portion of the governing equations presented in Equation 2.4.

$$\left(\rho C_p T\right)_{,t} + \rho C_p u_k T_{,k} = -p u_{k,k} + (K T_{,j})_{,j} + \lambda\left(u_{k,k}\right)^2 + \mu\left(u_{i,j} + u_{j,i}\right)u_{j,i} + \dot{Q}$$

By removing the velocity terms and assuming a one-dimensional heat flow, we can develop a simpler equation

$$\rho C_p \frac{dT}{dt} = K\frac{d^2T}{dx^2} + \dot{Q}$$

There is a closed form solution to this equation; however, an explanation of the solution procedure is beyond the scope of this book. Briefly, it is based on separating the time domain from the spatial domain. Then, the steady-state solution may be calculated and then combined with the time domain solution. A further simplification is that the temperature distribution is uniform in the spatial domain. Thus, we can concern ourselves solely with the time domain. On the heating cycle, the solution is

$$\left(T_{S-h} - T_0\right) = \left(T_F - T_0\right)\left[1 - \exp\left(-\frac{t}{RC}\right)\right]$$

and on the cooling cycle, the solution becomes

$$\left(T_{S-c} - T_0\right) = \left(T_{S-h} - T_0\right)\left[\exp\left(-\frac{t}{RC}\right)\right]$$

In this equation, subscript 0 refers to the initial condition and subscript F refers to the final or steady-state conditions, subscript S–h refers to the instantaneous temperature of the component in the heating cycle, and subscript S–c refers to the instantaneous temperature of the component in the cooling cycle. The variable t is time; R is the equivalent thermal resistance; and C, thermal capacitance, is defined as $C = W\,C_p$, where W is the mass of the heat sink in pounds or grams, and C_p is its specific heat. One may note that the product of thermal resistance and thermal capacitance (RC) has the units of time and, as a result this product is often referred to a system's thermal time constant. This product may be used to compare the performance of various heat sinks.

Temperature Rise of an Electronics Module

To calculate the time constant of the heat sinks used in the electronics module, we have already calculated its equivalent thermal resistance. The heat sink has a mass of 0.1239 lb and is aluminum (C_p=.22 BTU/[lb °F]). With an ambient temperature of 75°F, the time-dependent rectifier temperature equation is

$$RC = R_{\text{Equivalent}}\,W\,C_p = 1.534 \times 0.1239 \times 0.22 = 0.0418$$

$$\left(T_{S\text{-}h} - 75\right) = \left(99.17 - 75\right)\left[1 - \exp\left(-\frac{t}{0.0418}\right)\right]$$

$$T_{S\text{-}h} = 75 + \left(99.17 - 75\right)\left[1 - \exp\left(-\frac{t}{0.0418}\right)\right]$$

The solution to this equation is provided graphically in Figure 6.5 indicating that the rectifier reaches its steady-state temperature within 15 min (0.25 hr).

Now presume that the module is operated for 10 min and then shut down. We need to determine the temperature profile for the time needed for the system to cool down to room temperature.

After 10 min of operation the temperature of the heat sink is

$$T_{S\text{-}h} = 75 + \left(99.17 - 75\right)\left[1 - \exp\left(-\frac{\dfrac{10\,\text{min}}{60\,\text{min}/\text{hr}}}{0.0418}\right)\right]$$

$$T_{S\text{-}h} = 98.73\,°\text{F}$$

at this time the module is shut down, so the following equation applies to the cooling cycle:

$$T_{S\text{-}c} = 75 + \left(98.73 - 75\right)\left[\exp\left(-\frac{t}{0.0418}\right)\right]$$

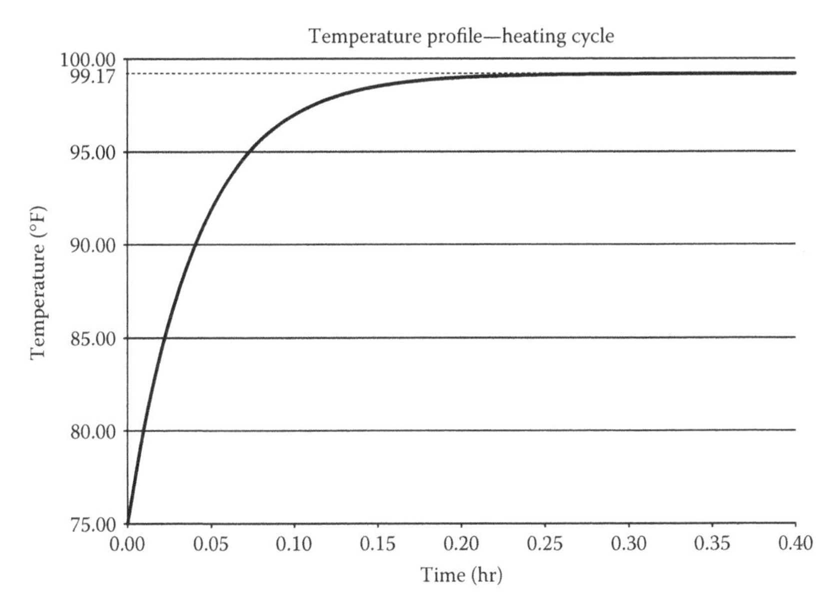

FIGURE 6.5
Time dependence of rectifier temperature.

Note that in this equation, t is not the accumulated time but the time from the start of cooling. Figure 6.6 shows the heating and cooling temperature profiles. It takes nearly 15 min for the system to cool down to room temperature.

Advanced Materials and Technologies

As the functional density of many electronics packages increases, so does the dissipated heat densities. Considering that this increase in functionality is generally accompanied with a reduction of the physical size of the package, the traditional means of removing the generated heat may no longer be sufficient for a reliable operation of the product. For the sake of completeness, in this segment, we will review a few of the more advanced materials and technologies used in thermal management of these denser electronics packages.

Solid-State Air/Liquid Pump Cooling

In some advanced technologies, the coolant is delivered directly to the chip via microfluidic channels. In particular, the pumping mechanism is conducted via a valve-less diaphragm micropump for forced convection within

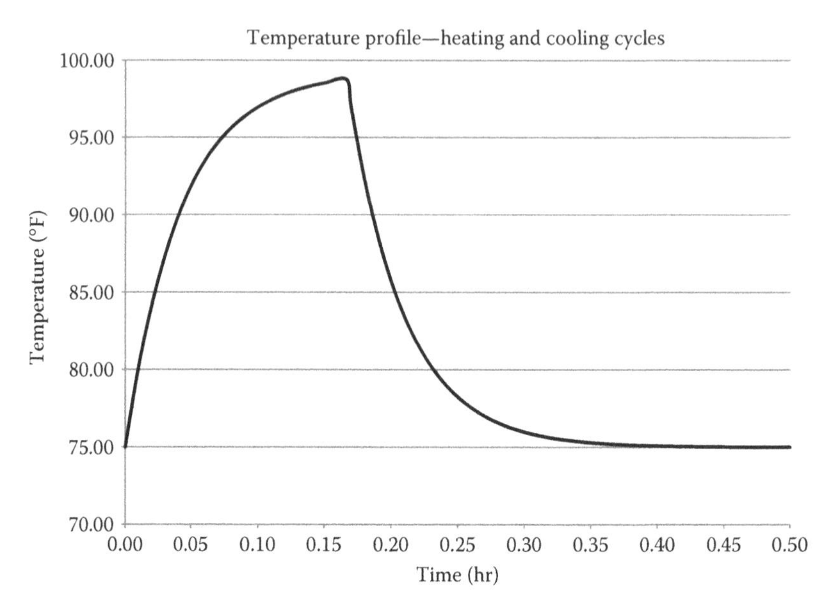

FIGURE 6.6
Temperature profile of the rectifier in heating and cooling cycles.

a small volume (Staley et al. 2008; Tay and Li 1997). Typically, these pumps consist of a solid-state diffuser/nozzle assembly driven by piezoelectric crystals. This technology is able to deliver relatively high coolant flow rates directly to the chip. Another example of this type of technology is the so-called "synthetic jet" which is based on period suction and ejection of fluid out of an orifice (Mahalingam et al. 2007) again using piezoelectric crystals.

Because this new pumping technology has no moving parts, it is generally held that they are more reliable and have a longer life (by nearly a factor of 6) than traditional fans or pumps. (Nuventix 2007). Their disadvantage is that each valve-less pump needs to be custom-made for their specific application.

Graphite Foam

The use of an open-cell graphite foam is to employ a more efficient "heat sink" material in the thermal management design. Traditional heat sinks are blocks of aluminum or other types of metals with fins, either machined or otherwise attached to them. Then, air or other fluids are forced over these fins in order to facilitate the removal of heat. As it was indicated in Chapter 5, there is a direct relationship between the level of heat transfer and the surface area in contact with the fluid. An open-cell foam material provides the largest possible surface area. Gallego and Klett (2003) have studied this material in some detail and have reported that carbon foams are more efficient than typical heat sinks; a 53% higher efficiency with power densities up to 100 W/cm^2

at temperatures less than 100°C. A potential drawback of this technology is that added protection against corrosion should be designed in the system.

Metal Matrix and Composite Materials

A number of advanced materials have been developed on the basis of combining two or more different but compatible materials in order to achieve higher performance. While, the concept of developing composite materials is not new, effort is being made to tailor these activities to electronics thermal management. The first advantage of using composite materials is that the electronic component may be directly soldered to the heat sink without the need of an interface material. Another advantage is that it is possible to match the coefficients of thermal expansion between the two materials. Further advantages include having extremely high thermal conductivities, ability to develop designs suitable for the used environment, as well as reductions in size and weight. For a review of these class of advanced materials, see Zweben (2001, 2010, 2012).

Graphene and Other Interface Materials

Another area of focus for developing novel materials is at the interface of heat producing component and the substrate conducting heat away. In Chapter 3, the concept of heat spreading was discussed. Traditional materials typically exhibit spread angles between 20° to 45°. By taking advantage orthotropic material behaviors, it is possible to increase the spread angle to approach values as high as 90°. Materials such as diamond and graphene exhibit such properties. For a review of these class of interface materials, see Zweben (retrieved on December 29, 2014) and Yan (2013).

7

Basics of Vibration and Its Isolation

Introduction

So far, we have been concerned with the thermal performance of electronics enclosures. There are other issues in their design that can be just as (or even more) important. Mechanical and electromagnetic issues are two major factors that must be managed. In general, once the enclosure is designed and the prototypes are developed, a few samples are tested in environmental chambers as well as vibration (shake) tables for compliance with various standards. A few may also be tested for drop and/or impact tests. Designs are only modified to pass the given test criteria. As a result, the designers, in general, know nothing about the behavior of their system in the field and any relationship that their data have to either failure rates or repair/maintenance scheduling.

In this chapter, we review one of the factors involved in the mechanical analysis of electronics packaging, namely, vibration management and possible approaches to its isolation. The term "management" is used because most of the time—and at least for new designs—we as designers must anticipate the impact of vibration by selecting proper components without full knowledge of the system's response. Once the design is complete and prototypes are made, various flaws may be identified through testing.

The study of vibration is concerned with the oscillatory motions of bodies and the forces associated with them. All bodies possessing mass and elasticity are capable of vibration. Sources of vibration may be categorized as follows:

- In stationary systems: unbalanced loads
- In road vehicles: rough surfaces of the roads
- In sea vehicles: fluid/structure interaction
- In air vehicles: aerodynamic loads

Vibration can be broadly characterized as linear or nonlinear, free or forced. In free vibration, the system will vibrate at one or more of its natural frequencies, which are properties of the dynamic system established by its mass and stiffness distribution. In forced vibration, if the excitation is oscillatory, the system is forced to vibrate at the excitation frequency. If this frequency coincides with one of the natural frequencies of the system, resonance occurs, which may potentially lead to large oscillations causing catastrophic failures. It is noteworthy to consider that all vibrating systems exhibit damping because friction and other factors dissipate energy.

Periodic and Harmonic Motions

When an oscillatory motion "repeats" itself regularly in equal intervals of time as shown in Figure 7.1a and b, it is called a "periodic" motion. Each interval of time denoted as τ is called the period of the oscillation and its reciprocal $f = 1/\tau$ is called the frequency. Customarily τ is measured in seconds and f is given in cycles per second. The simplest form of periodic motion is harmonic motion (Figure 7.1a) and is depicted as

$$x = A \sin\left(2\pi \frac{t}{\tau}\right)$$

where A is amplitude of oscillation.

To develop a simpler mathematical form, let us adopt the following relationship:

$$\omega = \frac{2\pi}{\tau} = 2\pi f$$

Circular frequency, ω, is generally measured in radians per second. Now consider the mathematical expression for velocity and acceleration of a point whose displacement is governed by x:

$$\text{Displacement: } x = A \sin \omega t$$

$$\text{Velocity: } \frac{dx}{dt} = x_{,t} = \omega A \cos \omega t$$

$$\text{Acceleration: } \frac{d^2x}{dt^2} = x_{,tt} = -\omega^2 A \sin \omega t$$

Thus, by combining these equations, the differential equation describing the motion of a single degree-of-freedom system (1 DOF) may be obtained:

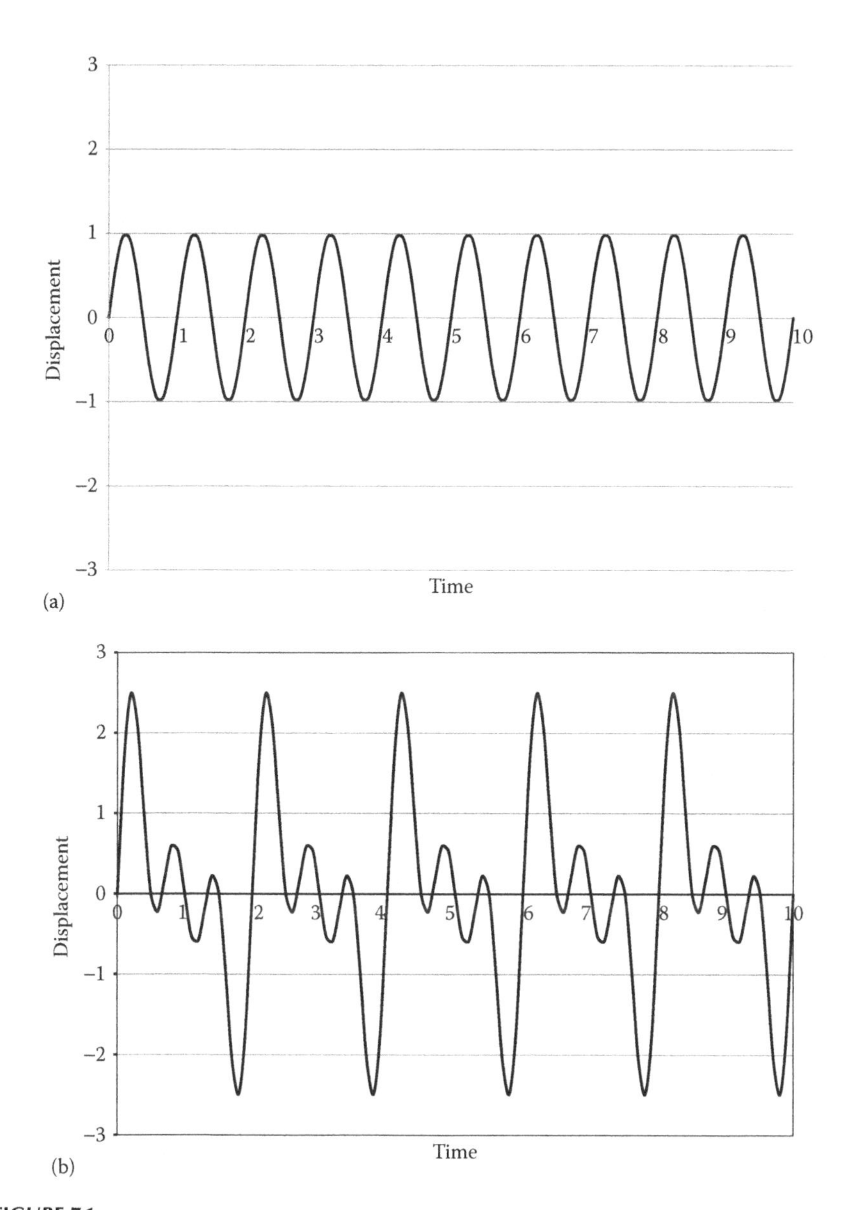

FIGURE 7.1
Time response of (a) typical harmonic and (b) period motions.

$$\omega^2 x + x_{,tt} = 0 \tag{7.1}$$

While the basis of a simple harmonic motion describes the motion of one point with only one frequency, it is conceivable to imagine a series of connected points whose movements are interconnected. Therefore, to describe the motion of one point, contributions from other points need to be considered as well. As a result, we develop a complex motion that is nevertheless periodic as shown in Figure 7.1b. An example of this periodic motion and the interdependence of various points is that of a guitar string, where there are theoretically an infinite number of natural frequencies corresponding to the infinite number of points on the string. When the string is struck, the vibration at each natural frequency contributes to the overall free vibration of the string.

Mathematically, periodic motion is described as

$$x = A\sin\omega_1 t + B\sin\omega_2 t + C\sin\omega_3 t + \cdots$$

The mathematical approach to solving more complex periodic motions is to understand the contribution of each frequency and the overall mode shape (*A*, *B*, *C*, etc.) to the overall displacement. Once this task is completed, it is quite possible to assemble the contributing frequencies and mode shapes to describe the overall behavior of a vibrating system. A full discussion of this subject, however, is beyond the scope of this book.

Free Vibration

Oftentimes, free vibration is demonstrated through the use of a spring–mass system as shown in Figure 7.2, with no other loads applied except gravity. Though this constitutes a very simple analogy, it leads to modeling simplifications for a large class of vibration problems.

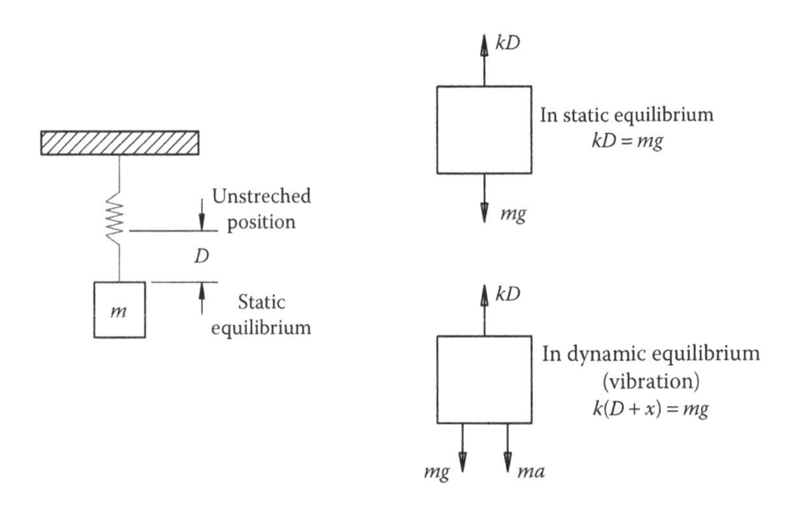

FIGURE 7.2
A single degree of freedom system shown in static and dynamic equilibrium.

For dynamic equilibrium, forces may be balanced as follows:

$$ma = mg - k(D + x)$$

From static equilibrium, we have $mg = kD$. Also a (acceleration) may be written as $x_{,tt}$. Thus we can write

$$mx_{,tt} = -kx$$

Now we define

$$\omega^2 = \frac{k}{m}$$

and obtain the following equation:

$$x_{,tt} + \omega^2 k = 0 \tag{7.2}$$

This is the same as Equation 7.1. Therefore, the solution to this equation (i.e., motion or displacement) is harmonic. The natural period of the oscillation and frequency are

$$\tau = 2\pi\sqrt{\frac{m}{k}} \text{ and } = \frac{1}{2\pi}\sqrt{\frac{k}{m}}$$

The expression for the frequency f may be written in a different way. Recall that $mg = kD$; thus,

$$\frac{k}{m} = \frac{g}{D}$$

Or

$$f = \frac{1}{2\pi}\sqrt{\frac{g}{D}}$$

In other words, the natural frequency of a spring–mass system is a function of applied load (g) and its static deflection (D).

First Application

Considering that any physical system deflects because of an applied load, theoretically, it is possible to develop a force-deflection relationship for such a system under such applied loads. In fact, in the field of strength of materials, many solutions to such problems as beam and/or plate deflections have been developed (Baumeister et al. 1979). If we were to limit our interest to only one location in the system and the corresponding applied load, we could convert a complex system into a simple spring–mass system and calculate its natural frequency. This may be referred to as the 1 DOF approximation. Later, it will be shown how we can employ this calculated frequency and applied dynamic loads to approximate maximum vibration amplitude at the same location.

As an example of this technique, let us calculate the natural frequency of a 10 lb block mounted at the free end of a cantilever beam as shown in Figure 7.3. Neglect the mass of the beam.

Based on strength-of-material assumptions, the expression for the maximum deflection of the free end of a cantilever beam due to a concentrated load at that end is

$$\delta = \frac{PL^3}{3EI}$$

where:

P is the load at the end of the beam
L is beam length
E is the Young's modulus of elasticity
I is the area moment of inertia

By rewriting this equation, we obtain

$$\left(\frac{3EI}{L^3}\right)\delta = P$$

Notice that this has the same format as a spring–mass force deflection relationship (i.e., $kx = F$), where

$$k \Rightarrow \frac{3EI}{L^3}, x \Rightarrow \delta, F \Rightarrow P(= mg)$$

Therefore, the expression for the natural frequency may be employed:

$$f = \frac{1}{2\pi}\sqrt{\frac{K}{m}} \Rightarrow f = \frac{1}{2\pi}\sqrt{\frac{3EI}{mL^3}} \text{ or } \omega = \sqrt{\frac{3EI}{mL^3}}$$

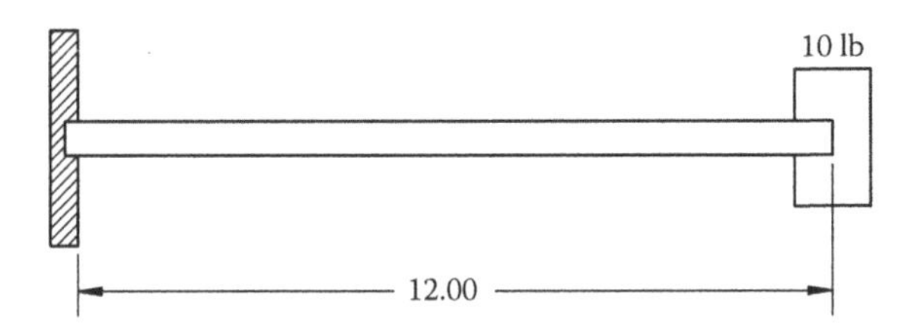

FIGURE 7.3
A cantilever beam with a 10 lb block at free end.

Suppose that the beam in Figure 7.3 is made of aluminum ($E = 10 \times 10^6$ psi) and has a moment of inertia $I = 0.03$ in^4, the equivalent mass for the 10 lb weight is

$$m = \frac{10}{32.2} = 0.31\frac{\text{lb}-\text{s}^2}{\text{ft}}$$

After substituting appropriate values, the natural frequency using the 1 DOF approach is calculated as

$$f = 6.5 \text{ or } \omega = 41$$

The approximation used here is only valid for a cantilever beam under a point load (or concentrated mass) at its free end. For this problem, this approximation is identical to the exact solution; however, we assumed that the weight of the beam was negligible compared to the mass of the 10 lb block. For other types of beams, the appropriate force-deflection relationship must be used. Clearly, equations developed in this manner provide natural frequencies of equivalent spring–mass systems (1 DOF), but the actual shape of the system (i.e., the vibration mode shape) is not readily available through this approach.

Similarly, by making proper assumptions, a printed circuit board assembly (PCBA) may be modeled as a spring and mass system for its first mode of vibration. Considering that fundamentally, a printed circuit board (PCB) is a plate, theoretically, it is possible that, based on this approach, one would use the force-deflection relationships readily available in various handbooks to calculate this fundamental frequency.

Example

Consider the system shown in Figure 5.6. Its PCBAs weigh one pound and the PCBs have the following dimensions: $10 \times 6.1 \times 0.063$ in. PCBs are glass-epoxy with copper planes: $E = 3 \times 10^6$, and $\nu = 0.18$. Components are uniformly distributed. Find the natural frequency of the PCBA for the following two conditions: first, simply supported (SS) on all edges, and second, clamped on all edges.

The deflection at the center of a SS and/or clamped–clamped (CC) plate of length (a) and width (b), ($a > b$), is (Baumeister et al. 1979)

$$\text{SS (hinged) plate: } y = 0.1106\frac{wb^4}{Et^3}$$

$$\text{CC plate: } y = 0.0277\frac{wb^4}{Et^3}$$

where:
E is Young's modulus
t is plate thickness
w is weight per unit area

The frequency of the equivalent spring–mass system is calculated from

$$f = \frac{1}{2\pi}\left(\frac{g}{y}\right)^{1/2}$$

Now substitute for y and obtain the 1 DOF frequency:

$$f = \frac{1}{2\pi}\sqrt{\left(\frac{g}{0.1106(wb^4/Et^3)}\right)}$$

for SS plates, and

$$f = \frac{1}{2\pi}\sqrt{\left(\frac{g}{0.0277(wb^4/Et^3)}\right)}$$

for CC plates.

Fortunately, formulae to calculate exact natural frequencies of plates under a variety of boundary conditions are readily available. For a SS and/or CC plate of length (a) and width (b), ($a > b$), the fundamental frequency is (Baumeister et al. 1979, Steinberg, 1988)

$$\text{SS plate: } f_{\text{Exact}} = \frac{\pi}{2}\left(\frac{gD}{w}\right)^{1/2}\left[\frac{1}{a^2}+\frac{1}{b^2}\right]$$

$$\text{CC plate: } f_{\text{Exact}} = \frac{\pi}{1.5}\left(\frac{gD}{w}\right)^{1/2}\left[\frac{3}{a^4}+\frac{2}{a^2b^2}+\frac{3}{b^4}\right]^{1/2}$$

where:
$$D = \frac{Et^3}{12\left(1-v^2\right)}$$
D is called flexural rigidity
v is Poisson's ratio

The frequency of the plate for the two boundary conditions as well as for the exact and approximate (1 DOF) solutions is provided and compared in Table 7.1. In a way, one should expect this level of discrepancy in the results. The reason is that in the 1 DOF approximation, we assume that the entire plate geometry (including its mass distribution) contributes equally to vibration, whereas in reality, different segments of the plate—depending on their location relative to the boundaries—undergo different displacements. Although a longer explanation is beyond the

TABLE 7.1

A Comparison of Frequency Values; $a = 10$, $b = 6.1$, $t = 0.063$, $w = 0.0164$, $E = 3 \times 10^6$, $\nu = 0.18$, $D = 64.6$

Boundary Conditions	Exact	1 DOF	% Error
SS frequency	71.44	54.05	24.34
CC frequency	141.58	108.01	23.72

scope of this book, we can say that the reason the frequencies are lower in the spring–mass simplification in general is because we are employing a larger mass than what is actually participating in the vibration. In fact, if we were to use a 55%–65% of the PCBA's mass as suggested by Crede (1951), we would get almost identical results with the exact equations.

Even though the 1 DOF approach loses its appeal because of its inherent deficiency in calculating a flexible body's natural frequency, it may still be very useful in providing us with important information about our system. For this purpose, let us look at a second application.

Second Application

In many vibration problems, a system undergoes a given acceleration level (e.g., 2*g*s). We may take advantage of this information to develop a means of calculating the maximum deflection that the vibrating part undergoes using a simple spring–mass analogy. Recall that for a simple harmonic vibration, the displacement is defined as

$$x = A \sin \omega t$$

and the acceleration is

$$x_{,tt} = -\omega^2 A \sin \omega t$$

A is the maximum deflection of the mass, and $x_{,tt}$ is its acceleration. The maximum acceleration happens when $\sin \omega t = 1$. Thus

$$\max x_{,tt} = \omega^2 A$$

Note that the negative sign is dropped because we are interested in the magnitude and not the direction of acceleration; therefore, the maximum deflection is:

$$A = \frac{\max x_{,tt}}{\omega^2} \tag{7.3}$$

or

$$A = \frac{\max x_{,tt}}{4\pi^2 f^2} \tag{7.4}$$

This equation in and of itself is not useful, but as mentioned earlier, in many cases an acceleration level is specified. For example, it may be that the cantilever system in Figure 7.3 undergoes a 2*g* acceleration. Thus, the maximum dynamic displacement is

$$A = \frac{2(32.2)}{\omega^2}$$

Recall that $\omega = 41$. Therefore, the maximum deflections is

$$A = \frac{2(32.2)}{41^2} = 0.038\,\text{ft or } 0.456\text{ in}$$

Interestingly, we may still use this equation to estimate the dynamic deflection of the PCBA used in the previous example. Should the PCBA be subjected to the same *g*-load, we would obtain the following deflections:

$$\text{SS (hinged) plate: } A = \frac{2(32.2)}{4\pi^2 84.95^2} = 2.26\times10^{-4}\,\text{ft or } 2.71\times10^{-3}\,\text{in}$$

$$\text{CC Plate: } A = \frac{2(32.2)}{4\pi^2 174.21^2} = 5.38\times10^{-5}\,\text{ft or } 6.45\times10^{-4}\,\text{in}$$

In Chapter 9, we will discuss how dynamic loads may be evaluated based on the maximum dynamic displacements. These loads are needed to calculate the dynamic stresses used to evaluate the enclosure vibration worthiness.

Mode Shapes

A mode shape is the deformation of a system corresponding to a particular frequency. Although it is possible to calculate the frequency of a system, only the relative shape of the corresponding mode can be calculated. The exact magnitude is only obtained after the equation of motion is integrated (or solved for every time point). To illustrate the reason behind this inability, once again, consider Equation 7.2 for a spring–mass system:

$$x_{,tt} + \omega^2 k = 0$$

Recall that

$$x = A\ \sin\omega t$$

$$x_{,tt} = -\omega^2 A\ \sin\omega t$$

Now, substitute these relationships into Equation 7.2:

$$-m\omega^2 A\ \sin\omega t + kA\sin\omega t = 0$$

This relationship may be further reduced to

$$(-m\omega^2 + k)\ A\ \sin\omega t = 0$$

Here, the product of three quantities, namely, $(-m\omega^2 + k)$, A, and $\sin\omega t$, is zero. A is defined as the amplitude of vibration in a 1 DOF system (i.e., a spring–mass system) and thus corresponds to the mode shape. If A were to be set to zero, it would imply that the amplitude of vibration is zero. So A cannot be zero and similarly $\sin\omega t$ is nontrivial. Therefore, $(-m\omega^2 + k)$ must be zero leading to the means by which the natural frequency of the system may be calculated. Thus, the exact mode shape of vibration may not be defined by any equations without the knowledge of applied excitations to the system.

Nowadays, most engineers employ finite element methods to determine the natural frequencies and mode shapes of complex systems. Generally, the mode shapes are stored in the same memory location as actual displacements. Thus, it would be easy to mistake the relative mode shapes for the actual deformations.

Figure 7.4 depicts an array of PCBAs—the enclosure is not shown. The first natural frequency of vibration is 64 Hz. The general mode shape is shown, but the magnitude of displacement is not calculated. Notice how some of the PCBAs have crossed each other. This obviously does not happen in real life but clapping of boards may indeed happen. Evaluation of the mode shapes enables us to pinpoint such problem areas.

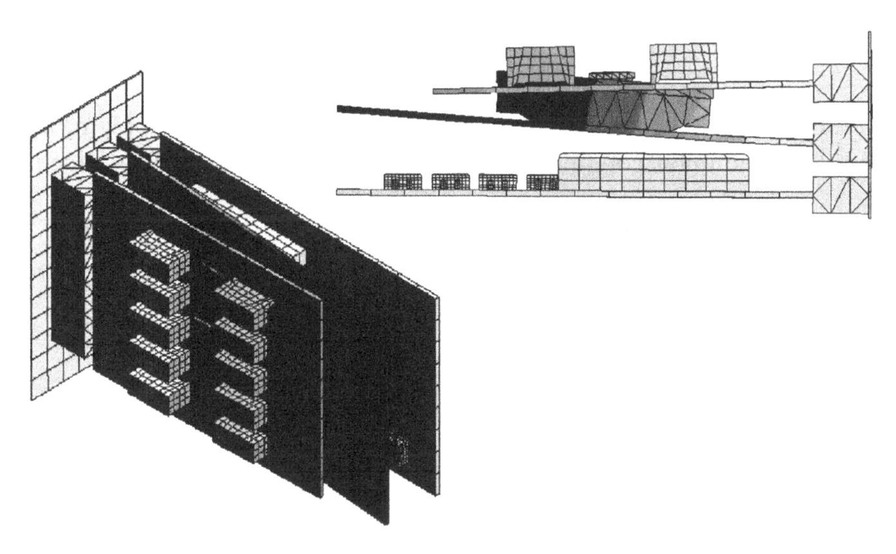

FIGURE 7.4
The first mode shape of an array of PCBAs.

Damped Vibration

Damping acts as a force, which opposes motion. When a linear system is excited, its response will depend on the type of excitation and the damping present in the system. For an undamped system, the amplitude of vibration is constant and does not change with time, and its natural frequency is a function of stiffness and mass. Damping affects both the amplitude as well as the natural frequency.

The governing equation for a 1 DOF free vibration spring–mass system is:

$$mx_{,tt} + f_d + kx = 0$$

where f_d represents the damping force. There are two classes of damping forces: one depends on friction called Coulomb damping and the other one depends on velocity and is called viscous damping. Coulomb damping depends on surface properties and the level of applied force normal to the surface and is difficult to quantify. Viscous damping, however, is expressed as

$$f_d = cx_{,t}$$

Thus, the equations of motion becomes:

$$mx_{,tt} + cx_{,t} + kx = 0 \tag{7.5}$$

Impact of Damping

The solution to Equation 7.5 is as follows

$$x(t) = A\mathrm{e}^{\alpha t}$$

By substituting this solution into the governing Equation 7.5 and applying the boundary and initial conditions, coefficients A and α may be determined:

$$\alpha_{1,2} = \frac{-c}{2m} \pm \sqrt{\mathfrak{A}}$$

where

$$\mathfrak{A} = \left(\frac{c}{m}\right)^2 - \frac{k}{m}$$

On the one hand, $\mathfrak{A}$ may be less than zero, leading to imaginary values of α. Thus, the system is called "underdamped"; the solution is oscillatory and vibration exists. On the other hand, $\mathfrak{A}$ may be greater than zero, leading to real values of α. This means that the solution is nonoscillatory and that vibration does not exist; the system is called "overdamped." At the interface of these

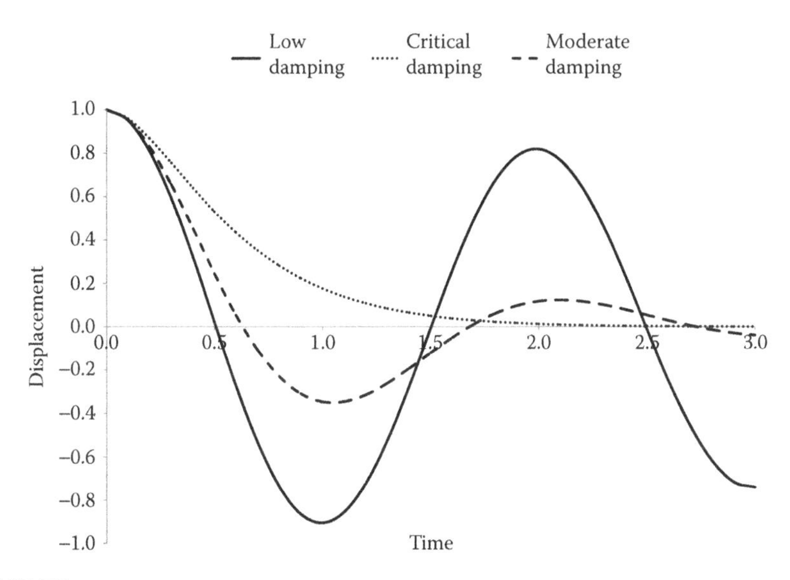

FIGURE 7.5
Impact of damping on vibration. Notice the signal frequency has decreased with increased damping.

two conditions, $\mathfrak{A} = 0$ and this condition is referred to as "critical damping"; however, vibrations do not exist. This is depicted in Figure 7.5.

Vibration only exists when the system is underdamped. However, the amplitude of vibration as well as the system's natural frequency are no longer constant. For a spring–mass system, the amplitude of the oscillations is described as

$$x_A = \left[x_\circ^2 + \frac{(v_\circ + x_\circ \xi \omega)^2}{\omega_D} \right]^{1/2} e^{-\xi \omega t}$$

where:

$x_\circ$ and $v_\circ$ are the initial position and velocity, respectively
ω is the undamped natural frequency
$\xi = c/(2m\omega)$ is defined as the damping ratio
ω_D the natural frequency of the damped system:

$$\omega_D = \omega\sqrt{1-\xi^2}$$

Note that the "natural" damping present in most materials hardly exceeds 15%, that is, $\xi = 0.15$. If we calculate the damped frequency, we get $\omega_D = 0.99\,\omega$. Therefore in evaluating natural frequencies and mode shapes—as long as artificial damping does not exist—the impact of damping is ignored.

Forced Vibration

Figure 7.6 depicts a spring–mass system under a loading *P(t)*. Similar to free vibration, the displacement of the mass is denoted by *x*, the spring stiffness by *k*, and the damping by *c*. The general governing equation for a 1 DOF system with damping is as follows:

$$mx_{,tt} + cx_{,t} + kx = P(t) \tag{7.6}$$

If *P(t)* is harmonic, it may be expressed as $P(t) = P_{\circ} \sin \bar{\omega} t$, thus

$$mx_{,tt} + cx_{,t} + kx = P_{\circ} \sin \bar{\omega} t$$

The solution to this equation is a combination of general and particular solutions:

$$x(t) = \left(A\cos(\omega_D t) + B\sin(\omega_D t)\right)e^{-\xi\omega t} + \frac{X_{\circ}}{\sqrt{\left(1-r^2\right)^2 + (2r\xi)^2}}\sin(\varpi t + \theta) \tag{7.7}$$

where

$$X_{\circ} = \frac{P_{\circ}}{K}, \tan\theta = \frac{-2\xi r}{1-r^2}, \text{ and } r = \frac{\varpi}{\omega}$$

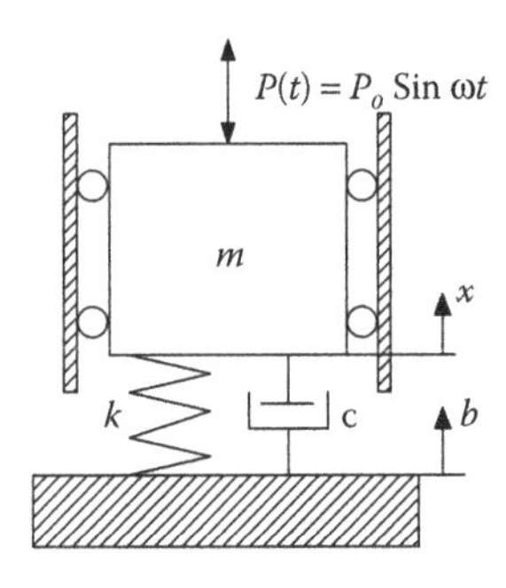

FIGURE 7.6
Depiction of a spring–mass system under dynamic loads.

Equation 7.7 has the following characteristics:

1. There is a startup "transient segment" that diminishes exponentially. The decay rate depends linearly on both the frequency as well as damping.
2. Under resonance conditions, that is, the forcing frequency equal to the natural frequency, $r = 1$, the magnitude of vibration is finite and approaches $X_{\circ}/2\xi$.
3. Systems response [i.e., $x(t)$] lags the forcing function by θ. This lag, called phase angle, is related to both damping present in the system as well as the ratio of the forcing frequency to undamped natural frequency.

Engineering Applications

From a design point of view, we are interested in a few guidelines to assess the behavior of our system. Particularly, as the excitation frequency nears the natural frequency, we need to know the following:

1. Will the displacements be magnified?
2. Will the excitation force be magnified?

Dynamic Magnification Factor

The dynamic magnification factor is a measure of dynamic displacements of a system compared to input or static values. It is an indicator of how displacements are magnified in a system's vibration. It is the ratio of steady-state response amplitude to the static response amplitude. For a spring–mass system,

$$D_m = \frac{X}{X_{\circ}} = \frac{1}{\sqrt{\left(1 - r^2\right)^2 + \left(2r\xi\right)^2}} \tag{7.8}$$

where:

$r = \varpi/\omega$ is the ratio of forcing frequency to natural frequency

$\xi = \left(c/2m\omega\right)$ is the damping ratio

The dynamic magnification factor is depicted in Figure 7.7. Note that D_m depends only on ξ and r. We can make the following observations:

1. For r much less than one, the magnification factor is nearly equal to one. This means that the dynamic displacements are nearly equal to static displacements. In other words, the problem can be solved as a static problem.

2. For r greater than 1.45, the magnification factor is much less than unity. This implies that when the frequency of the external load is about 45% greater than the natural frequency, the amplitude of the steady-state response is less than the static displacement. In other words, the impact of dynamic displacements is less severe than it is for the static displacements.
3. For r in the neighborhood of unity, D_m is equal to $1/(2r\xi)$. This shows that in resonance, the dynamic magnification factoris only a function of damping. For a typical value of ξ, say $\xi = 0.01$ (1% damping), $D_m = 50$. In other words, dynamic displacements will be 50 times greater than static deflections. Because of this amplification, we need to take any precaution to calculate the natural frequencies accurately. For instance, assume that ω_{exact} was not calculated accurately, and $\omega_{Calculated} = 0.9\,\omega_{exact}$, that is, 10% error in calculation. The erroneous frequency leads to a magnification factor of nearly 5 instead of 50. Clearly, large errors in displacement may be introduced if the natural frequencies are not calculated accurately.

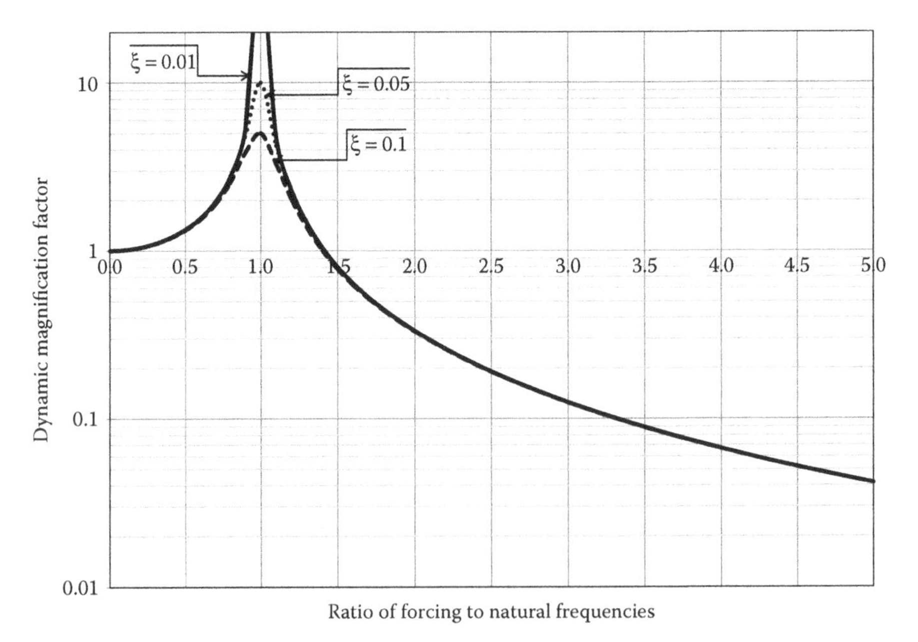

FIGURE 7.7
Graphical representation of dynamic magnification factor.

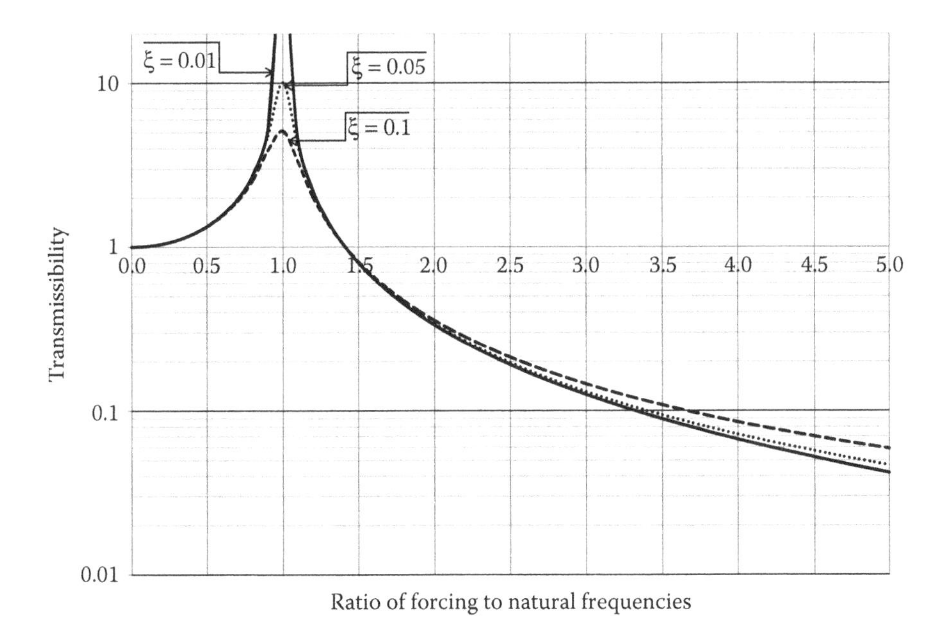

FIGURE 7.8
Graphical representation of transmissibility.

Transmissibility

Transmissibility is a measure of the magnification of input forces throughout the system. It is defined as the ratio of transmitted force through the spring and damper to the amplitude of the applied force.

For a single spring–mass system, it may be shown that

$$T_m = \frac{F_T}{F_o} = \frac{\sqrt{1+(2r\xi)^2}}{\sqrt{(1-r^2)^2+(2r\xi)^2}} \tag{7.9}$$

Transmissibility is shown in Figure 7.8. Note that T_m depends only on ξ and r. Similar to dynamic magnification, we can make the following observations:

1. For r much less than one, transmissibility is nearly equal to one. This means that the dynamic forces are nearly equal to static forces. In other words, the problem can be solved as a static problem.
2. For r greater than 1.50, transmissibility is much less than unity. This implies that when the frequency of the external load is about 50% greater than the natural frequency, the amplitude of the steady-state transmitted force is less than the applied static forces. In other words, the impact of dynamic forces is less severe than it is for the static forces.

3. For r in the neighborhood of unity, transmissibility is only a function of $\xi\left(T_m = \sqrt{1+\xi^2}/(2\xi)\right)$. For a typical value of ξ, say $\xi = 0.01$ (1% damping), $T_m = 50$. In other words, transmitted forces will be 50 times greater than the applied static loads. Similar to dynamic magnification, we need to take any precaution to calculate the natural frequencies accurately.

If we were to graph the dynamic magnification factor and transmissibility on the same graph, we would learn that the two variables have very similar numerical values particularly for small damping ratios. Figure 7.9 shows these two factors for a damping ratio of 0.15. Considering that inherent damping in a given material is rarely larger than 0.13, we notice that there is a good agreement between the two curves. However, the difference between the two curves must be clear. On the one hand, transmissibility reflects the ratio of the output excitation (whether force or displacement) to the input excitation. For instance, in an unbalanced rotary system, transmissibility reflects the ratio of the forces transmitted to the support to the unbalanced forces, or, in the case of vibrating foundations, the ratio of member displacement to that of the foundation. The dynamic magnification factor, on the other hand, reflects the ratio of the system's deformation under dynamic loads compared to static loads. Thus, there is a fundamental difference between how the two numbers must be interpreted, even though their numerical values may be similar.

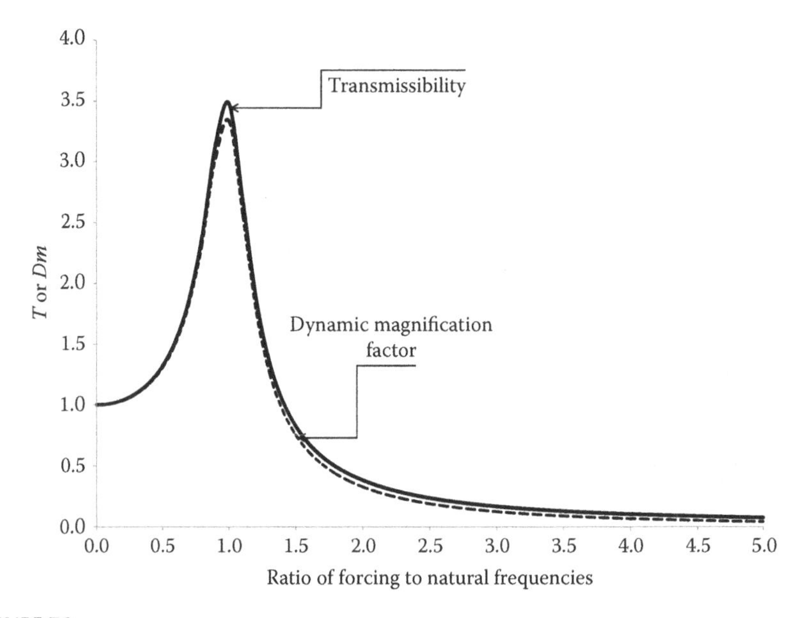

FIGURE 7.9
A comparison of dynamic magnification factor and transmissibility. Damping ratio (ξ) is 0.15 for this graph.

Phase Angle

Figure 7.10 depicts the behavior of the phase angle as a function of both frequency ratio (r) as well as damping. What this figure relays is that the system's response [i.e., $x(t)$] always lags the forcing function by θ. As the frequency ratio (r) grows and becomes much larger than unity, the phase angle (θ) approaches 180°. In other words, the applied force pushes down on the mass while it is still moving up and vice versa. In fact, the lower the damping, the faster the phase angle approaches 180°.

Vibration through Base Excitation

In the previous approach, we considered a spring–mass system that was excited through a sinusoidal force. In many systems, the source of excitation is the movement of the base. Refer to Figure 7.6 once more. Assume that the forcing function $P(t)$ is no longer present, but instead, the base is excited. The motion of the base is denoted as b; thus, we can write the following equation:

$$m(x-b)_{,tt}+c(x-b)_{,t}+k(x-b)=mb_{,tt}$$

If we were to substitute $y = x-b$, we obtain the following

$$my_{,tt}+cy_{,t}+ky=mb_{,tt} \tag{7.10}$$

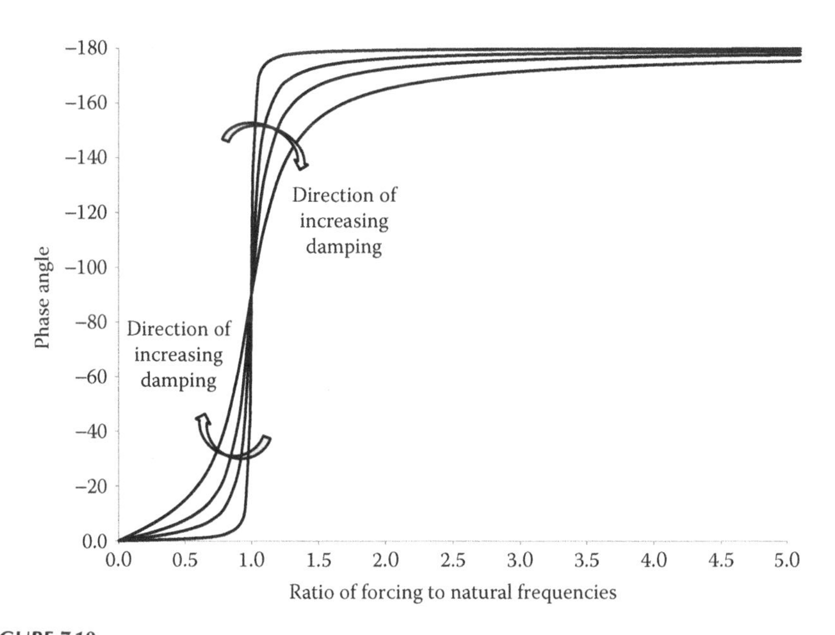

FIGURE 7.10
Phase angle as a function of frequency ratio and damping. Damping ratio (ξ) varies between 0.01 and 0.2.

Notice that this equation is identical in format to Equation 7.6. Therefore, we may conclude that should the base excitation be sinusoidal, the resulting set of equations would be identical to the excitations created by the applied forces.

Example

Recall the cantilever beam shown in Figure 7.3. Now assume that the 10 lb block is replaced with an oscillatory force of 10 lb with the following frequencies: ϖ = 30, 40, 50, and 60. Assuming a damping ratio (ξ) of 1%, we like to calculate the dynamic magnification factors and displacements as well as transmissibility values for this system (Figure 7.11).

Dynamic magnification factor and transmissibility are defined by Equations 7.8 and 7.9, respectively,

$$D_m = \frac{X}{X_\circ} = \frac{1}{\sqrt{\left(1-r^2\right)^2+\left(2r\xi\right)^2}}$$

$$T_m = \frac{F_T}{F_\circ} = \frac{\sqrt{1+\left(2r\xi\right)^2}}{\sqrt{\left(1-r^2\right)^2+\left(2r\xi\right)^2}}$$

where

$$r = \frac{\varpi}{\omega} \text{ and } \xi = \frac{c}{2m\omega}$$

Static deflection was discussed earlier as

$$\delta = \frac{PL^3}{3EI}, E = 10\times10^6 \text{ psi and } I = 0.03 \text{ in}^4$$

So,

$$\delta_\circ = \frac{PL^3}{3EI} = \frac{10\times12^3}{3\times\left(10\times10^6\right)\times0.03} = 0.0192 \text{ in}$$

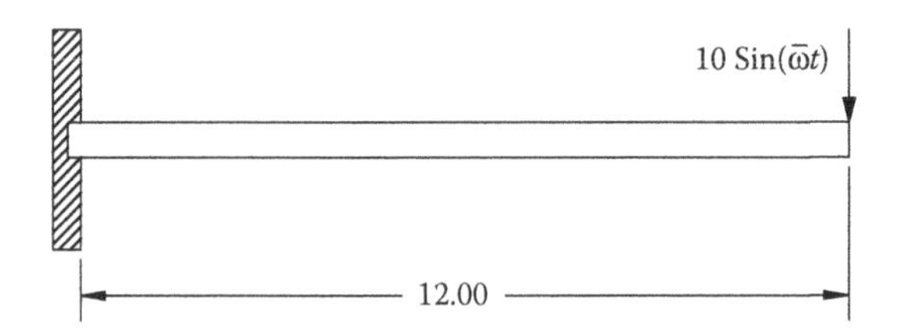

FIGURE 7.11
A cantilever beam under vibratory loads.

Previously, we calculated the natural frequency of this system to be $\omega = 41$ and $\xi = 0.01$:

For $r_1 = 30/41 = 0.7317$, thus

$$D_m = \frac{1}{\sqrt{\left(1-0.7317^2\right)^2 + \left(2\times 0.01\times 0.7317\right)^2}} = 2.15$$

and

$$\delta_1 = D_m\delta_\circ = 2.15\times 0.0192 = 0.041 \text{ in}$$

$$T_m = \frac{\sqrt{1+\left(2\times 0.01\times 0.7317\right)^2}}{\sqrt{\left(1-0.7317^2\right)^2 + \left(2\times 0.01\times 0.7317\right)^2}} = 2.15$$

Similarly,

R	D_m	δ	T_m
40/41 = 0.98	19.24	0.369″	19.24
50/41 = 1.22	2.05	0.039″	2.05
60/41 = 1.46	0.88	0.017″	0.88

Note: Calculations are made with fall significant figures and then the results are truncated.

Near resonance, this system will experience a theoretical displacement of 0.43 in compared to 0.019 in static displacement. In real life, the system may fall apart long before magnification factors of 19 are realized. Clearly, these vibrations need to be managed through selection of proper isolators.

Vibration Isolation

Physically, isolators store a portion of a system's energy and dissipate it on a longer timescale than that of forcing excitations. In general, an isolator reduces either the magnitude of motion transmitted from a vibrating foundation to the equipment, or the magnitude of forces transmitted from the equipment to its foundation. Furthermore, an isolator may be needed to maintain the amplitude of vibration below a desired value without any regard to the transmitted forces.

A system's energy may be stored in springs and dissipated through dampers. There are three classes of dampers; namely, viscous, Coulomb, and orifice dampers. The first relies on the viscosity and/or viscoelastic properties of the media to dissipate energy. The second relies on direct dissipation means such as friction. Orifice dampers rely on fluid properties as it is pushed through a small orifice. In general, viscous and orifice damping depends on the velocity and/or position of the mass being isolated, and Coulomb dampers are a constant force opposing the motion (Crede 1965).

From a design point of view, we may be able to combine spring elements with damper elements in a variety of ways. For instance, a damper may be attached between a vibrating mass and its support without the need of any elastic members, or, the same damper may be combined in series with a spring. All in all, there are four different, albeit similar, means of isolation. These are rigidly connected viscous dampers, rigidly connected Coulomb dampers, elastically connected viscous dampers, and elastically connected Coulomb dampers. See Crede (1951) and Harris (1996) for more details.

It is also important to note that there are a variety of materials that may be used in the manufacturing of isolators. These materials may be grouped into three categories: metals, elastomers, and plastics. Each category has unique properties that need to be carefully considered before being specified in a design. These properties are as follows.

Metal springs are used where large static deflections are required or harsh environmental conditions do not allow elastomers. On the one hand, they can be bulky and heavy and on the other hand, they do not dissipate any energy and there may still be a need for damping.

Elastomers provide the highest levels of energy storage and dissipation with an added benefit of being molded to custom shapes. The disadvantage of elastomers is that they may be adversely affected by environmental factors such as oxygen and/or oils and chemicals.

Plastic isolators are similar to elastomers but provide more rigidity.

From an analysis point of view, the stiffness and damping rations of isolators need to be calculated. To do so, we use the relationship for the dynamic magnification factor mentioned earlier to calculate the stiffness and damping ratio of isolators to properly reduce the deflection or deformation of critical components in the system below a specified value. In the case of transmitted forces, we calculate the needed stiffness and damping by using the relationship for transmissibility. From an engineering point of view, however, the approach to both problems is identical. It is finding a spring and damper combination so that the entire system including the added isolator would have a natural frequency that is outside of the forcing function frequency range (Crede 1951). Thus, the steps needed to specify an isolator are as follows.

1. Select the desired magnification factor or transmissibility
2. Select the damping ratio
3. Calculate the spring stiffness

Example 1

Find an appropriate spring to contain the magnification factors below 0.5 for the cantilever beam shown in Figure 7.7. Use 30% damping.

Notice that $\xi = 0.3$ and that $D_m \le (1/2)$. Thus, we write

$$D_m = \frac{1}{\sqrt{\left(1-r^2\right)^2 + \left(2r(0.3)\right)^2}} \le \frac{1}{2}$$

which leads to $\left(1-r^2\right)^2 + (2r(0.3))^2 \ge 4$. The solution to this inequality is $r \ge 1.65$.

This value is indicative of the new frequency ratio of the system with isolators. Recall that

$$r = \frac{\varpi}{\omega} \text{ and } \omega = \sqrt{\frac{k}{m}}$$

With $g = 32.2$ ft/sec^2 and $m = 10/32.2 = 0.31$ slugs $\Rightarrow \omega = 1.796\sqrt{k}$

$$1.65 = \frac{40}{1.796\sqrt{k}}$$

$$k = 182.2 \frac{\text{lb}}{\text{ft}}$$

This is the total stiffness. The stiffness of the cantilever beam is

$$k_{\text{beam}} = \frac{3EI}{L^3} = \frac{3\left(10 \times 10^6\right) \times 0.03}{12^3}$$

$$k_{\text{beam}} = 520.83 \frac{\text{lb}}{\text{in}} = 6249.96 \frac{\text{lb}}{\text{ft}}$$

For serial springs:

$$\frac{1}{k_{\text{total}}} = \frac{1}{k_{\text{beam}}} + \frac{1}{k_{\text{isolator}}}$$

From here $k_{\text{isolator}} = 187.67$ lb/ft

Example 2

A 3.0 lb electronics system can withstand a 5*g* sinusoidal vibration. The base will be loaded to 12*g* and frequency of 62 Hz. Design an isolation system with a damping factor of 20% for the mountings.

This problem is concerned with forces and their magnification or decay in the system. Therefore, the transmissibility must be taken into the calculations. Note that 20% damping means $\xi = 0.2$.

$$T_m = \frac{\sqrt{1 + \left(2r(0.2)\right)^2}}{\sqrt{\left(1-r^2\right)^2 + \left(2r(0.2)\right)^2}}$$

$$\frac{5g}{12g} = \frac{\sqrt{1+0.16\ r^2}}{\sqrt{\left(1-r^2\right)^2+0.16\ r^2}}$$

This expression leads to $r^4 - 2.762r^2 - 4.76 = 0$. The solution to this equation is $r = 1.99$. Thus, for $r > 1.99$ the response of the system will be less than $5g$. Note that r is related to the spring stiffness through natural frequency.

This value is indicative of the new frequency ratio of the system with isolators. Recall that

$$r = \frac{\bar{f}}{f} \text{ and } f = \frac{1}{2\pi}\sqrt{\frac{k}{m}}$$

With g =386 in/sec^2 and m =3/386 = 0.00777 $\Rightarrow f = 1.805\sqrt{k}$

$$\frac{62}{1.805\sqrt{k}} > 1.99$$

$$k \leq 297.8\frac{\text{lb}}{\text{in}}$$

Note that the softer the spring, that is, more isolation, the larger are the displacements. Furthermore, we use g = 386 in/sec^2 because, we wanted to calculate the stiffness in the units of lb/in.

Applications to Electronics Enclosures

Often, when there is a discussion of electronics equipment, the focus is generally on the electronics components and the PCBAs. One has to bear in mind that vibration management does solely focus on PCBAs and their components. In fact, the entire system structure must be evaluated. As shown in Figure 7.12, a typical electronics system is composed of a support structure, a chassis, PCBAs, and finally components.

As the support is subjected to vibration, it responds and begins to vibrate at the excitation frequency. Depending on the damping present as well as whether the excitation is near resonance conditions, vibration is transmitted to the chassis and similarly to the PCBAs and their components. Now, if the support structure has a transmissibility of 5 and chassis has a transmissibility of 10, the input vibration excitation to the PCBA is 50 times larger than the input to the support structure. If the PCBAs transmissibility is 4, then the response would have accelerations 200 times more than the input acceleration. Thus, the impact of vibration needs to be evaluated at all these levels even though, generally speaking, many failures happen at the PCBA level (Steinberg 2001).

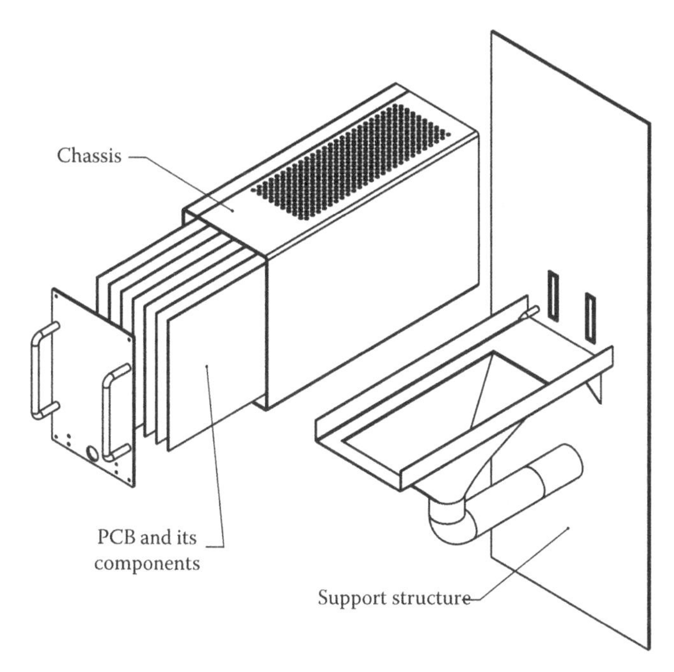

FIGURE 7.12
A typical electronics system exposed to vibration.

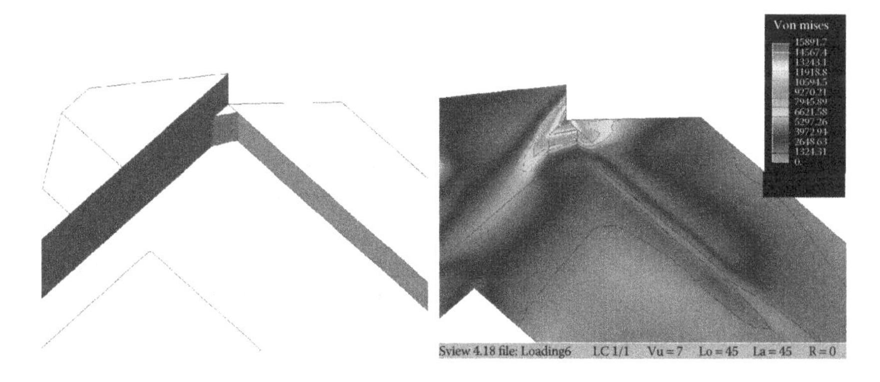

FIGURE 7.13
The stresses developing at the corner of supporting structure.

Ironically, it is possible that the response of one component would impact the response of the rest of the system. This coupling could potentially have a detrimental impact if the natural frequencies of the two components are close. Figure 7.13 shows a corner of the supporting structure of a PCBA containing relatively heavy transformers. This bracket fails the test conditions because of stresses that develop at the corner. A redesign is necessary for the system to remain functional under the test conditions.

Steinberg (2001) has shown that this coupling and feedback may be avoided if the following two conditions are avoided. First, the natural frequency of each PCBA in the system must be at least twice than the natural frequency of the empty chassis. Second, the weight ratio of the chassis to each PCBA must be at least 10 to 1. These conditions would provide favorable conditions such that vibration is not amplified by the chassis and the PCBAs would experience the same level of input vibration as the chassis. Should the need arise, ribs and stiffeners may be employed to increase the frequency of the PCBAs and separate it from the chassis frequency.

Furthermore, the response of fasteners, clamps, connectors, and wire harnesses must not be ignored. For instance, if the wire harnesses are not tied down, their vibration could potentially create fatigue failures near connectors where the motion of the conductors is restricted. Similarly, mating connectors may separate under vibration if the two parts are not tied together.

Maximum Deflection

Earlier, it was demonstrated that the primary natural frequency of a plate may be calculated from the maximum deflection and vice versa. Having the natural frequency enables one to calculate the maximum deflection. Furthermore, for a sinusoidal vibration, Equation 7.3 was developed to show the relationship between maximum displacement and the input accelerations. Let us rewrite this equation in the following format:

$$X_{\circ} = \frac{x_{,tt}}{\omega^2}$$

or

$$X_{\circ} = \frac{gG}{(2\pi f)^2}\text{in} \tag{7.11}$$

where $g = 386$ in/sec^2 is the gravitational constant and G is the level of acceleration experienced by the system. Clearly, $G = T_m G_{in}$. G_{in} is the input acceleration load and T_m is the system (or PCBAs) transmissibility; thus Equation 7.10 becomes

$$X_{\circ} = \frac{9.78\,T_m\,G_{in}}{f^2}. \tag{7.12}$$

Typical Transmissibility Values in Electronics Enclosures

Before we discuss typical values of transmissibility in electronics equipment, let us first review various factors impacting transmissibility. Recall that at resonance conditions both the dynamic magnification factor as well as transmissibility are inversely proportional to damping. Thus, we conclude that high levels of damping lead to low levels of transmissibility. In electronics equipment, the presence of conformal coatings, multilayer boards, and other features such as plastic housing and screw-in mounts tends to increase damping and thus reduce transmissibility (Steinberg 2001). Also as Steinberg (1988) points out, there is an inverse relationship between natural frequency and transmissibility. Furthermore, Steinberg (2001) suggests that for an electronics system

$$T_m = A\left[\frac{f}{\left(G_{in}\right)^{0.6}}\right]^{0.76} \tag{7.13}$$

where f is the fundamental frequency and the factor A is 1.0 for beams, 0.5 for plates (and by extension PCBAs which are treated as a plates), and 0.25 for boxes with aspect ratios higher than 2. It is noteworthy to consider that for plug-in PCBAs with edge guides, Steinberg (1988, 2001) further suggests that $T_m = B\sqrt{f}$, where B is 0.5 for boards with spring guides as well as boards with natural frequencies below 100 Hz. As the board frequency increases, so does B in such a way that at frequencies about 400 Hz, a value of two must be used.

Maximum Desired PCBA Deflection

Steinberg (1988) suggests that if the peak single-amplitude displacement of the PCBA is limited to X_{max}, the component can achieve a fatigue life of about 10 million stress reversals in a sinusoidal vibration environment. For sinusoidal vibrations

$$X_{max} = \frac{0.00022B}{Chr\sqrt{L}} \text{ in} \tag{7.14}$$

where:

B is the length of the PCB edge parallel to component (in)

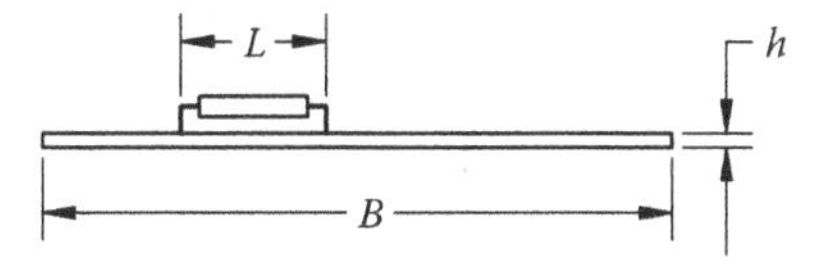

L is the length of the electronic component (in)

h is the thickness of the PCB (in)

C = 1.0 for standard DIP; = 1.26 for DIP with side-brazed lead wires; = 1.26 for pin grid array (PGA) with two parallel rows of wires; = 1.0 for PGA with wires around the perimeter; = 2.25 for leadless ceramic chip carriers; = 1.0 for leaded chip carriers where the lead length is about standard DIP

r = 1.0 when the component is at the center of the PCB; = 0.707 when the component is @ ½ point X and ¼ point Y on a PCB supported on 4 sides; = 0.5 when the component is @ ¼ point X and ¼ point Y on a PCB supported on 4 sides

Although this information is useful in determining the maximum allowable displacement of a vibrating PCBA, the reader should bear in mind that Steinberg's correlation was developed prior to the restriction of hazardous substances (RoHS) directive and lead-free requirements on solder. As such these data may be marginally applicable to new solders being developed, and further studies are needed to either verify this formulation or develop new ones.

Random Vibration

Earlier, we mentioned that in periodic motion, displacements follow a predictable pattern. However, in certain vibration cases, the amplitude of displacements or accelerations does not follow a particular pattern. Should we plot these variations as a function of time, we would develop a graph very similar to Figure 7.14. This figure may be considered to be a typical time history curve for random vibration.

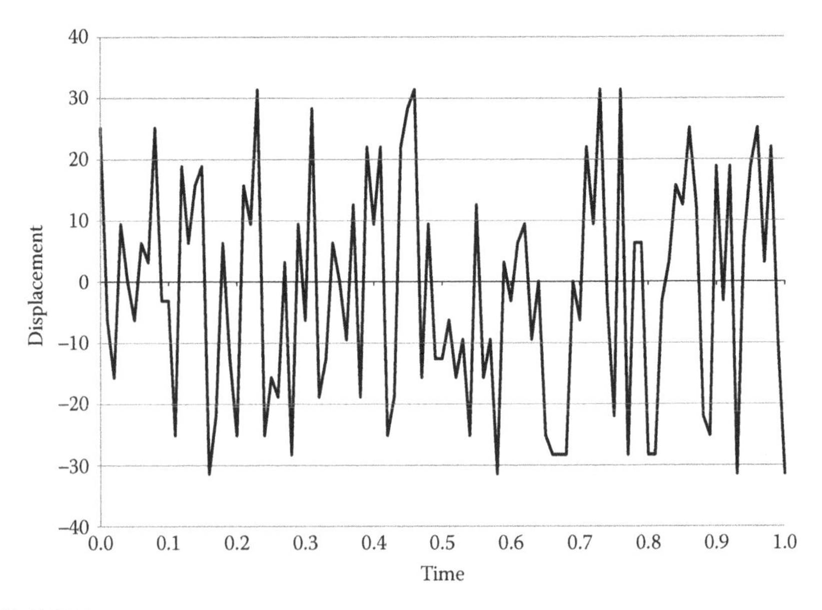

FIGURE 7.14
A typical displacement time history in random vibration.

Clearly, Figure 7.14 is much more complicated than Figure 7.1 where a periodic motion could easily be formulated. If we were to study Figure 7.14 carefully, we would begin to notice at least two features within this seeming chaos. First, in this particular case, values are within about +30 to −30. Second, the mean value of data appears to be around zero. Furthermore, although it is not readily evident, it may be shown that the majority of peaks and valleys are between +20 and −20. The lesson learned here is that although it may be difficult to identify the displacement of the mass [i.e., $x(t)$] exactly, we may be able to calculate its bounds and the probability of the displacement values.

This type of vibration is called random, simply because there are no exact ways of defining the relationship between the input excitation, and the location of the mass in a space–time continuum. In a way, one may say that modeling random vibration is an attempt to model nature. After all, there are no perfect sinusoidal excitations in real life. All true vibration are, in fact, random. Previously, it was pointed out that free and forced vibrations are periodic. However, random vibration is nonperiodic. Although the solution to periodic vibration is deterministic, probabilistic approaches must be used to find the displacements and/or stresses in a random vibration excitation. This constitutes the most important difference between sinusoidal and random vibrations. In other words, we can guess what maximum accelerations, forces, or displacements the system will undergo but we can never say with certainty at which time.

A detailed explanation of the mathematics to address the random vibration phenomenon is beyond the scope of this book; however, it should be mentioned that the mathematics is based on transforming the data set from the time domain into the frequency domain through the application of Fourier transforms. Effectively, this enables us to transform our fundamental variable from time to frequency through the use of a function called the transfer function (Wirsching et al. 1995, Curtis and Lust 1996).

An interesting aspect of this transformation is that while in the time domain, we encounter past, present, and future; in the frequency domain, all the frequencies within a given bandwidth are present and active. To illustrate this point, imagine a periodic step function. This function may be represented as its Fourier series expansion (Boyce and DiPrima 1977). An accurate model of the step function requires that an infinite number of frequencies be present. This requirement is graphically shown in Figure 7.15, where the step function is first approximated with one frequency and then second and third frequencies are added. Similarly, when a system is subjected to a random excitation, all natural frequencies in that system will be excited (or activated) to accurately produce the response to that excitation.

Since all frequencies are present in a random vibration, resonance conditions associated with each natural frequency will also be present. Therefore, there is an overamplification of displacements and/or stresses. As a result, failures may be produced that cannot be duplicated in a harmonic vibration sweep.

Next, we will examine certain terminology used in vibration analysis and measurement.

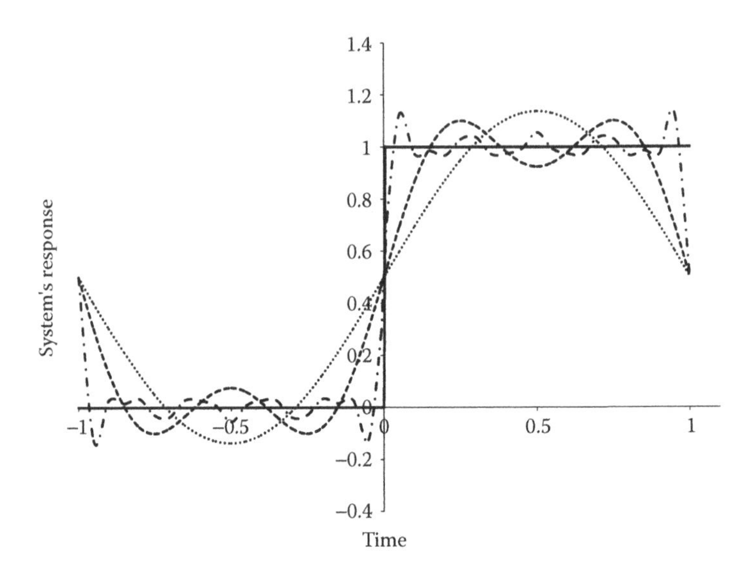

FIGURE 7.15
A Fourier series approximation of a step function.

Vibration Terminology

As pointed out, modeling random vibration is almost like simulating vibrations that take place in nature—almost. For that reason, we would need to treat our data as if we were considering test data. In that sense, it would become necessary to develop an understanding of various terms that are used in this field.

Peak Value

This generally indicates the maximum stress that the vibrating part is undergoing. It also places a limit on the "rattle space" requirement.

Average Value

The average value indicates a steady or static value, somewhat like the DC level of an electrical current. It can be evaluated by

$$\bar{x} = \lim_{T \to \infty} \frac{1}{T} \int_0^T x(t)\,dt$$

For example, the average value for a complete cycle of a sine wave [A $\sin(t)$] is zero, whereas its average value for a half cycle is

$$\bar{x} = \frac{A}{\pi} \int_0^{\pi} \sin(t)\,dt = \frac{2A}{\pi} = 0.637\,A$$

Mean Square Value

This is the average of the square values, integrated over some time interval T. This term is generally associated with the energy of the vibration.

$$\overline{x^2} = \lim_{T\to\infty} \frac{1}{T} \int_0^T x^2(t)\,dt$$

For a simple harmonic system, this definition leads to

$$\overline{x^2} = \frac{1}{2} A^2$$

Root Mean Square

A common measure of vibration, the root mean square (RMS) is defined as the square root of the mean square value:

$$\text{RMS} = \sqrt{\overline{x^2}}$$

It is generally understood that the RMS value corresponds to 68% of the data in the Gaussian distribution curve—or one standard deviation. Values within two standard deviations and three standard deviations are multiples two and three of the "RMS" value.

Decibel

The decibel is a unit of measurement that is frequently used in vibration. It is defined in terms of a power ratio:

$$dB = 10 \ \log \frac{\wp_1}{\wp_2}$$

However, power is proportional to the square of the amplitude or voltage, thus we have

$$dB = 20 \ \log \frac{A_1}{A_2}$$

Thus an amplifier with a voltage gain of 5 has a decibel gain of 14:

$$dB = 20 \ \log 5 = 14$$

Octave

When the upper limit of a frequency range is twice its lower limit, the frequency span is said to be an octave.

Spectral Density

Spectral density is defined as the following relationship:

$$S.D. = \lim_{\Delta f \to 0} \frac{\varphi^2}{\Delta f}$$

where φ may be the RMS of acceleration, velocity, or displacement. Spectral density may be classified as

1. Power spectral density (PSD)
2. Velocity spectral density (VSD)
3. Displacement spectral density (DSD)

PSD is commonly used and is available from various handbooks and/or standards:

$$PSD = \lim_{\Delta f \to 0} \frac{G^2}{\Delta f}$$

where G is the RMS of acceleration.

Solution Techniques for Random Vibration

Before we begin to solve a random vibration problem, there are two assumptions that we need to make. The main assumption is that the system under consideration is stable and that its response has a well-defined mean value. Furthermore, it is assumed that this mean value is reached within a relatively short number of cycles. Figure 7.14 is a good example of such a system. Once these assumptions are satisfied, the steps to solve random vibration problems are relatively straightforward; primarily, we need to understand the level of excitation that our system experiences and how it responds to this excitation.

To calculate an RMS of input excitation, we need to have a spectral density curve of that excitation. It may, then, be shown that the area under the spectral density curve is equal to the RMS of the excitation (Boyce and DiPrima 1977). In general, different industries require equipment to be tested to random excitations that are common to that industry. The required curves are generally published in various standards and handbooks such as MIL-STD-810F for avionics and SAE J1211 for surface vehicles.

The next step is to determine the equipment response. In other words, we should find out how the system behaves under such a load. Mathematically speaking, we have to calculate the RMS of the system spectral density. To do so, we need to know the following two characteristics of our system: its natural frequencies and its transmissibility values at these frequencies.

In Chapter 9, we will study how we can use this information to determine the RMS of the stresses in the system and, finally, the probability of failure.

Excitation Spectrum

As it was pointed out in random vibration, the exact location, velocity, and/or acceleration of a system under random vibration may not be known and/or calculated. The reason is because the excitation input to the system is not known exactly; rather, an excitation spectrum is known or given in a test standard. For instance, consider Figure 7.16, which depicts the power (or acceleration) spectral density vibration exposure in a rail cargo environment, which is provided in MIL-STD-810F (2000).

Recall that an RMS of the excitation may be calculated taking the square root of the area under the spectral density curve. Since the units of power (or acceleration) spectral density—also known as PSD—is G^2/Hz, the RMS of the excitation would have units of gravity. More specifically, G_{RMS} is the square root of the area under the PSD curve. Note that since G_{RMS} is the area under the PSD curve, various PSD curves may indeed produce the same G_{RMS}.

Let us explore the PSD curve in more detail. It is plotted on a log–log coordinate system and is a function of frequency. In other words, the x-axis shows the frequency and the y-axis shows the PSD, which has the units of G^2/Hz. The simplest PSD curve is a straight horizontal line known as white noise, as shown in Figure 7.17. In this figure all frequencies between 10 Hz and 2000 Hz are active with a spectral density value of 0.08 G^2/Hz.

The square root of the area under these curves provides a RMS of the acceleration loads (G_{RMS}). For example, to calculate the G_{RMS} for the white noise input PSD as shown in Figure 7.17, the following steps need to be taken:

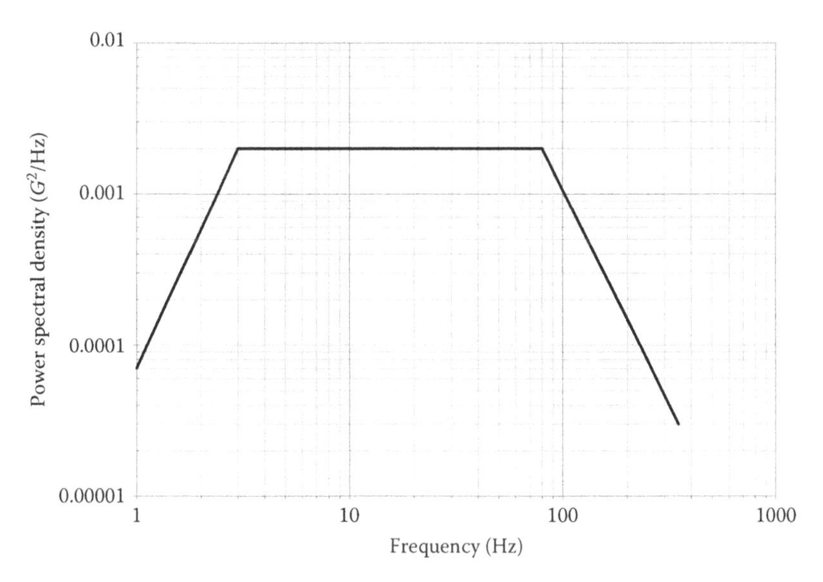

FIGURE 7.16
Power spectral density for rail cargo vibration exposure.

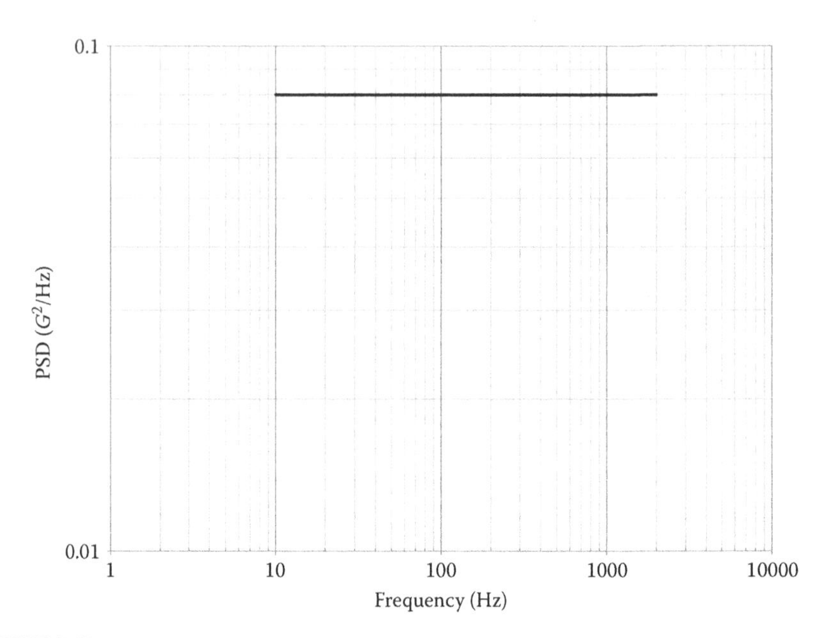

FIGURE 7.17
A white noise PSD curve.

$$G_{RMS} = \sqrt{PSD(f_2 - f_1)}$$

$$G_{RMS} = \sqrt{0.08(2000 - 10)}$$

$$G_{RMS} = 12.61$$

To evaluate the area under a more general PSD curve, such as the one shown in Figure 7.16, one should bear in mind that this is a log-log scale and not be fooled by "straight lines" and usage of the triangle rule—except if the lines are horizontal.

In general, to calculate the area under the sloping lines, such as the ones shown in Figure 7.18, special attention must be given to the basic area relations using log equations. For example, using the point slope equation from analytic geometry:

$$\ln PSD = m \ln f + \ln n$$

m is the slope of the line and In n is the intercept on the "PSD" axis. This equation may be rewritten as

$$PSD = mf^n$$

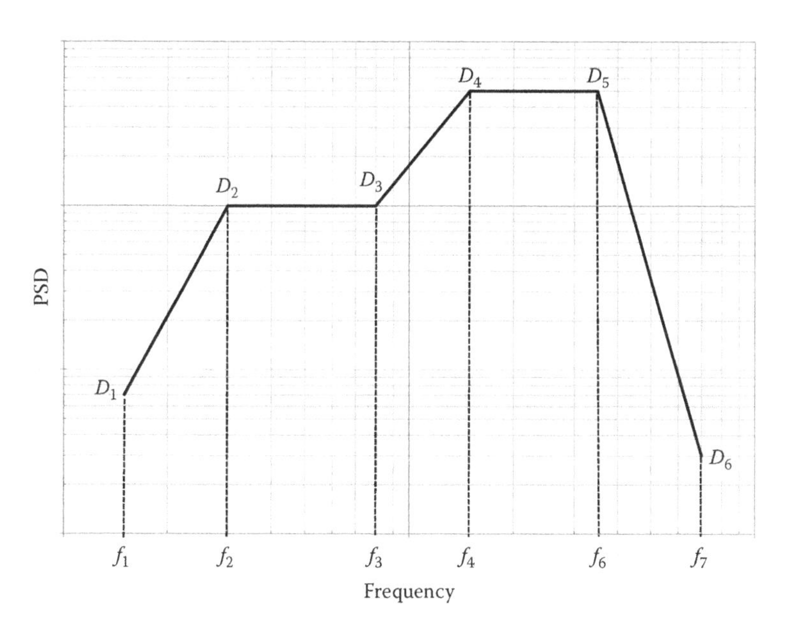

FIGURE 7.18
A typical PSD curve.

The area under the curve can then be determined by integration:

$$\text{Area} = \int_1^2 y dx = \int_1^2 PSD df = \int_1^2 m f^n df$$

Probability Distribution Functions

Earlier, it was mentioned that random vibration is probabilistic. In order to predict the probable acceleration levels, it is necessary to use the probability distribution function. Although providing these details is beyond the scope of this chapter, for a Gaussian (or normal) distribution curve, there are three bands of interest, namely, 1S, 2S, and 3S. S stands for standard deviation.

1. A 1S probability means that there is a 68% possibility that the accelerations will reach the G_{RMS} levels.
2. A 2S probability means that there is a 27% possibility that the accelerations will reach the $2 \times G_{RMS}$ levels.
3. A 3S probability means that there is a 4% (possibility that the accelerations will reach the $3 \times G_{RMS}$ levels.

The RMS acceleration of the white noise depicted in Figure 7.17 was shown to be 12.61. This means that there is 68% (1S) probability that the acceleration levels will reach 12.61 Gs, or there is 27% (2S) probability that the acceleration levels will reach 25.22 Gs, or there is 4% (3S) probability that the acceleration levels will reach 37.83 Gs.

Equipment Response

In order to calculate the equipment response, we need to know both the input PSD curve as well as the transmissibility in the system. Furthermore, if the input PSD curve is relatively smooth in the vicinity of the system's natural frequency (as shown in Figure 7.19) and the damping ratio is small so that $T_m = 1/2\xi$ holds true near resonance, it may be shown that the response of a 1 DOF system to random vibration input can be determined from the following equation (Steinberg 1988, Wirsching et al. 1995, Wang 2003):

$$G_{\mathrm{RMS}}^{\mathrm{out}} = \sqrt{\frac{\pi}{2} f T_m PSD} \tag{7.15}$$

where:

PSD is the spectral density at resonance (G^2/Hz)
fn is the natural frequency of the system (Hz)
T_m is the transmissibility at resonant frequency

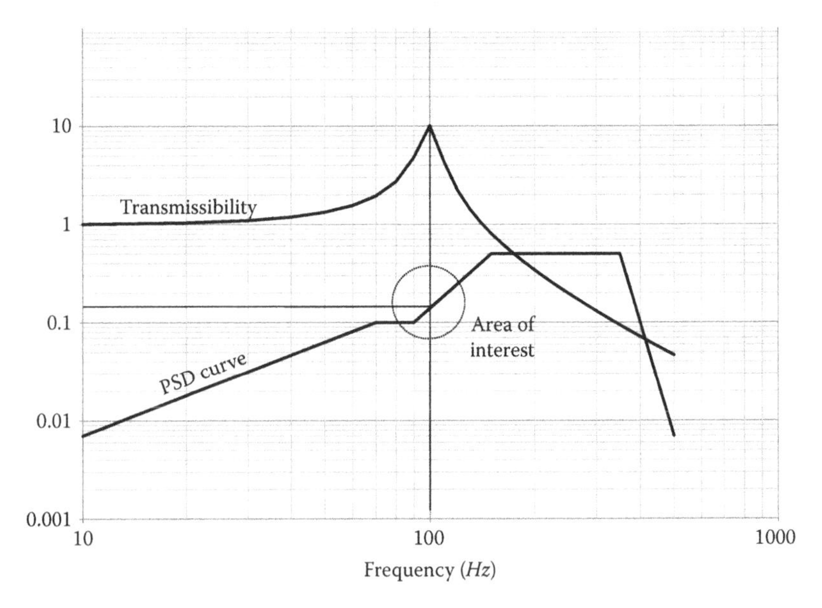

FIGURE 7.19
Transmissibility and PSD curves on the same graph.

Figure 7.19 shows these two curves on the same graph. Thus, $G_{\text{RMS}}^{\text{out}}$ for this configuration may be calculated as follows:

$$f = 100$$

$$T_m = 10$$

$$PSD_{\text{At Resonance}} = 0.11$$

$$G_{\text{RMS}}^{\text{out}} = \sqrt{\frac{\pi}{2} 100 \times 10 \times 0.11}$$

$$G_{\text{RMS}}^{\text{out}} = 13.14$$

This result indicates that the system represented in Figure 7.19 experiences a loading of 13.14 gs 68% of the time, a loading of 26.18 gs 27% of the time and a loading of 39.42 gs 4% of the time.

A frequently asked question is how a spring–mass system (a 1 DOF) would vibrate when subjected to a (flat spectrum) random vibration environment. The answer is that although the system can only vibrate at its natural or resonant frequency, it is the amplitude of vibration that varies.

In most practical cases, the magnification factor must be evaluated from available test data and in its absence must be estimated. The general rule is that low displacements and low strains lead to high transmissibility values and vice versa. Therefore, high frequencies, low-input gs or any other factor such as high-pressure interfaces between ribs, supports, connectors, and conformal coatings tend to lower transmissibility. Earlier, some guidelines were provided for periodic vibration environments. The same rules hold here as well.

Maximum Deflection

In the forced vibration section (Equation 7.11), we calculated the maximum deflection under a sinusoidal load. Interestingly enough, this equation may be used to calculate the RMS of the displacement. By substituting Equation 7.15 into Equation 7.11 we obtain

$$X_{\text{RMS}} = \sqrt{\frac{PSD\, T_m g^2}{32\pi^3 f^3}} \tag{7.16}$$

It should be mentioned that the same assumption that holds for calculating $G_{\text{RMS}}^{\text{out}}$, holds true here as well. Furthermore, while we offered a simplistic derivation of this equation, Crede and Ruzicka (1996) provide a much more rigorous derivation.

Maximum Desired Deflection of PCBAs

Similar to vibration, Steinberg (1988) suggests a maximum desired PCBA deflection for at least 10 million vibration cycles:

$$X = \frac{0.00022B}{Chr\sqrt{L}} \tag{7.17}$$

B, C, h, r, and L are defined for Equation 7.14.

Vibrations and Mechanical Stresses Caused by Acoustics and Noise

Most people's encounters with acoustics in electronics are when they notice a humming sound while operating a computer—a noise that is only too familiar and is just about taken for granted. Fans are probably the first source of noise that comes to mind; however, airflow—particularly through small opening and orifices may be another source of high noise.

Sound (or noise) is said to have been generated when pressure waves propagate through a material or medium such as air. Depending on the problem, there may be a standing wave, a transient wave, or a random pressure wave (noise) present in the medium. Pressure fluctuations will act as a pressure load on any structure in the medium and will cause resonance with any matching structural natural frequency. This problem may be quite acute for electronics systems operating near noisy equipment such as very loud machinery or engines and propulsion systems where noise levels of 130 dB to about 185 dB may cause fatigue failures, and levels above 185 dB could in fact cause static failures (Hubbard and Houbolt 1996). Earlier, a definition for decibel (dB) was proposed. We may write this equation in terms of sound pressure level P and a reference value P_{ref}:

$$dB = 20 \log \frac{P}{P_{ref}} \tag{7.18}$$

Generally, P_{ref} is set to 2.90×10^{-9} psi or 1.999×10^{-5} Pascal.

Considering that noise is a random phenomenon, its impact on the structure must be calculated using probabilistic techniques. To do so, we first need to calculate the pressure spectral density and then the system's response. Assuming a linear variation of dB levels with frequency (McKeown 1999), we have

$$dB = A + B \log(f) \tag{7.19}$$

By definition

$$\mathrm{PSD} = \lim_{\Delta f \to 0} \frac{P^2}{\Delta f} \approx \frac{P^2}{\Delta f} \tag{7.20}$$

It may be shown that for a banded frequency range, we have (Steinberg 1988)

$$\Delta f = \mathcal{F} f_c \tag{7.21}$$

where the bandwidth factor $\mathcal{F} = 0.2315$ for a typical one-third band, and f_c is the center frequency of the band. McKeown (1999) provides bandwidth factors for a different number of bands per octave. By combining Equation 7.18 with 7.19 and 7.20 into 7.21, we obtain the following relationship for the pressure spectral density (McKeown 1999):

$$\mathrm{PSD}_{sp} = \frac{P_{\mathrm{ref}}^2 10^{(A/10)}}{\mathcal{F}} f_c^{((B/10)-1)} \tag{7.22}$$

where the subscript *sp* stands for sound pressure.

If we were to assume a constant dB with frequency and a one-third octave band, we would develop the same relationship as Steinberg (1988):

$$\mathrm{PSD}_{sp} = \frac{\left(2.9 \times 10^{-9}\right)^2 10^{(dB/10)}}{0.231 f_c}$$

Calculating the system's response to acoustic excitations is relatively straightforward from here. It may be shown that a $P_{\mathrm{RMS}}^{\mathrm{out}}$ from the following equation:

$$P_{\mathrm{RMS}}^{\mathrm{out}} = \sqrt{\frac{\pi}{2} f T_m PSD_{sp}} \tag{7.23}$$

$P_{\mathrm{RMS}}^{\mathrm{out}}$ may now be used as a pressure load to calculate the stresses in the structural components subjected to the acoustic loads.

Multiple DOF Systems

When a multiple DOF system is subjected to a vibration input, the motion of each mass in the system will influence the motion of every other mass in the system, as shown in Figure 7.20. To evaluate the degree of coupling and the resulting system response to vibration, forces at each mass need to be balanced.

A detailed treatment of this subject is beyond the scope of this chapter. However, there are some practical points that we need to keep in mind. First,

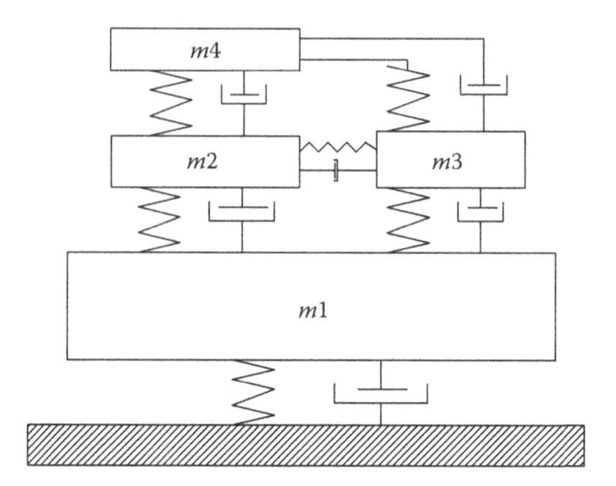

FIGURE 7.20
Multiple degrees of freedom systems.

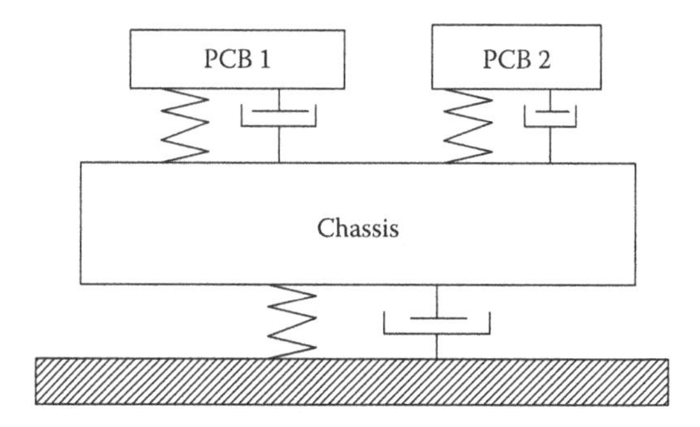

FIGURE 7.21
A representation of a chassis with two PCBAs.

if the natural frequency of the chassis is different from the natural frequency of PCBAs by a factor of two (forward or backward), the coupling between the chassis and the PCBAs is minimized (Crede 1951, Steinberg 1988). To avoid any vibration coupling between the PCBAs, it is recommended that either all PCBAs are directly mounted on the chassis. If a PCBA is to be mounted on another PCBA, then it is recommended that the aforementioned frequency rule be followed.

It should be noted that decoupling does not mean that the chassis would not influence the response of the PCBA and vice versa. For example, Figure 7.21 depicts a chassis with two mounted PCBAs. From Figure 7.22, we learn that the natural frequencies of the two PCBAs are 100 Hz and 150 Hz, respectively, and the natural frequency of the chassis is 400 Hz.

It stands to reason that if the chassis were excited precisely at 400 Hz, the inertia effects would act as an applied force on both PCBAs. In fact, the heavier the chassis (in relationship to the PCBA) the more severe this forcing function would be. In this regard, Figures 7.23 and 7.24 reflect the influence of the chassis on the

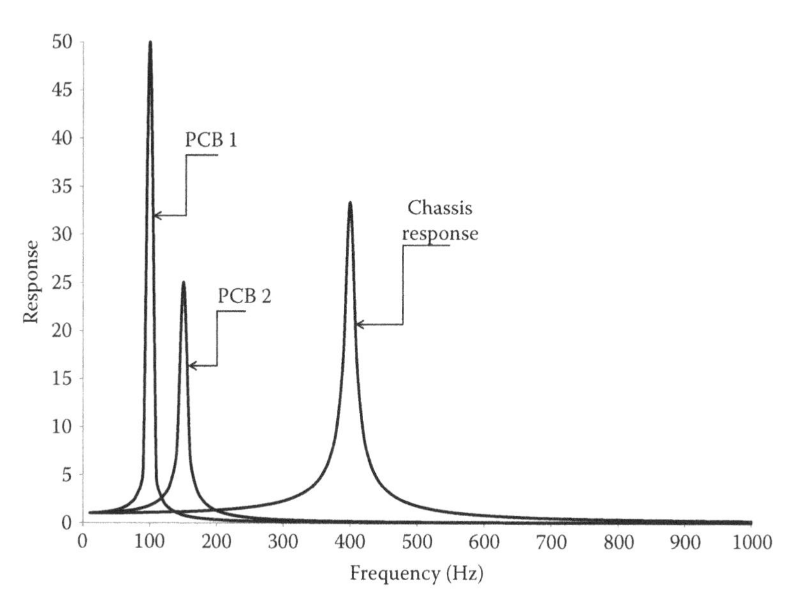

FIGURE 7.22
Frequency response of each component.

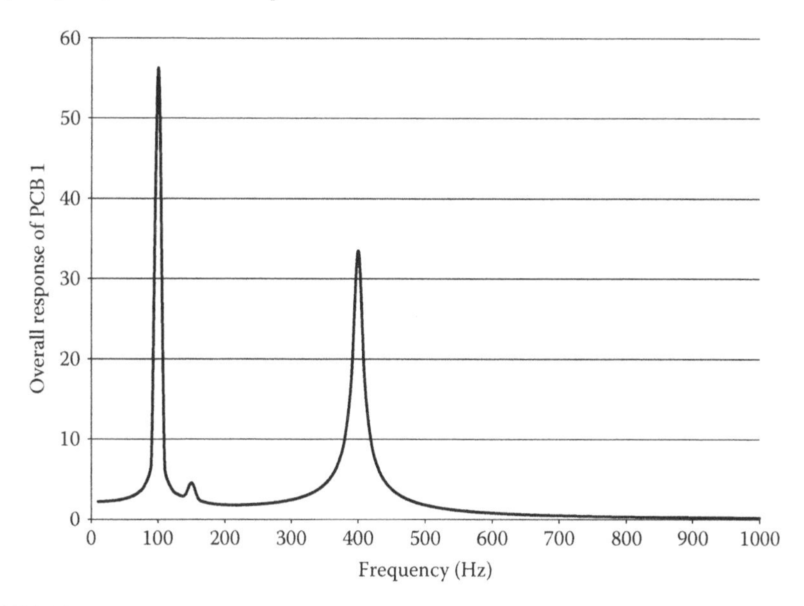

FIGURE 7.23
Frequency response of the first PCBA including the influence of the chassis.

individual PCBAs. Similarly, it may be argued that the PCBAs have a similar impact on the vibration of the chassis at lower frequencies, albeit, because of the typically lower mass, this influence is generally much less. Figure 7.25 shows the relative influence of the PCBA vibration on the chassis movement.

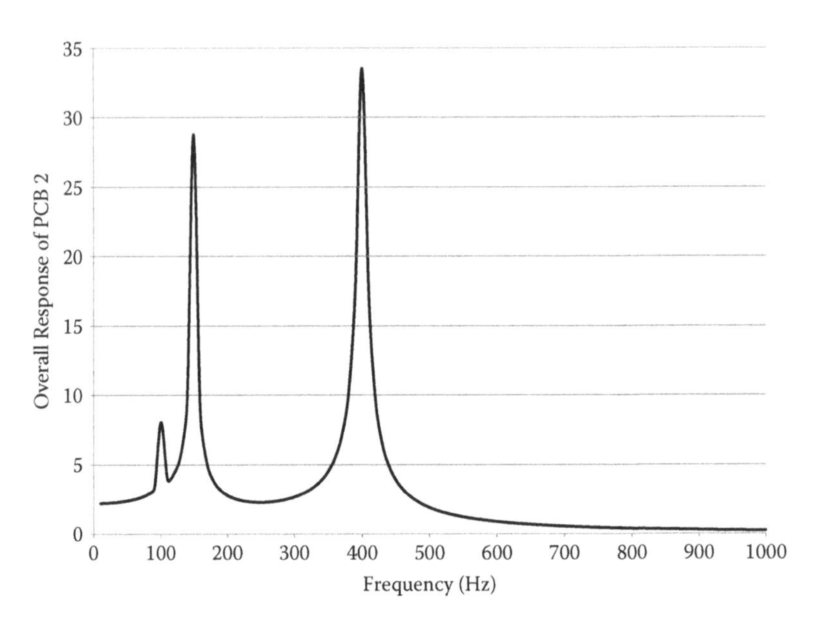

FIGURE 7.24
Frequency response of the second PCBA including the influence of the chassis.

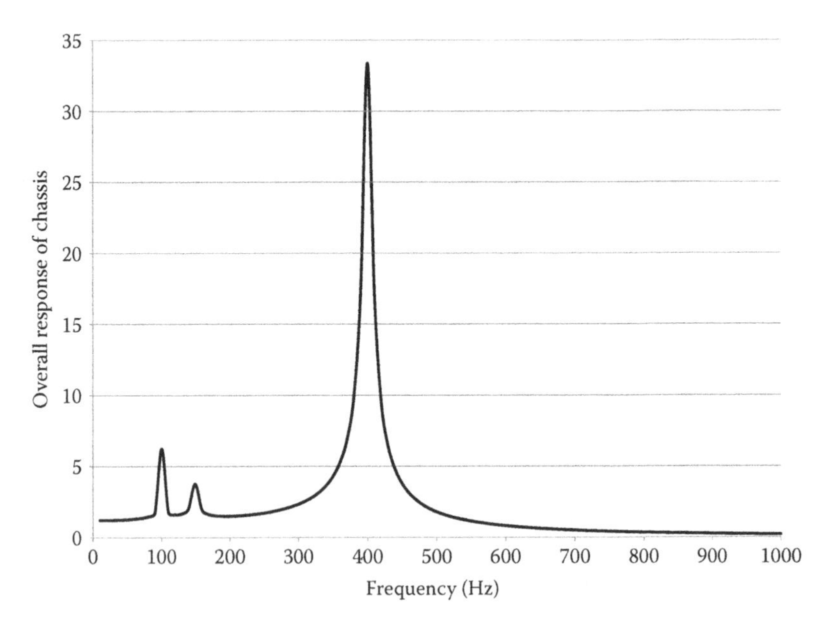

FIGURE 7.25
Frequency response of the chassis including the influence of the PCBAs.

There is one more point that needs to be discussed in the presence of a third small peak in Figure 7.23 as well as in Figure 7.24. These two peaks correspond to the resonance of each PCBA (PCBA 2 in Figure 7.23 and PCBA 1 in Figure 7.24) influencing the other through the connection with the chassis.

Existence of these resonance peaks is particularly important in random vibration and the induced stresses. Earlier, it was pointed out that in random vibration all frequencies would be present. Thus, in this case, both PCBAs as well the chassis would vibrate at 100-, 150-, and 400-Hz frequencies. The following procedure outlines the steps needed to calculate the system's response:

1. Calculate the natural frequency each component including the PCBA and the chassis.
2. Calculate the transmissibility of the chassis.
3. Calculate the frequency ratio of each PCBA to the chassis:

$$R_{\text{Chassis}} = \frac{f_{\text{Forcing}}}{f_{\text{Resonant}}} = \frac{f_{\text{PCB}}}{f_{\text{Chassis}}}$$

4. Calculate the contribution of the chassis movement to each PCBA:

$$T_m^{\text{Contri}} = \frac{1}{1 - R_{\text{Chassis}}^2} \text{ for each frequency.}$$

5. Calculate the transmissibility of each PCBA.
6. Calculate the frequency ratio of the chassis to each PCBA:

$$R_{\text{PCBA}} = \frac{f_{\text{Forcing}}}{f_{\text{Resonant}}} = \frac{f_{\text{Chassis}}}{f_{\text{PCBA}}}$$

7. Calculate the contribution of the PCBA movement to the chassis:

$$T_m^{\text{Contri}} = \frac{1}{1 - R_{\text{PCBA}}^2}$$

8. Calculate each component's response using

$$G_{\text{RMS}} = \sqrt{\frac{\pi}{2} \sum_i f_i T_{mi} PSD_i}$$

   where i refers to the frequency at each peak (Steinberg 1988).
9. G_{RMS} may now be used to evaluate deflections and/or induced stresses in each component.

Numerical techniques, in general, and finite element analysis (FEA) in particular, may be used to find "exact" solutions and are the tools of choice for many engineers. Although, our back-of-the-envelope assumptions would reduce a PCBA to a spring–mass system and would identify only the fundamental frequency, a finite element model of the same PCBA could potentially identify hundreds of natural frequencies associated with each mode shape. Numerical tools, although useful, must not be used blindly. To develop a better understanding of FEA, Chapter 10 covers this topic.

A Few Words on Advanced and Active Isolation Techniques

In this chapter, we primarily studied the impact of the vibration on the induced stresses and ultimately, whether the design could withstand the vibration or not. The implication was that if the product has weaknesses, those area would then be either redesigned or vibrations of "troublesome" frequencies isolated.

This approach to vibration isolation is called "passive isolation." Unlike passive isolation methods, there are "active isolation" techniques as well. These methods rely on feedback loops and a secondary means of counteracting the primary vibration inputs. One such technique is to attach a "dynamic absorber" to the device, which has the same frequency as the base unit; however, it provides its own driving force in such a way as to counteract the initial excitation. Aida et al. (1999) patented a dynamic absorber whose driving force was based on changes to a magnetic field. Dynamic absorbers are well studied. See, for instance, Malfa et al. (2000) or Sang-Myeong et al. (2008).

Some of the proposed dynamic absorbers could be rather bulky, as a result, others have looked into piezoelectric devices to provide the needed counterexcitation to isolate vibrations. Chomette et al. (2010) placed piezoelectric components on a PCBA to target specifically those modes of vibration that cause the most damage. He demonstrated the effectiveness of this approach experimentally by observing a reduction of PCBA damage via post excitation damage analysis.

Although active vibration isolation (or control) is a thriving scientific area of study, any further discussion is beyond the scope of this book.

8

Basics of Shock Management

Introduction

The physics of the shock environment is essentially different from that of vibration. Generally speaking, vibrations are steady-state phenomena caused by periodic excitations that have relatively long durations, whereas shocks are transient phenomena caused by very short nonperiodic excitations. Therefore, the absorbed energy in a shock environment must be released over a longer period of time—compared to the frequency of excitation—than required by vibration isolators.

There are two basic types of shock environments: one is caused by a sudden motion of a support or foundation, and the other is caused by applied or generated forces by the equipment and experienced by the support. In electronics packaging, the primary concern is with the sudden movement of the support structure. This support structure either suddenly moves under the influence of a large force in a very short time (impact) and goes back to its original position (here termed a pulse shock), or undergoes a sudden change in velocity (here called a velocity shock) as in a drop test. In both models, it is relatively straightforward to calculate the maximum acceleration, velocity, and displacement based on the pulse shape—generally a half sine, square wave, or saw tooth form—or the change in velocity.

It is also possible to calculate the structural response to the excitation. This last model is called shock response spectrum. It involves rigorous mathematics, and because of its complexity is not reviewed here. It is the most accurate and complete approach, but at the same time, the most difficult. Practical applications of this technique require the use of numerical modeling and will not be discussed here (Sloan 1985; Curtis and Lust 1996).

Our interest in shock management is to determine whether our system is capable of enduring a particular shock environment, and if not, how we may select (or design) an isolator. We need to keep in mind that a system experiencing a shock tends to move and vibrate. As with any vibration problem—albeit transitory—there are three main factors, namely, its natural frequency of vibration, the maximum amplitude, and the level of damping in the system. Therefore, in the design of a shock isolator, these three factors need to be considered.

Since vibration is a natural by-product of shocks, it may be easily concluded that the same isolator may be used in isolating vibrations as well. Although this is a logical expectation, it should be kept in mind that isolation for each phenomenon has different requirements. If we need to design an isolator that would be used in both shock and vibration environments, we need to be prepared to make compromises. Later, we will review an example of this type of isolator design that will outline the challenges involved.

The challenge in isolating a shock excitation is to reduce the induced displacements to a manageable level. It turns out that for high values of damping, the excitation forces can easily be transmitted (and amplified) to the critical components. It may be shown that if the natural frequency of the isolator is substantially less than that of the system (on the order of 10%–50%), the induced displacements may be reduced significantly (Curtis and Lust 1996).

Pulse Shock Isolation

The steps in isolating a pulse shock are relatively simplistic:

1. Identify shock strength as its pulse frequency. Generally, a pulse duration τ is given, for example, 5 ms. Pulse frequency is calculated from

$$f_p = \frac{1}{2\tau}$$

2. Assume an isolator frequency about 10%–15% of the system's first natural frequency
3. Calculate the ratio of isolator frequency f_i to the pulse frequency

$$R = \frac{f_i}{f_p}$$

4. Calculate shock transmissibility. Transmissibility curves are provided by Harris (1996); however, for $R \leq 0.5$, transmissibility may be calculated from the following relationship:

$$T_s = 2R$$

5. Next, using the following equation, calculate the maximum displacement induced by the shock excitation:

$$\delta_{max} = \frac{9.8 G_{Shock} T_s}{f_p^2}$$

6. Finally, ensure that δ_{max} is less than $\delta_{Allowable}$

Example 1

A piece of electronics equipment weighs 4.5 lb. It will be subjected to a 70*g* shock with a half sine pulse of 3.5 ms. Select a set of four springs that would limit the displacement of this unit to less than 0.5 in.

Start by calculating the pulse frequency:

$$f_p = \frac{1}{2 \times 0.0035} = 142.86$$

Now assume an isolator frequency of 10 Hz and calculate R and T_s:

$$R = \frac{10}{142.86} \simeq 0.07$$

$$T_s = 2 \times 0.07 = 0.14$$

From here, we calculate the maximum displacement:

$$\delta_{max} = \frac{9.8 G_{Shock} T_s}{f_p^2} = \frac{9.8 \times 70 \times 0.14}{10^2} = 0.96 \text{in}$$

Since this value is larger than the allowable limit, we need to choose a different spring. For the sake of argument, assume an isolator frequency of 20 Hz. We can repeat the calculations again:

$$R = \frac{20}{142.86} \simeq 0.14,\ T_s = 2 \times 0.14 = 0.28,\ \delta_{max} = \frac{9.8 \times 70 \times 0.28}{20^2} = 0.48 \text{in}$$

Clearly, this would be the correct spring selection. Now, we can calculate the spring stiffness. From

$$f_i = \frac{1}{2\pi}\sqrt{\frac{k}{m}}$$

we obtain

$$f_i = 20$$

$$k = 4\pi^2 f_i^2 m = 4\pi^2 \times 20^2 \times \frac{3\,\text{lb}}{386\dfrac{\text{in}}{\text{Sec}^2}}$$

$$k = 122.73\frac{\text{lb}}{\text{in}}$$

Obviously, we are assuming that this is an equivalent spring constant. Should our system look like a brick and we need to place four springs, each spring's stiffness is calculated as follows.

$$k_{\text{at each corner}} = \frac{122.73}{4} = 30.68\frac{\text{lb}}{\text{in}}$$

In the real world, calculations would not stop here. Chances are that such a spring does not exist as a standard item. Our next step would, therefore, be to identify the nearest available spring and repeat the calculations to make sure that we have the right selection.

Example 2

Let us once again consider the printed circuit board assembly (PCBA) used in the system depicted in Figure 5.6. It was shown that this PCBA has a fundamental natural frequency of 71.44 Hz if it is simply supported on its boundaries, and has a fundamental natural frequency of 141.58 Hz if it is clamped on four sides. This board and its critical component, as shown in Figure 8.1, must pass a 10*g* sinusoidal vibration test and a 50*g* shock with a half sine pulse of 0.005 s. Select a set of isolators that will permit these constraints.

First, conduct the analysis for the simply supported sides, and then repeat for the clamped boundary conditions.

System without Isolators

First, study the board's response in vibration without isolators using Equation 7.11:

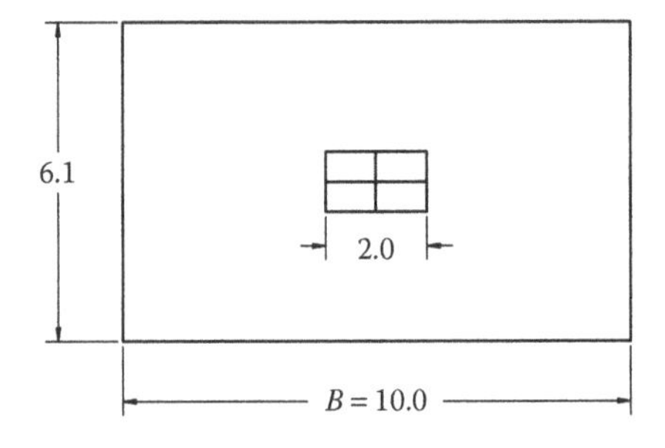

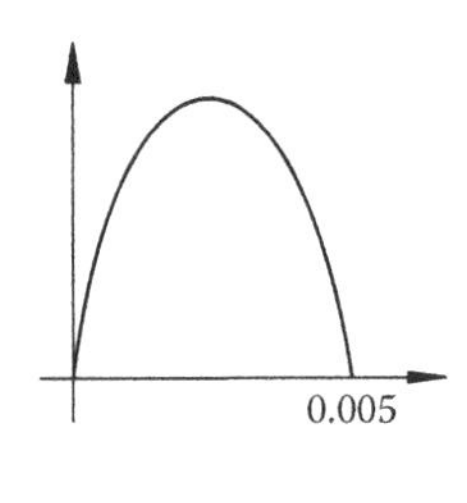

FIGURE 8.1
A PCB subjected to shock and vibration.

$$X_e = \frac{9.78 G_{in} T}{f^2} = \left\{ \begin{array}{c} G_{in} = 10 \\ f = 71.44 \\ T = \frac{1}{2}\sqrt{f} = \frac{1}{2}\sqrt{71.44} = 4.226 \end{array} \right\} \Rightarrow X_e = 0.0811 \text{ in}$$

The maximum allowable displacement based on Equation 7.13 is

$$X_{\max} = \frac{0.00022 \ B}{\text{Chr}\sqrt{L}} = \left\{ \begin{array}{c} B = 10 \\ C = r = 1 \\ h = 0.063 \\ L = 2.0 \end{array} \right\} \Rightarrow X_{\max} = 0.0247 \text{ in}$$

So there is a great chance that the system will fail under the vibration condition; thus, we need an isolator to work properly under both the shock and vibration conditions.

System with Isolators

Design 1

Assume a set of isolators with a resonant frequency of 5 Hz to provide vibration isolation for the 71.44 Hz PCBA. For the shock environment the input factors are

$$G_{\text{Shock}} = 50$$

$$f_i = 5$$

$$f_p = \frac{1}{2 \times 0.005} = 100$$

$$R = \frac{5}{100} = 0.05$$

$$A = 2R = 2 \times 0.05 = 0.1$$

$$\delta_{\max} = \frac{9.78 G_{\text{Shock}} T_s}{f_p^2} = \frac{9.78 \times 50 \times 0.10}{5^2} = 1.96 \text{ in}$$

This dynamic displacement value is too large for this application. A more appropriate value should be in the range of 0.3 or 0.4 in.

Design 2

Assume a set of isolators with a resonant frequency of 20 Hz:

$$f_i = 20$$

$$R = \frac{20}{100} = 0.2$$

$$A = 2R = 2 \times 0.2 = 0.4$$

$$\delta_{\text{max}} = \frac{9.78 G_{\text{Shock}} T_s}{f_p^2} = \frac{9.78 \times 50 \times 0.40}{20^2} = 0.49 \text{ in}$$

This value may still be too large.

Design 3

Assume a set of isolators with a resonant frequency of 35 Hz:

$$f_i = 35$$

$$R = \frac{35}{100} = 0.35$$

$$A = 2R = 2 \times 0.35 = 0.7$$

$$\delta_{\text{max}} = \frac{9.78 G_{\text{Shock}} T_s}{f_p^2} = \frac{9.78 \times 50 \times 0.70}{35^2} = 0.28 \text{ in}$$

This value is acceptable; however, we need to make sure that the displacements under vibration conditions will be acceptable as well. The way we would go about this is to imagine that the PCBA in resonance has become the forcing function for the isolators. As such, we can calculate the dynamic magnification factor for a system that has a natural frequency f_p and a forcing frequency equivalent to that of the PCBA.

For small levels of damping, we can assume $\xi = 0$ and write the equation for dynamic magnification as

$$D_m = \frac{1}{1 - r^2}$$

where

$$r = \frac{f_{\text{PCBA}}}{f_{\text{isolator}}}$$

Again, keep in mind that here, D_m is the dynamic magnification factor of the isolators and not of the PCBA:

$$r = 71.44 / 30 = 2.0411 \Rightarrow D_m = -0.3158$$

A negative sign only means that the system response is out of phase with the input:

$$X_i = X_e D_m \Rightarrow X_i = 0.0811 \times 0.3158 \Rightarrow X_i = 0.0256 \text{ in}$$

This is larger than the allowable displacement and therefore this set of isolators is not acceptable.

Design 4

Assume a set of isolators with a resonant frequency of 30 Hz:

$$f_i = 30$$

$$R = \frac{30}{100} = 0.30$$

$$A = 2R = 2 \times 0.30 = 0.6$$

$$\delta_{\text{max}} = \frac{9.8 G_{\text{Shock}} T_s}{f_p^2} = \frac{9.8 \times 50 \times 0.6}{25^2} = 0.33 \text{ in}$$

Now let us check for vibration:

$$r = 71.44 / 30 = 2.3813 \Rightarrow D_m = -0.2141$$

$$X_i = X_e D_m \Rightarrow X_i = 0.0811 \times 0.2141 \Rightarrow X_i = 0.0174 \text{ in}$$

This value is below the allowable limit, and therefore the design is acceptable if the simply supported design is used. Should the sides be clamped, the calculations would be as follows:

$$X_e = \frac{9.78 G_{\text{in}} T}{f^2} = \left\{ \begin{array}{c} G_{\text{in}} = 10 \\ f = 141.58 \\ T = \sqrt{f} = \sqrt{141.58} = 11.87 \end{array} \right\} \Rightarrow X_e = 0.0582 \text{ in}$$

Note that the relationship

$$T = \sqrt{f}$$

instead of

$$T = \frac{1}{2}\sqrt{f}$$

was used because the frequency of the clamped board is now higher than 100 Hz.

Design 4 suggested that an isolator with a frequency of 30 Hz would be acceptable. Now, let us examine this isolator for the clamped board:

$$r = 141.58 / 30 = 4.7167 \Rightarrow D_m = -0.0471$$

$$X_i = X_e D_m \Rightarrow X_i = 0.0582 \times 0.0471 \Rightarrow X_i = 0.0027 \text{ in}$$

This value is much less than the allowable level and thus the same isolators are acceptable regardless of the condition of the sides. The decision we need to make here is whether we like to isolate the PCBA or the chassis. Depending on that decision, we would then calculate the spring constant of the isolators.

Velocity Shock Isolation

Implicitly, when we considered the pulse shock, we had assumed that the entire system behaves as a spring-mass system. In fact, this is a very good assumption. If shock or vibration isolation is done properly, the system (or the electronics package) would in fact be much more rigid than the isolators and generally have much higher frequencies. A similar assumption holds true here, particularly if the velocity shock is a drop test. Thus, we can assume a spring-mass system's behavior at the moment of impact.

This assumption enables us to equate the kinetic energy of the system with the energy required to deflect the spring a distance Δ (or to the stored energy in the spring). By doing so, we obtain the maximum deflection:

$$\Delta = V\sqrt{\frac{m}{k}} = \frac{V}{2\pi f_i} \tag{8.1}$$

where f_i is the natural frequency of the system (without isolators) or of the isolators if they exist. V is the change in velocity.

For a drop test, V may be calculated by equating the potential energy of the mass (here, electronics package) at a height h to its kinetic energy at the moment of impact:

$$V = C\sqrt{2gh} \tag{8.2}$$

g is gravitational constant, h is the drop height, and C is the coefficient of rebound. It is equal to two (2) for a full rebound and one (1) for no rebound at all.

Now, we need to relate this information to the transmitted shock acceleration. Again, by equating the maximum kinetic energy at the moment of impact and the stored energy in the elastic element (i.e., springs) it may be shown that shock acceleration is given by the following relationship (Crede 1951; Steinberg 1988):

$$G_{max} = \frac{2\pi f_i V}{g} \tag{8.3}$$

By combining Equations 8.2 and 8.3 and rearranging the terms, an expression for the height may be developed:

$$h = \frac{G_{max}^2 g}{8\pi^2 C^2 f_i} \tag{8.4}$$

Recall that in this equation, C is the rebound factor. Therefore, if the electronics box is dropped from any heights above h, it will be damaged.

An interesting aspect of Equation 8.4 is this: suppose that the unit that we are designing had to survive a drop from a height more than h. What type of isolators should we choose? The steps to do this are as follows:

1. Calculate the change in velocity from Equation 8.2
2. From Equation (8.3), calculate the natural frequency of the isolators:

$$f_i = \frac{G_{max} g}{2 V}$$

3. Calculate the dynamic deflection of the isolators from Equation 8.1
4. Finally, calculate the spring constants from the calculated natural frequency of the isolators and the mass of the unit

Example

A 20-lb electronics box with a natural frequency of 125 Hz has a fragility level of 12gs (i.e., it cannot withstand more than 12gs). This unit has to survive a drop from a height of 12 in with no rebound. We need to design a set of isolators that would make this drop possible.

Noting that there is no rebound (C = 1), from Equation 8.3, we have

$$h = \frac{G_{max}^2 g}{8\pi^2 C^2 f_i} = \frac{12^2 \times 386}{8 \times 3.14^2 \times 1^2 \times 125}$$

$$h = 5.64 \text{ in}$$

Now calculate the velocity from a 12-in drop from Equation (8.2):

$$V = \sqrt{2 \times 386 \times 12}$$

$$V = 96.25 \frac{\text{in}}{\text{Sec}}$$

From Equation (8.3), calculate the natural frequency of the isolators:

$$f_i = \frac{G_{\max} g}{2\pi V} = \frac{12 \times 386}{2 \times 3.14 \times 96.25}$$

$$f_i = 7.66\,\text{Hz}$$

The dynamic deflection of the isolator is [Equation 8.1]:

$$\Delta = \frac{96.25}{2\pi \times 7.66}$$

$$\Delta = 1.999\text{ in}$$

The displacement of this box will be approximately 2 in. Isolator spring constant (stiffness) is easily computed next:

$$f_i = \frac{1}{2\pi}\sqrt{\frac{k}{m}}$$

From here

$$k = 4\pi^2 f_i^2 m$$

$$k = 4 \times 3.14^2 \times 7.66^2 \times \frac{20}{386}$$

$$k = 120.0 \frac{\text{lb}}{\text{in}}$$

This is the stiffness of the dynamic system to ensure survival from a 12-in drop test.

Maximum Desired PCBA Deflection

Similar to vibration, Steinberg (1988) suggests that if the peak single-amplitude displacement of the PCBA is limited to $X_{\max}$, the component can achieve a fatigue life of about 10 million stress reversals in a sinusoidal vibration environment. For a shock condition,

$$X_{\max} = \frac{0.00132\,B}{Chr\sqrt{L}}\,\text{in}$$

Chr and *L* are defined by Equation 7.14. Again, as mentioned when Equation 7.13 was introduced, the accuracy of this equation with new lead-free solders needs to be verified.

Equipment Design

Aside from the PCBA, the electromechanical engineer must be mindful of the design aspects of other parts of the packaging. These include the chassis design, which may be formed from sheet metal, cast from a light metal and machined, or even injected molded plastic or a variety of metals; the cabinet design and its structural strength, as well as its shock and, vibration issues; and, the vibration of wires and cables as well as their interaction with other components, which may lead to failure.

Methods of construction of the chassis or the cabinet play an important role in the survival of the system as a whole in vibration and shock environments. For example, the friction between bolted or riveted joints dissipates energy and as a result represents a great level of damping in the system. Therefore, these types of structures are superior to welded structures in vibration environments. Furthermore, in welded joints, stress concentration is a common defect leading to low fatigue life. Reducing the number of welded joints as well as their heat treatment to reduce the residual stresses may eliminate this problem.

As for wires and cables, it is prudent to tie wires that extend in the same direction, support the harness length, minimize the lead length, and clamp the wires near the termination to a structure.

In short, we cannot afford to place all of our attention on one aspect of the package and ignore the impact of the rest of the system.

9

Induced Stresses

Introduction

In calculating natural frequencies and mode shapes, the applied loads (either forces or acceleration, e.g., gravity g loads) do not play a role. However, when we speak of a system's failure in a shock or vibration environment, we are making a reference to the deflection (and by necessity to stress) at a critical point in the system. These values cannot be evaluated without knowledge of the applied loads on the system. This necessitates our understanding and evaluation of the dynamic loads acting on as well as the behavior of the deformable structure. Here, the assumption of a spring-mass may no longer be accurate.

Fortunately, calculating a system's response under dynamic loads is not difficult at all. The principles of strength of material still hold as one realizes that the components used in an electronics enclosure are beams, plates, and frames. The question remains, how should one treat the "dynamic" loads? We need to keep in mind that there are three categories of dynamic loads each belonging to forced vibration, random vibration, and shock loading. We need to calculate the dynamic loads somewhat differently for each condition.

Forced Vibration

Recall that transmissibility is the ratio of the transmitted forces to the static forces. Thus, dynamic loads are nothing but the product of transmissibility and the static loads. As a result, one may enumerate the steps to calculate the induced stresses as follows:

1. Calculate the natural frequencies.
2. For a given load, calculate transmissibility of the system.
3. Calculate the dynamic load (a product of transmissibility and static loads).
4. Calculate the stresses from the dynamic loads.

Although engineers must develop a solid foundation of the analysis tools that they use, the techniques described here would be best used for developing a "sense" for the system. Nowadays, finite element analysis (FEA) tools can be readily used to calculate the loads and stresses on the electronics enclosures. Thus, the following example is provided for the sake of completeness, because in reality, a computational tool will be used and the steps in calculating the stresses and strains will be transparent to the user.

Sample Problem

As an example, let us consider the beam problem that we used previously to calculate the dynamic magnification factors. Now we need to understand the stress levels so an appropriate material may be selected (Figure 9.1).

Based on strength of materials theories, the stresses in a beam are given by the following relationship:

$$\sigma = \frac{Mc}{I}$$

where M is the bending moment, I is moment of inertia, and c is the distance from the outer fibers to the neutral axis. For more details, the reader is encouraged to read any textbook on strength of materials. The bending moment is evaluated from the shear and moment diagram, which may be developed by using a free body diagram. Note that the dynamic bending moment is the static value multiplied by transmissibility.

Recall that for this system, $\omega = 41$, $\xi = 0.01$ and $c = 0.5$. From a free body diagram (not shown here), we obtain

$$M = 120\,\text{lb-in}$$

Therefore, a static stress value is

$$\sigma_{\text{static}} = \frac{Mc}{I} = \frac{120 \times 0.5}{0.03}$$

$$\sigma_{\text{static}} = 2000\,\text{psi}$$

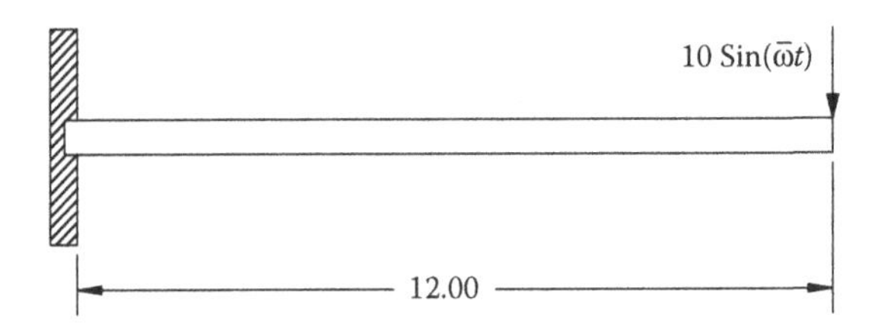

FIGURE 9.1
A cantilever beam under a vibratory load.

Previously, we calculated for $r_1 - 0.73$, T_m –2.13; therefore,

$$M = 120 \times 2.13 = 256.8 \text{ lb-in}$$

leading to

$$\sigma_1 = \frac{256.8 \times 0.5}{0.03}$$

$$\sigma_1 = 4280 \, \text{psi}$$

Similarly,

R	T_m	M	σ
30/41 = 0.73	2.13	256.8	4280
40/41 = 0.98	19.24	2308.7	38,479
50/41 = 1.22	2.05	246.1	4101
60/41 = 1.46	0.88	105.1	1767

Note that moment is in lb-in and stresses are in psi. Furthermore, static bending stress is 2000 psi, while dynamic stress may rise to as much as 38,479 psi. Clearly, at dynamic loads, failure may occur!

Random Vibration

In the section on random vibration in Chapter 7, we discussed how the system response (acceleration) to a given excitation spectrum may be calculated (Equation 7.15). Dynamic loads (forces and moments) may be evaluated from the output acceleration. In the British system, dynamic force is the product of the system's weight and root mean square of the output acceleration (G_{RMS}). Whereas in other measurement systems, G_{RMS} must be converted into an appropriate unit. One has to exercise caution not to ignore stress concentration factors if there are holes or notches in the system and to take proper units into account if strength of material equations is used. Otherwise, if we use an FEA package, G_{RMS} value may be used as an input body force value in a static analysis to calculate stresses and deflections.

Probability of Failure

There are three levels of stresses corresponding to the three levels of the probable G_{RMS} levels. As an example, assume that 1S stress is 4624 psi. Then, 2S and 3S stresses are 9248 psi and 13,872 psi, respectively. If the material is aluminum with a yield stress of 12,000 psi, then there is a 4% chance that the part will undergo yielding.

Example

Calculate the stresses for the beam shown in Figure 9.1 using the previous white noise power spectral density (PSD) shown in Figure 7.17. Determine the probability of failure for 6061-T6 aluminum and stainless steel. Assume a 10% damping in the system because of coatings, wiring, and so on. The natural circular frequency is 41. For aluminum the yield stress is about 12 ksi and for steel it is 40 ksi.

Calculation of stresses starts with Equation 7.15 as follows:

$$G_{\mathrm{RMS}}^{\mathrm{out}} = \sqrt{\frac{\pi}{2} f T_m \mathrm{PSD}}$$

$$G_{\mathrm{RMS}}^{\mathrm{out}} = \sqrt{\frac{1}{4} \omega T_m \mathrm{PSD}}$$

$$T_m = \frac{1}{2\xi} = \frac{1}{2 \times 0.1} = 5 \text{ at resonance}$$

$$G_{\mathrm{RMS}}^{\mathrm{out}} = \sqrt{\frac{1}{4} \times 41 \times 5 \times 0.08} = 2.02$$

$$F = lbs \times G_{\mathrm{RMS}}^{\mathrm{out}} = 10 \times 2.02 = 20.2\,\mathrm{lb}$$

$$M = F \times l = 20.2 \times 12 = 242.4\,\mathrm{lbs\text{-}in}$$

$$\sigma_{\mathrm{RMS}} = \frac{KMc}{I} = \frac{1 \times 242.4 \times 0.5}{0.03} = 4040\,\mathrm{psi}$$

K is the stress concentration factor. Since there are no holes or notches in the system, $K = 1$. $\sigma_{\mathrm{RMS}} = 4040\,\mathrm{psi}$ is the 1S stress level compared to a static value of 2000 psi. The 2S and 3S are 8080 (= 2 × 4040) psi and 12,120 (= 3 × 4040) psi, respectively. Clearly, there is a 4% chance that aluminum will fail, and thus it may be not a good material choice.

Shock Environment

For loads in the shock environment, we adapt a method similar to the one used in the random vibration. In the pulse shock, generally we have the shock acceleration (e.g., 20 or 50*g*'s). If we are to include the impact of the isolators, we need to modify this value by the transmissibility of the isolators. In the case of velocity shock, we can employ Equation 8.3 to compute this acceleration. Once this value is known, we can easily compute dynamic loads as follows.

$$F = lbs \times G_{\text{Shock}}$$

$$M = F \times l$$

Again, in an FEA, we can use G_{Shock} as an input body force.

10

The Finite Element Methods

Introduction

In previous chapters, we familiarized ourselves with tools for developing an understanding of heat transfer, shock, and vibration issues. Although these tools may suffice in solving design issues, they are not appropriate for conducting in-depth analyses. For realistic problems of today, a numerical—and mainly finite element—method is used. Nowadays, it is possible to use a computer-aided design (CAD) software to "design" a system and then by click of a button, solve for stresses and strains in various components without the slightest notion of any degree of accuracy.

Since a CAD model is closest to the "real thing," an interesting question arises: What is the difference between a real-world system, an engineering model, and a finite element model?

As engineers, we transform real-world systems into engineering models by making "engineering" assumptions. This enables us to simplify the physics of the system to a level where realistic solutions may be sought. By further transforming these models into idealized mathematical models, finite element models are developed and are characterized by nodes, elements, and boundary conditions.

The purpose of this chapter is to provide a basic understanding of finite element analysis (FEA) and its proper applications in the context of heat transfer and vibration.

The basic idea in FEA is to find the solution of a complicated problem by replacing it with a simpler one. The earliest application of FEA may be considered to be the ancient technique of finding the lower and upper bounds to the circumference of a circle. However, FEA as it is known today was presented by Turner and his colleagues in 1956 for the analysis of aircraft structures (Turner et al. 1956).

Some Basic Definitions

The following are basic definitions of some terms commonly used in the finite element formulation:

Node: A node is a location in the model where variables—such as displacements and temperatures are calculated. A node contains degrees of freedom

Element: A building block for the model, it contains and dictates the relationship between nodes

Element connectivity: This is a set of nodes that make up an element. Furthermore, element connectivity allows information to be shared in between elements

Higher order elements: Recall that the finite element approximations make use of shape functions (also called trial functions or interpolation polynomial). Should this shape function be of the order of two or more, the element is known as a "higher order element." In higher order elements, some interior nodes are introduced in addition to the corner nodes in order to match the number of nodal degrees of freedom with the number of generalized coordinates in the interpolation polynomial

Isoparametric elements: One of the strong points of the finite element methods is its ability to model curved geometries—such as round boundaries. If the same shape function is used to define both the field variables and the geometry, then the formulation is said to be isoparametric

Subparametric elements: If the interpolations function used to define the field variables has a lower order than that of the geometry, then the formulation is said to be subparametric

Superparametric elements: If the interpolations function used to define the field variables has a higher order than that of the geometry, then the formulation is said to be superparametric

The FEA Procedure

To take advantage of any FEA approach, one must develop a basic understanding of the physical and engineering issues, a basic understanding of the fundamental concepts of finite element method, and knowledge of the capabilities and limitations of the approach used. In general, there are six steps that need to be followed in a FEA procedure. These are as follows:

Step 1. From the physical to the FEA representation: In this step, one has to decide what the true concerns are, that is, what is it that needs to be determined, overall deflections, localized stresses, and so on. By knowing what variables are needed, appropriate elements can be chosen that lead to an overall knowledge of how to make the transition from the physical to the FEA model

Step 2. Discretization: Once the FEA model is known, it is then subdivided into a number of elements. The analyst must take into account changes in geometry, material properties, loading, and so on. Furthermore, questions on the size and the number of elements used as well as simplifications afforded by the physical configuration of the body and loading must be addressed

Step 3. Application of loads: In general, this step is simple unless loads depend on geometry or other variables

Step 4. Application of boundary conditions: Constraints must be applied where physical boundaries are constrained or when modeling simplifications, such as symmetry, are used

Step 5. Assembly of element equations and the solution phase: In today's sophisticated FEA programs, this step is done automatically

Step 6. Review of the solution and validation: This is the time that the analyst would study the field variables. Care must be exercised here to not treat the results as absolute. The results must be validated in order to *gain* confidence in the solution

Steps 1 through 4 are generally called the preprocessing phase, Step 5 is the solution phase, and Step 6 is the post-processing phase.

Finite Element Formulation

By way of an example, the following is an illustration of how an engineering problem may be solved using finite element formulation. This problem is a bar with two different cross-sectional areas, clamped at one end, and loaded with a lateral force at the free end, as shown in Figure 10.1. Our goal is to calculate the displacements of this bar at various locations, particularly where the two cross sections meet and at the end where the load is applied. Note that in this formulation displacements are calculated first. They are called the primary variables. If stress or strain field is needed, it must be developed from the calculated displacements.

Consider a generic line element of length l (as in Figure 10.2) and assume that the displacement varies linearly from one end to the other—each end is called a node. Then, a simple finite element model of this system is created as shown in Figure 10.3.

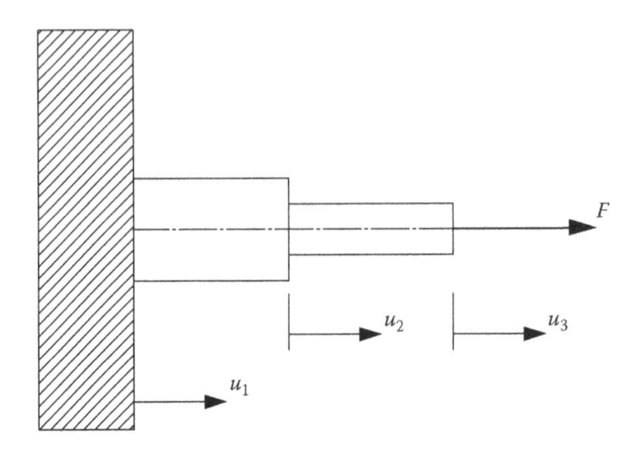

FIGURE 10.1
Two bars loaded at one end.

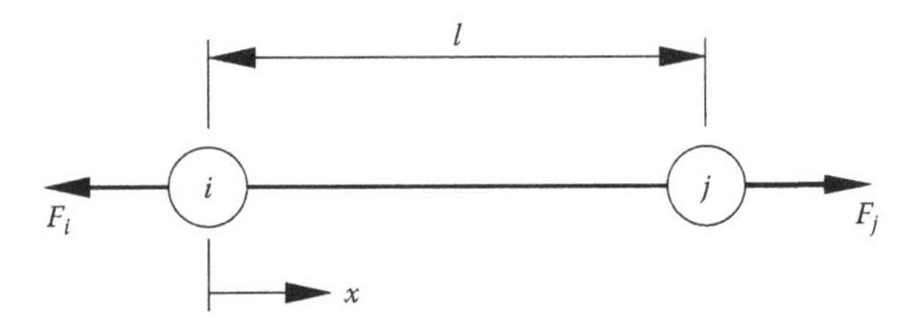

FIGURE 10.2
A simple line element of length l.

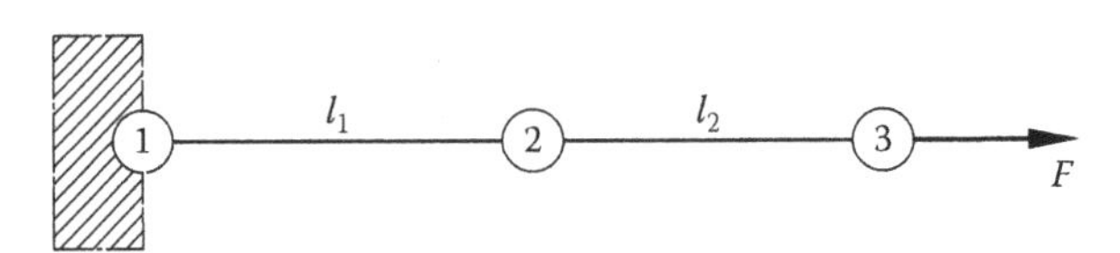

FIGURE 10.3
The finite element model of the original system.

Now, we assume that the displacement varies linearly from one node to the other. In other words:

$$u(x) = a + bx \tag{10.1}$$

Let us apply the boundary condition:

$$\text{at } x = 0 \quad u(0) = u_i$$

$$\text{at } x = l \quad u(l) = u_j$$

Thus, one concludes that

$$a = u_i$$

$$b = \left(\frac{u_j - u_i}{l}\right)$$

By substituting these relationships into Equation (10.1), we develop the elemental equation for displacement based on nodal values:

$$u(x) = \left(1 - \frac{x}{l}\right)u_i + \left(\frac{x}{l}\right)u_j \qquad 0 \le x \le l \tag{10.2}$$

In general this equation may be written as

$$u(x) = \sum_{i=1}^{n} \varphi_i u_i$$

where φ_i s are called the shape or trial functions. In this illustration, there are two nodes, thus $n = 2$, and

$$\varphi_1 = \left(1 - \frac{x}{l}\right)$$

$$\varphi_2 = \left(\frac{x}{l}\right)$$

The goal of this formulation is to find the displacements. Thus, from a stress–strain relationship, we can drive a force–displacement relationship as follows:

$$\sigma = E\varepsilon$$

where σ is stress, E is Young's modulus of elasticity, and ε is strain. Multiply each side by the cross-sectional area, and we obtain

$$A\sigma = EA\varepsilon$$

Note that the right-hand side is equivalent to force and strain is the derivative of displacement with respect to length, thus,

$$F = EA\frac{du}{dx}$$

or

$$F \cong EA\frac{\Delta u}{\Delta x} = EA\frac{u_j - u_i}{x_j - x_i} = EA\frac{u_j - u_i}{l_{ij}}$$

For each element, we balance the force at each node. For the first element:

$$\begin{bmatrix} \left(\frac{EA}{l}\right)_1 & \left(\frac{EA}{l}\right)_1 \\ \left(\frac{EA}{l}\right)_1 & \left(\frac{EA}{l}\right)_1 \end{bmatrix} \begin{Bmatrix} u_1 \\ u_2 \end{Bmatrix} = \begin{Bmatrix} F_1 \\ F_2 \end{Bmatrix} \tag{10.3}$$

For the second element:

$$\begin{bmatrix} \left(\frac{EA}{l}\right)_2 & \left(\frac{EA}{l}\right)_2 \\ \left(\frac{EA}{l}\right)_2 & \left(\frac{EA}{l}\right)_2 \end{bmatrix} \begin{Bmatrix} u_2 \\ u_3 \end{Bmatrix} = \begin{Bmatrix} -F_2 \\ F_3 \end{Bmatrix} \tag{10.4}$$

Notice that in a general formulation, the stiffness and the length of the two elements are different. Now combine the two elements:

$$\begin{bmatrix} \left(\frac{EA}{l}\right)_1 & -\left(\frac{EA}{l}\right)_1 & 0 \\ -\left(\frac{EA}{l}\right)_1 & \left(\frac{EA}{l}\right)_1 + \left(\frac{EA}{l}\right)_2 & -\left(\frac{EA}{l}\right)_2 \\ 0 & -\left(\frac{EA}{l}\right)_2 & \left(\frac{EA}{l}\right)_2 \end{bmatrix} \begin{Bmatrix} u_1 \\ u_2 \\ u_3 \end{Bmatrix} = \begin{Bmatrix} F_1 \\ 0 \\ F_2 \end{Bmatrix} \tag{10.5}$$

In matrix notation, this equation is written as

$$[\mathbf{K}]\{\mathbf{u}\} = \{\mathbf{F}\}$$

In general, if **[K]**—defined as the characteristic or stiffness matrix—depends on **{u}** the problem is nonlinear and if **[K]** does not depend on **{u}** the problem is said to be linear. The size of the matrix **[K]** for practical problems easily exceeds several million entries. Thus, a great deal of research has been dedicated and continues to be dedicated to solving these equations most efficiently.

To obtain the displacements, first we need to apply the boundary conditions to Equation 10.5 (i.e., zero displacement at node 1 and force equal to F at node 3) and specify the values of element length as well as stiffness. Then, we can solve for u_2 and u_3.

Formulation of Characteristic Matrix and Load Vector

In the previous example, we developed the finite element matrices based on our understanding of the physics of the problem. In general, there are three approaches to this formulation:

1. Direct approach
2. Variational (energy) approach
3. Weighted residual approach

Direct Approach

In this approach, as in the previous example, basic physics is used to form the elemental matrices. As an example consider the bar problem in Figure 10.4.

From physics, we know that force and displacement have the following relationship:

$$[\mathbf{K}]\{\mathbf{u}\} = \{\mathbf{F}\}$$

or

$$\begin{bmatrix} K_{ii} & K_{ij} \\ K_{ji} & K_{jj} \end{bmatrix} \begin{Bmatrix} u_i \\ u_j \end{Bmatrix} = \begin{Bmatrix} F_i \\ F_j \end{Bmatrix}$$

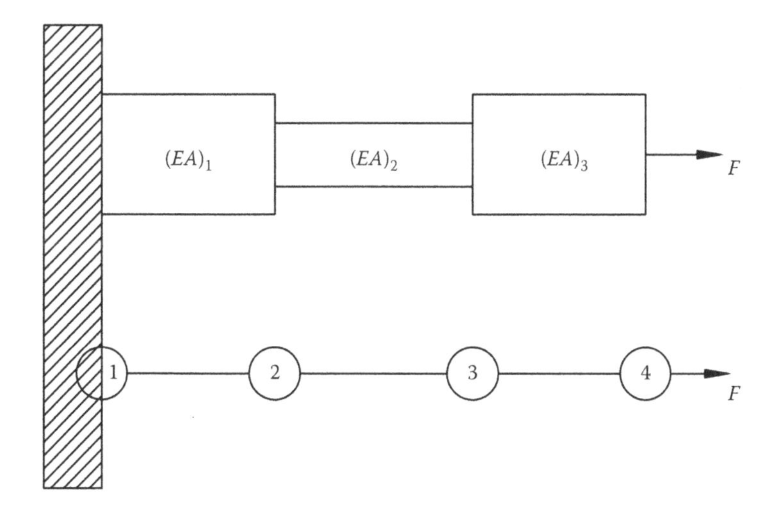

FIGURE 10.4
Bar with different cross-sectional areas.

To calculate the **[K]** matrix, we need to take advantage of influence function or coefficient. The stiffness influence coefficient is defined as the force needed at node i (in the direction of u_j) to produce a unit displacement at node j (i.e., $u_i = 1$) while all other nodes are restrained. Therefore,

$$K_{ii} = F_i$$

However from strength of materials,

$$F = \sigma A$$

And

$$K_{ii} = F_i = E\varepsilon A = EA\frac{\Delta l}{l}$$

where l is element length and Δl is the change in length. Recall that

$$\Delta l = u_j - u_i$$

$$\Delta l = 1 - 0 = 1$$

Thus,

$$K_{ii} = \frac{EA}{l}$$

Note that for calculating K_{ij} a negative force must be used because of the direction of the coordinate system:

$$K_{ij} = -\frac{EA}{l}$$

The matrix equation may now be assembled easily:

$$\begin{bmatrix} \left(\frac{EA}{l}\right)_1 & -\left(\frac{EA}{l}\right)_1 & 0 & 0 \\ -\left(\frac{EA}{l}\right)_1 & \left(\frac{EA}{l}\right)_1 + \left(\frac{EA}{l}\right)_2 & \left(\frac{EA}{l}\right)_2 & 0 \\ 0 & -\left(\frac{EA}{l}\right)_2 & \left(\frac{EA}{l}\right)_2 + \left(\frac{EA}{l}\right)_3 & -\left(\frac{EA}{l}\right)_3 \\ 0 & 0 & -\left(\frac{EA}{l}\right)_3 & \left(\frac{EA}{l}\right)_3 \end{bmatrix} \begin{Bmatrix} u_1 \\ u_2 \\ u_3 \\ u_4 \end{Bmatrix} = \begin{Bmatrix} F_1 \\ F_2 - F_2 \\ F_3 - F_3 \\ F_4 \end{Bmatrix}$$

Variational (Energy) Approach

In this approach, one takes advantage of the fact that nature would take the path of least resistance, that is, minimum energy expenditure to change from one state into another. Thus, by expressing the energies involved and then by minimizing them, not only are the equations of "motion" defined but the boundary conditions are also determined. The advantages of variational formulation are as follows:

1. Generally, the energy functional (I) has a clear physical meaning.
2. The functional contains lower order derivatives of the field variables.
3. Sometimes, both upper and lower bounds may be found.
4. Complicated boundary conditions—either natural or free are satisfied. Only geometric—or forced—boundary conditions need to be imposed.

For our example:

$$I = \text{strain energy} - \text{work done by external forces}$$

$$I = \sum_k \Pi_k - W_p$$

$$\Pi_k = \left[A \int_0^l \frac{1}{2} \sigma \cdot \varepsilon \right]_k, \quad \sigma = E\varepsilon$$

but

$$\varepsilon = \frac{u_j - u_i}{l}$$

substitute back in the strain energy relationship and obtain

$$\Pi_k = \frac{EA}{2l}\left(u_i^2 + u_j^2 - 2u_i u_j\right)$$

or in matrix form

$$\Pi_k = \left(\frac{EA}{2l}\right)_k \begin{bmatrix} u_i & u_j \end{bmatrix} \begin{bmatrix} 1 & -1 \\ -1 & 1 \end{bmatrix} \begin{Bmatrix} u_i \\ u_j \end{Bmatrix}$$

The work done by external forces is the sum of applied forces in the direction of displacements:

$$W_p = \sum_i F_i u_i$$

The functional may now be formulated as follows:

$$I = \sum_{k=1}^{3} \left\{ \left(\frac{EA}{2l}\right)_k \begin{bmatrix} u_i & u_j \end{bmatrix} \begin{bmatrix} 1 & -1 \\ -1 & 1 \end{bmatrix} \begin{Bmatrix} u_i \\ u_j \end{Bmatrix}_k - \left(\sum_i F_i u_i\right)_k \right\}$$

This energy functional must be minimized with respect to displacements, thus

$$\frac{\partial I}{\partial u_i} = 0 \text{ for } i = 1,2, \text{and } 3$$

This differentiation leads to the following set of elemental equations, which may be assembled for the overall equation:

$$\sum_{k=1}^{3} \left\{ \left(\frac{EA}{l}\right)_k \begin{bmatrix} 1 & -1 \\ -1 & 1 \end{bmatrix} \begin{Bmatrix} u_i \\ u_j \end{Bmatrix}_k - F_k \right\} = 0$$

Weighted Residual Approach

Another approach to develop the elemental equations is the weighted residual technique. The first step of this approach is based on the idea that in many practical problems, the governing differential equation may be obtained much easier than the energy functional. The next step is to find an approximate solution—such as finite element approximation—and minimize the error. This step is expressed mathematically as

$$[\mathbf{K}]\{\mathbf{u}\} = \{\mathbf{F}\}$$

Or

$$[\mathbf{K}]\{\mathbf{u}\} - \{\mathbf{F}\} = \mathbf{0}$$

If $\{\tilde{\mathbf{u}}\}$ is an approximation to $\{\mathbf{u}\}$ then

$$[\mathbf{K}]\{\tilde{\mathbf{u}}\} - \{\mathbf{F}\} = \{\mathbf{R}\}$$

Needless to say, the best approximation provides the lowest error, that is, lowest $\{\mathbf{R}\}$. To achieve this, the method of weighted residual approach dictates that we multiply the error by a "weight" function, integrate the product over the domain of interest, and set the outcome equal to zero.

This technique is best suited for areas of physics such as fluid dynamics, where development of the differential equations is much easier than the energy functional.

Finite Element Formulation of Dynamic Problems

Recall that the finite element matrix equation has the following general form:

$$[\mathbf{K}]\{\mathbf{u}\} = \{\mathbf{F}\}$$

This has the same form as the force deflection relationship in a spring-mass system. One may say that to develop the equations for a FEA, a continuous (engineering) system is broken into a series of interconnected spring-mass-damper systems.

There are three classes of dynamic response problems:

1. Rigid body dynamics where although bodies move in space, they maintain their original shape. This class of problems is not of concern to us.
2. Wave propagation where a stress (or shock) wave travels through a system. Stresses and deflection vary with time but no periodic behavior may be identified.
3. Vibration problems, where a periodic/harmonic response is expected.

Wave Propagation Type

Let us look at the vibration and numerical solution of this problem. The equation of motion for a spring-mass system is:

$$mu_{,tt} + c\,u_{,t} + k\,u = f(t) \tag{10.6}$$

In this equation, $f(t)$ is a general time-dependent forcing function. Now, time derivatives may be replaced by an approximation similar to the following:

$$u_{,t} = \frac{u_{t+\Delta t} + u_t}{\Delta t} \tag{10.7}$$

$$u_{,tt} = \frac{u_{t+2\Delta t} - 2u_{t+\Delta t} + u_t}{\Delta t^2} \tag{10.8}$$

Equations 10.7 and 10.8 indicate that to calculate the velocity at time t, the location at time $t + \Delta t$ must be known. Furthermore, to find the acceleration at the same time, the location at time $t + 2\Delta t$ must also be known. By substituting this approximation in Equation 10.6, we obtain:

$$\frac{m}{\Delta t^2} u_{t+2\Delta t} + \left(\frac{c}{\Delta t} - \frac{2m}{\Delta t^2}\right) u_{t+\Delta t} + \left(\frac{m}{\Delta t^2} - \frac{c}{\Delta t} + k\right) u_t = f(t)$$

At time $t = 0$, the initial conditions, that is, the location of the mass as well as its velocity, must be known. Then, by taking advantage of Equation 7.7, the location of the mass at time Δt may be calculated. The calculation may begin in earnest at time 2 Δt by

$$u_{2\Delta t} = \frac{\Delta t^2}{m}\left(f(0) - \left(\frac{c}{\Delta t} - \frac{2m}{\Delta t^2}\right)u_{\Delta t} - \left(\frac{m}{\Delta t^2} - \frac{c}{\Delta t} + k\right)u_0\right)$$

Once $u_{2\Delta t}$ is evaluated, $u_{3\Delta t}$ is evaluated as follows:

$$u_{3\Delta t} = \frac{\Delta t^2}{m}\left(f(\Delta t) - \left(\frac{c}{\Delta t} - \frac{2m}{\Delta t^2}\right)u_{2\Delta t} - \left(\frac{m}{\Delta t^2} - \frac{c}{\Delta t} + k\right)u_{\Delta t}\right)$$

This approach is called time marching and may be continued to evaluate the displacement at any time. However, there is an inherent inaccuracy in this approximation, and as time progresses, errors accumulate and eventually the approximate solution becomes completely erroneous. The task of finding proper time approximations to alleviate this problem rests on the shoulders of mathematicians, and some sophisticated strategies have been developed. A discussion of these strategies is beyond the scope of this chapter.

Despite the inherent inaccuracies, this approach is applicable to any dynamic system provided that the mass, damping, and stiffness matrices are available. Furthermore, $f(t)$ may be any function of time including periodic functions. However, this is not a good approach to vibration problems because

1. The time step required to model several periods accurately will be prohibitively small for any realistic problem.
2. Natural frequencies may not be readily calculated. The time response is a combination of all frequencies involved and discrete values may not be extracted.

Vibration Type

Another mathematical approach for solving time-dependent problems is to use Fourier transform and construct a time response curve using frequencies and mode shapes. To do that, the first step is to calculate the natural frequencies and mode shapes. Earlier, it was shown that damping has little effect on the value of natural frequencies, therefore, only the undamped equations will be considered:

$$[\mathbf{m}]\{\mathbf{U}_{,tt}\} + [\mathbf{k}]\{\mathbf{U}\} = \mathbf{0},$$

Assume displacements as follows:

$$\{\mathbf{u}\} = \{\chi\}\ \mathbf{e}^{-\mathbf{iwt}}$$

Thus, we obtain

$$\left[[\mathbf{k}]-\omega^2[\mathbf{m}]\right]\{\chi\}=\mathbf{0}.$$

This means that the determinant of $[[\mathbf{k}]-\omega^2\,[\mathbf{m}]]$ must be zero. Therefore, values of ω must be found to satisfy this condition.

Methods of Solving This Equation

Two general types of methods are available:

1. Transformation methods such as those given by Jacobi, Givens, and Householder; these methods should be considered when all of the frequencies are needed (Rao 1982, Bathe 1982)
2. Iterative methods such as power methods; these methods must be considered when few frequencies and mode shapes are required

It should be pointed out that there are two possible ways of formulating the mass matrix **m**. The first is called consistent mass matrix. This matrix is fully populated. It is also possible to diagonalize this matrix (i.e., only the terms on the diagonals are nonzero). This is effectively an assumption that the mass of the systems is lumped at the nodes. There are advantages and disadvantages in using either of the methods. This will be illustrated in the following example.

Example

Find the natural frequencies of longitudinal vibration of the unconstrained stepped bar shown in Figure 10.5:

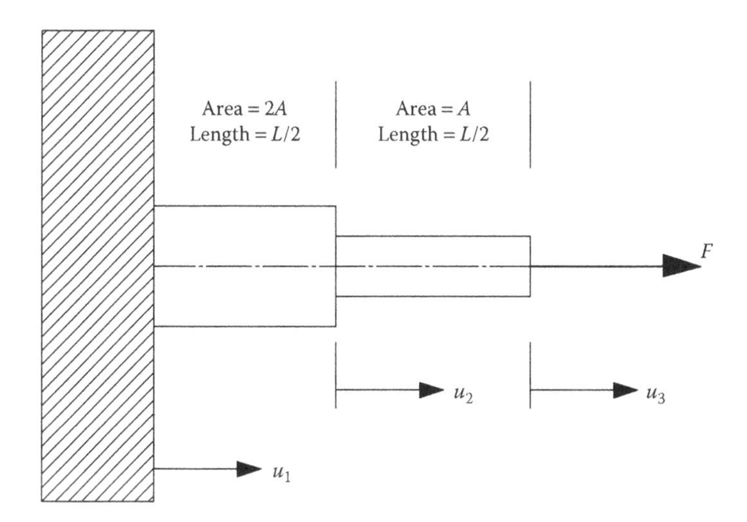

FIGURE 10.5
Longitudinal vibration; the first rod has a cross-sectional area twice the second one.

Suppose that we only use two elements:

$$[K]_{(1)} = \left(\frac{EA}{l}\right)_1 \begin{bmatrix} 1 & -1 \\ -1 & 1 \end{bmatrix} = \frac{4EA}{L} \begin{bmatrix} 1 & -1 \\ -1 & 1 \end{bmatrix}$$

$$[K]_{(2)} = \left(\frac{EA}{l}\right)_2 \begin{bmatrix} 1 & -1 \\ -1 & 1 \end{bmatrix} = \frac{2EA}{L} \begin{bmatrix} 1 & -1 \\ -1 & 1 \end{bmatrix}$$

Consistent Mass Matrix

To use this formulation, one must use the energy approach to develop the mass matrix. The result is as follows:

$$[m]_{(1)} = \left(\frac{\rho Al}{6}\right)_1 \begin{bmatrix} 2 & 1 \\ 1 & 2 \end{bmatrix} = \frac{\rho AL}{6} \begin{bmatrix} 2 & 1 \\ 1 & 2 \end{bmatrix}$$

$$[m]_{(2)} = \left(\frac{\rho Al}{6}\right)_2 \begin{bmatrix} 2 & 1 \\ 1 & 2 \end{bmatrix} = \frac{\rho AL}{12} \begin{bmatrix} 2 & 1 \\ 1 & 2 \end{bmatrix}$$

where ρ is density. The assembled matrices are:

$$[k] = \frac{2EA}{L} \begin{bmatrix} 2 & -2 & 0 \\ -2 & 3 & -1 \\ 0 & -1 & 1 \end{bmatrix}$$

$$[m] = \frac{\rho AL}{12} \begin{bmatrix} 4 & 2 & 0 \\ 2 & 6 & 1 \\ 0 & 1 & 2 \end{bmatrix}$$

The determinant of **[[k]–ω^2 [m]]** must be zero. Define

$$\beta^2 = \frac{\rho L^2 \omega^2}{24E}$$

and form the determinant:

$$\left(\frac{2EA}{L}\right) \begin{vmatrix} 2(1-2\beta^2) & -2(1+\beta^2) & 0 \\ -2(1+\beta^2) & 3(1-2\beta^2) & -(1+\beta^2) \\ 0 & -(1+\beta^2) & (1-2\beta^2) \end{vmatrix} = 0$$

This equation may be expanded to provide the following equation.

$$18\beta^2\left(1-2\beta^2\right)\left(\beta^2-2\right)=0$$

Notice that there are only three degrees of freedom in this system, that is, u_1, u_2, and u_3. Therefore, there are only three possible solutions to this equation:

$$\beta^2 = 0 \Rightarrow \omega_1^2 = 0 \Rightarrow \omega_1 = 0$$

A zero frequency corresponds to a rigid body motion, that is, the entire system moves together. This is because no boundary conditions have been applied. The other two modes of vibration have the following frequencies:

$$\beta^2 = \frac{1}{2} \Rightarrow \omega_2^2 = \frac{12E}{\rho L^2} \Rightarrow \omega_2 = 3.46\sqrt{\frac{E}{\rho L^2}}$$

$$\beta^2 = 2 \Rightarrow \omega_3^2 = \frac{48E}{\rho L^2} \Rightarrow \omega_3 = 6.92\sqrt{\frac{E}{\rho L^2}}$$

Lumped Mass Matrix

To use this formulation, one must "lump" the mass at each node. For this example we obtain

$$[m]_{(1)} = \left(\frac{\rho A l}{2}\right)_1 \begin{bmatrix} 1 & 0 \\ 0 & 1 \end{bmatrix} = \frac{\rho A L}{2}\begin{bmatrix} 1 & 0 \\ 0 & 1 \end{bmatrix}$$

$$[m]_{(2)} = \left(\frac{\rho A l}{2}\right)_2 \begin{bmatrix} 1 & 0 \\ 0 & 1 \end{bmatrix} = \frac{\rho A L}{4}\begin{bmatrix} 1 & 0 \\ 0 & 1 \end{bmatrix}$$

This leads to the assembled mass matrix as follows:

$$[m] = \frac{\rho A L}{4}\begin{bmatrix} 2 & 0 & 0 \\ 0 & 3 & 0 \\ 0 & 0 & 1 \end{bmatrix}$$

Define

$$\beta^2 = \frac{\rho L^2 \omega^2}{8E}$$

and form the determinant:

$$\left(\frac{2EA}{L}\right)\begin{vmatrix} 2\left(1-\beta^2\right) & -2 & 0 \\ -2 & 3\left(1-\beta^2\right) & -1 \\ 0 & -1 & \left(1-\beta^2\right) \end{vmatrix} = 0$$

This equation may be expanded to provide the following equation:

$$6\beta^2\left(1-\beta^2\right)\left(\beta^2-2\right)=0$$

This leads to the following frequencies:

$$\beta^2 = 0 \Rightarrow \omega_1^2 = 0 \Rightarrow \omega_1 = 0 \text{ a rigid body motion}$$

$$\beta^2 = 1 \Rightarrow \omega_2^2 = \frac{8E}{\rho L^2} \Rightarrow \omega_2 = 2.83\sqrt{\frac{E}{\rho L^2}}$$

$$\beta^2 = 2 \Rightarrow \omega_3^2 = \frac{16E}{\rho L^2} \Rightarrow \omega_3 = 4\sqrt{\frac{E}{\rho L^2}}$$

The results of this formulation are tabulated in Table 10.1. We will discuss the differences later.

Impact of Boundary Conditions

Now we need to include the effect of boundary conditions. To do so, we need to go back to the assembled equation $[[\mathbf{k}]-\omega^2\,[\mathbf{m}]]$ and set the rows and columns associated with the zero displacements equal to zero. This in effect reduces the size of the matrix by the number of degrees of freedom set to zero. For this example, the first row and column must be set to zero, and the resulting equations for the stepped bar are as follows.

TABLE 10.1

A Comparison of Calculated Frequencies Using Different Formulations

	ω_1	ω_2	ω_3
Consistent mass formulation	0	$3.46\sqrt{\frac{E}{\rho L^2}}$	$\omega_3 = 6.92\sqrt{\frac{E}{\rho L^2}}$
Lumped mass formulation	0	$2.83\sqrt{\frac{E}{\rho L^2}}$	$4\sqrt{\frac{E}{\rho L^2}}$
% difference	0	18.2%	42.2%

Consistent Formulation

$$\begin{vmatrix} 3\left(1-2\beta^2\right) & -\left(1-\beta^2\right) \\ -\left(1-\beta^2\right) & \left(1-2\beta^2\right) \end{vmatrix} = 0, \quad \beta^2 = \frac{\rho L^2 \omega^2}{24E}$$

This equation leads to

$$\omega_1 = 1.96\sqrt{\frac{E}{\rho L^2}}$$

$$\omega_2 = 5.16\sqrt{\frac{E}{\rho L^2}}$$

Lumped Formulation

$$\begin{vmatrix} 3\left(1-\beta^2\right) & -1 \\ -1 & \left(1-\beta^2\right) \end{vmatrix} = 0, \quad \beta^2 = \frac{\rho L^2 \omega^2}{8E}$$

This equation leads to

$$\omega_1 = 1.84\sqrt{\frac{E}{\rho L^2}}$$

$$\omega_2 = 3.55\sqrt{\frac{E}{\rho L^2}}$$

Similarly, Table 10.2 depicts the results when proper boundary conditions have been applied.

TABLE 10.2

A Comparison of Calculated Frequencies Using Different Formulations and Applied Boundary Conditions

	ω_1	ω_2
Consistent massformulation	$1.96\sqrt{\frac{E}{\rho L^2}}$	$\omega_3 = 5.16\sqrt{\frac{E}{\rho L^2}}$
Lumped mass formulation	$1.84\sqrt{\frac{E}{\rho L^2}}$	$3.55\sqrt{\frac{E}{\rho L^2}}$
% difference	6.1%	31.2%

Uniform Bar

In this example, we assumed that the bar has two different cross-sectional areas. Now, let us carry the same calculations for a bar of uniform cross section (as the smaller one) and study the impact that the cross section may have.

$$[K]_{(1)} = \left(\frac{EA}{l}\right)_1 \begin{bmatrix} 1 & -1 \\ -1 & 1 \end{bmatrix} = \frac{2EA}{L} \begin{bmatrix} 1 & -1 \\ -1 & 1 \end{bmatrix}$$

$$[K]_{(2)} = \left(\frac{EA}{l}\right)_2 \begin{bmatrix} 1 & -1 \\ -1 & 1 \end{bmatrix} = \frac{2EA}{L} \begin{bmatrix} 1 & -1 \\ -1 & 1 \end{bmatrix}$$

Consistent Mass Matrix

The result is as follows:

$$[m]_{(1)} = \left(\frac{\rho Al}{6}\right)_1 \begin{bmatrix} 2 & 1 \\ 1 & 2 \end{bmatrix} = \frac{\rho AL}{12} \begin{bmatrix} 2 & 1 \\ 1 & 2 \end{bmatrix}$$

$$[m]_{(2)} = \left(\frac{\rho Al}{6}\right)_2 \begin{bmatrix} 2 & 1 \\ 1 & 2 \end{bmatrix} = \frac{\rho AL}{12} \begin{bmatrix} 2 & 1 \\ 1 & 2 \end{bmatrix}$$

where ρ is density. The assembled matrices are

$$[k] = \frac{2EA}{L} \begin{bmatrix} 1 & -1 & 0 \\ -1 & 2 & -1 \\ 0 & -1 & 1 \end{bmatrix}$$

$$[m] = \frac{\rho AL}{12} \begin{bmatrix} 2 & 1 & 0 \\ 1 & 4 & 1 \\ 0 & 1 & 2 \end{bmatrix}$$

The determinant of $[[\mathbf{k}] - \omega^2 [\mathbf{m}]]$ must be zero. Define

$$\beta^2 = \frac{\rho L^2 \omega^2}{24E}$$

and form the determinant:

$$\left(\frac{2EA}{L}\right) \begin{vmatrix} (1-2\beta^2) & -(1+\beta^2) & 0 \\ -(1+\beta^2) & 2(1-2\beta^2) & -(1+\beta^2) \\ 0 & -(1+\beta^2) & (1-2\beta^2) \end{vmatrix} = 0$$

This equation may be expanded to provide the following equation:

$$6\beta^2\left(1-2\beta^2\right)\left(\beta^2-2\right)=0$$

As before, there are only three degrees of freedom in this system, that is, u_1, u_2, and u_3. Therefore, there are only three possible solutions to this equation:

$$\beta^2=0 \Rightarrow \omega_1^2=0 \Rightarrow \omega_1=0$$

$$\beta^2=\frac{1}{2} \Rightarrow \omega_2^2=\frac{12E}{\rho L^2} \Rightarrow \omega_2=3.46\sqrt{\frac{E}{\rho L^2}}$$

$$\beta^2=2 \Rightarrow \omega_3^2=\frac{48E}{\rho L^2} \Rightarrow \omega_3=6.92\sqrt{\frac{E}{\rho L^2}}$$

Lumped Mass Matrix

For this example we obtain

$$[m]_{(1)}=\left(\frac{\rho Al}{2}\right)_1\begin{bmatrix}1 & 0\\ 0 & 1\end{bmatrix}=\frac{\rho AL}{4}\begin{bmatrix}1 & 0\\ 0 & 1\end{bmatrix}$$

$$[m]_{(2)}=\left(\frac{\rho Al}{2}\right)_2\begin{bmatrix}1 & 0\\ 0 & 1\end{bmatrix}=\frac{\rho AL}{4}\begin{bmatrix}1 & 0\\ 0 & 1\end{bmatrix}$$

This leads to the assembled mass matrix as follows:

$$[m]=\frac{\rho AL}{4}\begin{bmatrix}1 & 0 & 0\\ 0 & 2 & 0\\ 0 & 0 & 1\end{bmatrix}$$

Define

$$\beta^2=\frac{\rho L^2\omega^2}{8E}$$

and form the determinant:

$$\left(\frac{2EA}{L}\right)\begin{vmatrix}\left(1-\beta^2\right) & -1 & 0\\ -1 & 2\left(1-\beta^2\right) & -1\\ 0 & -1 & \left(1-\beta^2\right)\end{vmatrix}=0$$

This equation may be expanded to provide the following equation:

$$2\beta^2\left(1-\beta^2\right)\left(\beta^2-2\right)=0$$

This leads to the following frequencies:

$$\beta^2=0 \Rightarrow \omega_1^2=0 \Rightarrow \omega_1=0 \text{ a rigid body motion}$$

$$\beta^2=1 \Rightarrow \omega_2^2=\frac{8E}{\rho L^2} \Rightarrow \omega_2=2.83\sqrt{\frac{E}{\rho L^2}}$$

$$\beta^2=2 \Rightarrow \omega_3^2=\frac{16E}{\rho L^2} \Rightarrow \omega_3=4\sqrt{\frac{E}{\rho L^2}}$$

The results of this formulation are tabulated in Table 10.3. We will discuss the differences later.

These results indicate that as long as the boundary conditions have not been applied (in other words, the bar has free–free boundaries), the cross-sectional area has no impact on the frequencies.

Now we need to include the effect of boundary conditions. As we have done once before, we need to go back to the assembled equation **[[k]–**ω^2 **[m]]** and set the rows and columns associated with the zero displacements equal to zero. We obtain the following.

Consistent Formulation

$$\begin{vmatrix} 2\left(1-2\beta^2\right) & -\left(1+\beta^2\right) \\ -\left(1+\beta^2\right) & \left(1-2\beta^2\right) \end{vmatrix}=0, \quad \beta^2=\frac{\rho L^2\omega^2}{24E}$$

This equation leads to

$$\omega_1=1.61\sqrt{\frac{E}{\rho L^2}}$$

TABLE 10.3

A Comparison of Calculated Frequencies of a Uniform Bar Using Different Formulations

	ω_1	ω_2	ω_3
Consistent mass formulation	0	$3.46\sqrt{\frac{E}{\rho L^2}}$	$\omega_3=6.92\sqrt{\frac{E}{\rho L^2}}$
Lumped mass formulation	0	$2.83\sqrt{\frac{E}{\rho L^2}}$	$4\sqrt{\frac{E}{\rho L^2}}$
% difference	0	18.2%	42.2%

$$\omega_2 = 5.63\sqrt{\frac{E}{\rho L^2}}$$

Lumped Formulation

$$\begin{vmatrix} 2\left(1-\beta^2\right) & -1 \\ -1 & \left(1-\beta^2\right) \end{vmatrix} = 0, \quad \beta^2 = \frac{\rho L^2 \omega^2}{8E}$$

This equation leads to

$$\omega_1 = 1.52\sqrt{\frac{E}{\rho L^2}}$$

$$\omega_2 = 3.70\sqrt{\frac{E}{\rho L^2}}$$

Similarly, Table 10.4 depicts the results when proper boundary conditions have been applied and compared to exact results.

It was clearly shown—by way of the example—that the lower the frequency to be calculated, the more accurate it is expected to be. The inherent inaccuracies in using FEA could lead to gross failures. For instance, assume that a 10% error in natural frequency calculation has taken place. This error leads to a dynamic magnification factor of 5 instead of 50. Thus, we may design a system that may not be able to withstand loads experienced in the field.

Theoretically, one may calculate as many frequencies and mode shapes as there are nodes in the system. However, from a practical point of view, the accuracy of calculations drops very rapidly, as shown in the example above. To increase accuracy, three general rules may be cited:

TABLE 10.4

A Comparison of Calculated Frequencies Using Different Formulations versus Exact Results

	Exact	Consistent		Lumped	
ω_1	$1.57\sqrt{\frac{E}{\rho L^2}}$	$1.61\sqrt{\frac{E}{\rho L^2}}$	Error = 2.5%	$1.52\sqrt{\frac{E}{\rho L^2}}$	Error = −3.2%
ω_2	$4.71\sqrt{\frac{E}{\rho L^2}}$	$5.63\sqrt{\frac{E}{\rho L^2}}$	Error = 19.5%	$3.70\sqrt{\frac{E}{\rho L^2}}$	Error = −21.4%

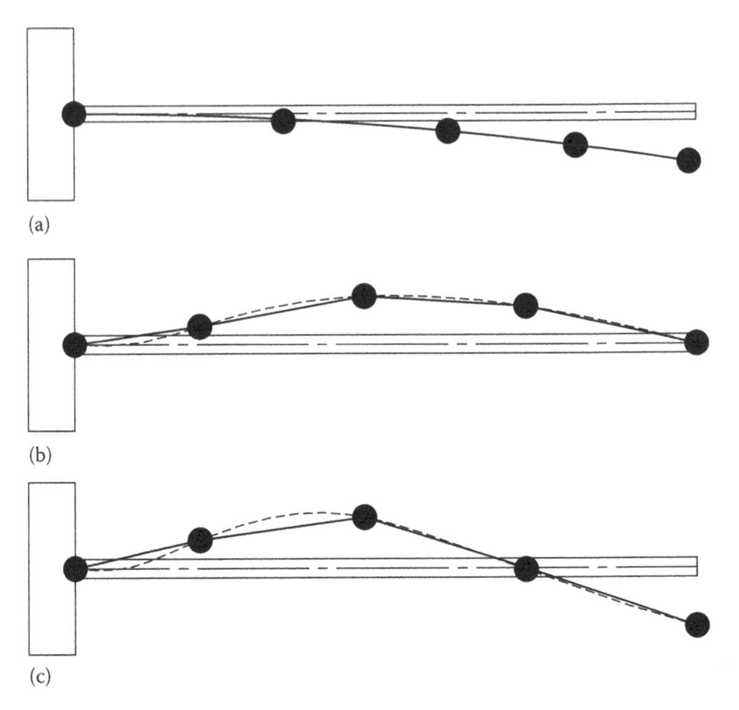

FIGURE 10.6
Approximations for different mode shapes: (a) good approximation to the first mode of vibration of a cantilever beam, (b) relatively good approximation to the second mode of vibration of a cantilever beam, and (c) poor approximation to the third mode of vibration of a cantilever beam.

1. Apply the boundary conditions correctly. Application of boundary conditions is not always straightforward. An example of ambiguity in this application is presence of friction
2. Use much more nodes in the system than the required number of frequencies. Generally speaking, there should be enough nodes to describe the mode shape accurately (Figure 10.6)
3. The lumped mass matrix gives a very good solution if only the first mode shape and frequency needs to be calculated. The advantage of consistent formulation is that stress and deformations are calculated with more precision

Finite Element Formulation of Heat Conduction

For the sake of completeness, let us review the finite element formulation of heat conduction in a bar with two different cross-sectional areas clamped at one end and at a temperature of T_1, as shown in Figure 10.7. Also, heat is

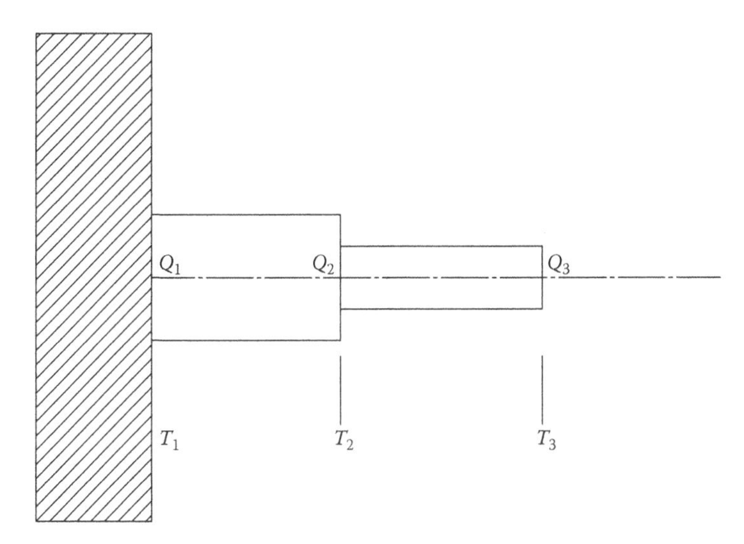

FIGURE 10.7
Heat conduction in a bar with different cross-sectional areas.

being generated at each end of the bar as well as where the two cross sections meet. Our goal is to calculate the temperature distribution in this bar, particularly where the two cross sections meet and at the free end. In this particular situation, we assume that the problem is steady state and that no other mode of heat transfer exists.

Consider a generic line element of length L and assume that temperature varies linearly from one end to the other—each end is called a node. Then, a simple finite element model of this system is as shown in Figure 10.8.

For a one-dimensional heat conduction problem, we have

$$Q = \frac{KA}{L}\left(T_{\text{hot}} - T_{\text{cold}}\right)$$

We use this equation and balance the heat equation (i.e., heat in less heat out equal to zero) for each node as shown below. To maintain consistency, assume that the temperature at each node i is lower than its surroundings.

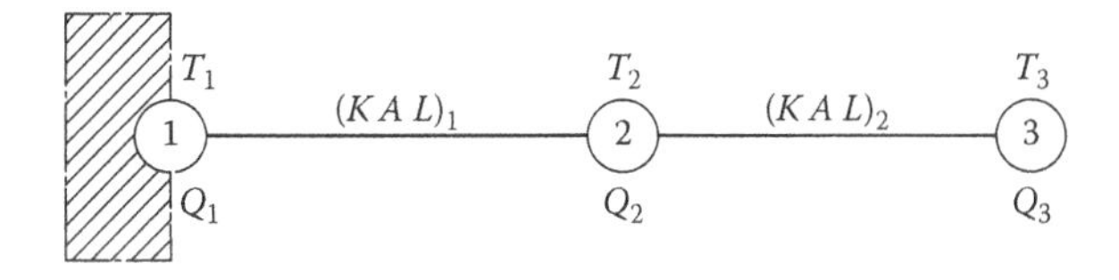

FIGURE 10.8
A finite element model of the bar.

Node 1:

$$\frac{K_1A_1}{L_1}(T_2 - T_1) + Q_1 = 0 \tag{10.9}$$

Node 2:

$$\frac{K_2A_2}{L_2}(T_3 - T_2) + \frac{K_1A_1}{L_1}(T_1 - T_2) + Q_2 = 0 \tag{10.10}$$

Node 3:

$$\frac{K_2A_2}{L_2}(T_2 - T_3) + Q_3 = 0 \tag{10.11}$$

Now, we may assemble the matrix equation:

$$\begin{bmatrix} \left(\frac{KA}{L}\right)_1 & -\left(\frac{KA}{L}\right)_1 & 0 \\ -\left(\frac{KA}{L}\right)_1 & \left(\frac{KA}{L}\right)_1 + \left(\frac{KA}{L}\right)_2 & -\left(\frac{KA}{L}\right)_2 \\ 0 & -\left(\frac{KA}{L}\right)_2 & \left(\frac{KA}{L}\right)_2 \end{bmatrix} \begin{Bmatrix} T_1 \\ T_2 \\ T_3 \end{Bmatrix} = \begin{Bmatrix} Q_1 \\ Q_2 \\ Q_3 \end{Bmatrix} \tag{10.12}$$

One may easily recognize the similarity of this equation and Equation 10.5.

CAD to FEA Considerations

One must differentiate between the function of CAD programs and FEA software. CAD packages are design tools developed to give a "touchy-feely" taste to the designer. Furthermore, they are used to develop engineering drawings and manufacturing tools. On the one hand, in today's culture, it is very easy to design a part and then press a button to conduct an analysis of the part or system without any consideration of the implications of the solutions that we find. On the other hand, FEA software is designed to model nature and make calculations based on exact data. The solution is only as good as the data that we provide and the discretization that we allow to take place. Thus, the two have very different roles.

CAD to FEA, Do's and Do Not's

The following is a list of do's and do not's in CAD to FEA file translations.

1. Do not just bring your CAD over to FEA
2. Create a file to be modified for FEA
3. Do you really need to simulate the entire model?
4. Do you really need the degree of detail in regions of relatively no importance?
5. Take out the extras before translating

Criteria for Choosing Engineering Software

Choosing engineering software for today is also choosing software for tomorrow, and like toothpaste out of the tube, it is a difficult decision to reverse if it is wrong. There are many choices. How can one make an intelligent estimate and choose the right package for an application when such a large number of products are available? And what are some of the related issues once the code is chosen and paid for?

With the advent of the new generation of computers and the speed with which they operate, it is only natural for this tool to become commonplace in the world of engineering design and analysis. Up to about the mid-1980s, most engineers wrote their own specialized programs to do certain calculations. General-purpose commercial engineering programs were not commonly used. They were usually used by highly educated engineers in research and development areas for highly complicated tasks. The most effective means for marketing of these programs at that time was word of mouth!

Engineering applications and computers have come a long way. On the one hand, personal computers and low-end workstations have made it easy for design and manufacturing engineers to access these sophisticated programs more readily, and on the other hand, the software vendors have recognized the potential of this market and have attempted to make their programs more user-friendly. In 2014, the global engineering software market was valued to be around $20 billion with a growth potential of over 12%. Many new companies have entered this market, and right now there are hundreds of programs available for the engineering community.

The following is a methodology for making software choices—take the points that work for you. These are summarized in a six-step formula at the end of this section.

What Types of Engineering Programs Are There?

By and large, there are four types of engineering software available today. These general classification are CAD and manufacturing (CAD/CAM), FEA of stress and heat conduction problems, computational fluid dynamics (CFD), and finally specialized areas such as electromagnetic or combustion. I will examine each category briefly.

CAD and CAM

These software programs aid the engineer in developing drafts and blueprints. They are usually designed to produce three-view and isometric representations of objects with the given dimensions and special considerations. These drawings can then be used in manufacturing the piece. The same CAD programs may be used in construction and architecture in order to convey a sense of the building to be designed. Although traditionally, these packages are for graphics alone, the new release of these packages offer very limited analysis capabilities. More recently, however, these programs offer solid modeling capabilities that can be used to transfer model information to FEA packages for analysis.

Solid Mechanics and Stress Analysis FEA Software

These programs actually do the various and necessary calculations to evaluate whether or not a structural design would maintain its integrity through its life cycle. Historically, this is where engineering applications of computers and programming began on a wide scale. The matrix analysis techniques of structural engineering opened the doors to the applications of finite difference and finite element techniques. In return, these techniques allowed engineering simplifications of realistic systems, and before long, FEA became a common technique. Soon after, it was realized that these very techniques are also applicable in other areas of engineering and physics. Nowadays, FEA applications in solid mechanics range anywhere from linear static to transient nonlinear analyses, to thermal stress and conduction calculations, and from simple impact problems to crash analysis.

Fluid Mechanics and Heat Transfer CFD Software

Once adequate experience was gained in solid mechanics, attention was focused on fluid flow, heat, and mass transfer types of problems. CFD software may employ one of several analysis techniques available, namely, control volume techniques, finite difference techniques, and finite element methods. To this day, each formulation has shown its own unique characteristics and to my knowledge, none is shown to be superior to the others in general. There is a range of problems that are solved using CFD. Examples may be cited from simple incompressible viscous flow through pipes and pressure drop calculations to heating, ventilating, and air conditioning to complicated supersonic flows.

General Physics CFD and/or FEA

As analysis becomes a more integral segment of the industrial design cycle, and as the computing power of modern computers increases almost exponentially, engineering simplifications and assumptions begin to take new forms. Although in the past, coupling of various disciplines of physics were ignored, they can now be considered and their effects can be taken into account with the advent of new computational techniques and engineering software programs. By no means is the application of numerical methods limited to obvious applications of solid mechanics, fluid mechanics, or heat transfer. Various disciplines such as magnetics, chemical reactions, combustion, and molding have found their way into various computer programs as well.

There is a price to pay with these engineering advances. True, many complicated calculations may be done easily now; however, as the physics becomes more difficult, the numerical treatment becomes more delicate. For example, CFD problems are by far much more difficult to solve than their FEA counterpart. It is estimated that 75% of all solid mechanics, that is, stress analysis problems and only 10%–15% of CFD problems are linear. In a linear problem the mathematical equations do not pose any difficulty, and they can be solved in a straightforward manner. However, if a problem is nonlinear special considerations needs to be made to it so that these mathematical equations can be solved. There are other added complications. As the physics becomes more complicated, smaller elements are needed in order to capture and resolve changes in the primary variables such as velocity components, turbulence energies, or magnetic fluxes. These naturally result in much larger problem sizes requiring larger computer disk spaces and CPU time allocations.

Which Software Should I Choose?

First determine your general area of application, that is, solid modeling, solid mechanics, fluid dynamics, and so on. Then, to answer this question adequately, one has to research the following six areas: applications, the completeness of the package, manuals, customer support, software quality assurance (SQA), and, finally, interfacing to other programs.

Application

Back in late 1990s, a client once asked me to help him choose the proper package for his application. My first thoughts were that he should go with package A because it was the leader in the industry in stress analysis—with a yearly lease price of $10,000 at that time—as well as package B, another industry leader in fluid flow analysis—with a price tag of $17,000. Upon examining his application, I realized that he needed a nonlinear stress analysis solver with occasional applications of fluid mechanics and heat transfer. I suggested that, based on his applications, he should choose package C, because this company offered their product on a modular basis and he only had to pay for what he was going to use. True, companies A and B were industry leaders, but the user would end up paying for many features—such as crack tip propagation, stamping, nonlinear plastic flow, and so on—that would not have been used at all. The necessary modules of software C cost this client about $7000.

When choosing a software package, it is important to know the extent of your applications and the physics involved. This, I believe, is one of the most important factors for a decision. And it is important to remember that some software vendors sell their package as a whole and others in modules. Application, however, is not the *only* issue to be considered.

Complete Package

It is a normal assumption to think that the program you are about to buy or lease comes all in one package. That is not so! There are many "solvers" out there that can solve some very difficult problems: however, they do not have any modeling capabilities or any tools to look at the results once the analysis is done. These programs generally depend on information provided by third-party software in order to work, without which, they are useless. One has to be mindful of this fact particularly when comparing prices.

User Interface, Manuals, and Training

As general-purpose commercial engineering codes are becoming more and more available to non-expert users, user interfaces play more important roles, and code developers dedicate larger portions of their resources to user-friendly interfaces. If we were to use the word "interface" as any way that code developers interact with users, then four levels of user interface exist:

1. Graphical user interface (GUI), where information is input through the usage of some picking device such as the mouse. The more sophisticated the GUI, the less usage of the keyboard
2. Textual user interface, where the information is input primarily through the keyboard
3. Manuals, training courses, and notes such as tutorials
4. Customer support

These factors are intertwined and need to be looked at closely. The following are only a sample of questions to be entertained. The question of user support will be treated separately.

1. How user-friendly is the GUI (text-based interfaces have become a past trend)? How well is the information and logical flow of modeling laid out on the screen? How much access to manuals does one need to start using the program?

2. How user-friendly are the manuals? Are they written as a reference or as tutorials? Would the vendor send copies for evaluations to your site?
3. Are there specific manuals to teach the new user how to get started and come up to speed in a short period of time?
4. How inexpensive and accessible are the training classes? One point to keep in mind is that some "Introductory Classes" can be so simple that one would wonder "why pay several hundreds of dollars just to sit in the class!" Look at the course syllabus before actually registering. Another issue is that some companies do not allow for cancellation once the money is paid. Or they have a surcharge for rescheduling

Customer Support

Support is an issue that is difficult to evaluate before buying the code. Yet, it is an extremely important issue because users—even pros—come across obstacles at times that they cannot resolve. These obstacles range from simple negligence on the part of the user or not knowing portions of the code to discovering an undetected bug. In a way, support engineers are the backbone of any software (or even hardware) company. The questions to research in this area are as follows:

1. The actual number of support engineers—Are the code developers also the ones who support it?
2. Their level of competence—To be a good support engineer requires not only the ability to "see" the user's model and what causes the obstacle, but also the technical competence to be knowledgeable in all of the disciplines that the code has to offer. Many times a user calls for support with an issue that has to do with the physical nature of the simulation and not the software. The support engineer must be knowledgeable of the assumptions and limitations of the code
3. Friendliness—Rapport eases communications
4. Timeliness—How quickly do they respond to phone calls, faxes, and e-mails? Is there any type of procedure that oversees these activities and their timely responses?

The best technique to find the answers to these questions is to talk to other clients and users of that software and network with them.

Quality Assurance

Quality assurance (QA) seems to be the "buzzword" these days. Many companies pride themselves for having a QA department or a set of QA procedures in place. Some software companies have even taken the initiative to become

ISO certified. It may be that for the consumer, it is more important to talk about SQA than QA. SQA depends on *maintenance, error reports,* and *bug fixes, user interfaces,* and *developing new* features. Software performance testing (SPT) may be a better term for what is generally known as QA.

In general, the user's encounter with SQA is mainly through maintenance, error reports, and bug fixes. Furthermore, the majority of users make the assumption that SPT has been successful and never ask for verification of the code prior to purchase.

Maintenance

From a user's point of view, maintenance means upward and platform compatibility, and life cycle. It implies that once a feature is developed, it works equally the same across any computer platform and would produce identical results. Moreover, updates should support the capabilities available in previous versions. Life cycle, here, refers to the number of updates that are supported in one major release. As an example, suppose that version 1.0 of a program has just been released, that is V1.0. The number 1 refers to a major release and the number 0 refers to the updates. Say, after six months, new updates—either features or bug corrections—are released. This new program may be called V1.1 because nothing major has been done to the program and all changes are considered to be minor. The next release might be V1.2, and the next V1.3, and so on. Then, the software company may decide to make major changes and come out with version V2.0. Usually, the input files used for V1.*x* versions will not be accepted by V2.0 major release. The time span from V1.0 to V2.0 is called a life cycle.

Error Reports and Bug Fixes

Each software developer must have a comprehensive error reporting system with different classes of error identifications. This is a measure of how well the program developers and SQA keep track of bugs. If such a system is not in place, the same errors are likely to happen time and again. These errors and their implications must be reported to the users after a formal investigation and verifications. And for specific cases, solutions and/or a work-around should be offered. Ask about it, if the marketing people have not clarified this point.

Software Performance Testing

From a user's point of view, SPT is a set of problems that should check every aspect of the program for the following:

1. Backward compatibility. The new version of the code should easily accept the previously defined input files and produce the same results—with the exception of error corrections.

2. If the program allows for two ways of solving the same problem, such as different techniques for solving equations, they produce the same results.
3. The mathematical limits and assumptions of various capabilities are tested.

Validation

At times, validation problems that are presented as QA results taken for SPT validation issues are different than software testing in that they try to show that the program produces results that are the same as those obtained by other means. There are three types of validation studies:

1. Comparison with exact results. Unfortunately, exact results are only available for a very small group of simple problems. Usually one-dimensional problems belong to this category
2. Comparison with experimental results. Traditionally, this has been the most reliable and most effective approach to validating any theory. But it, too, has its drawbacks for use as a technique for validating numerical works. Errors are almost always expected, and most numerical techniques show "good" agreement with the experiments and not necessarily "exact" agreement. One has to exercise care not to denounce any code or algorithm on this basis, because many mathematical models have been developed to resemble the physical model under specific conditions. If these models are used in a way contrary to their basic assumptions, the results will not be dependable
3. Comparison with other numerical solutions. This is probably one of the most common methods of validating any one code on different types of physics. Somehow it is believed that if the code compares with some other code, it renders the code validated! In reality, it only proves that there are no programming errors

In the final analysis, the one who truly validates the code is the user. The user validates the code by using the program in his or her real-life engineering problems time and again. And the user trusts that it produces reliable results for use in his or her analysis.

Interfacing to Other Programs

In today's engineering environment, and with the recognition of concurrent engineering, it is essential that various engineering software—even competing ones—have the ability to exchange data. It is quite common that a designer would create a model using a CAD program and would then send it to the analyst for further work. Generally, neutral files such as Initial

Graphics Exchange Specification (IGES) would allow for such information to be transferred. However, other companies have collaborated to make this transfer much easier. It is important to know of such capabilities, if any. The following is a set of probing questions to ask:

1. In case files are imported from CAD programs: Does the translator recognize both the wire frame and solid models?
2. Does the user have a choice to indicate which entities are translated? This is an important option particularly when CAD models are exported to FEA and CFD programs. CAD models usually contain symbols (such as dimension lines as well as dimensions) that if brought into the analysis package introduce meshing difficulties. These "extras" have to be deleted before any finite element discretization can take place
3. Do the data flow in one direction or is it both ways? In other words, if software A is capable of recognizing files from other programs, how many other programs would recognize files produced by software A?

Summary

It is not easy to decide which software to buy. However, one has to review three levels. First, one has to define the scope of the work to be done by the software; second, one has to define the duration of the work; and finally, one has to define the extent of usage. Considering these three and combining them with the available budget and knowing what to look for in each product, making the decision may become easier. So, here is the six-step formula:

1. Determine the extent of your application.
2. Decide whether you need a complete package or your application can be satisfied with certain modules.
3. Research the type of training that the vendor provides to help you get started. Next, try to develop a feel for the clarity of the manuals. Then, find out how easily the program can be used.
4. Contact other users and ask them how professional the customer support team is.
5. Research the QA techniques and procedures used by the vendor and the extent of the tested capabilities. Does QA monitor even the manuals and the training courses?
6. Finally, find out how easily the software interfaces with other programs. Are there specific interfaces or should general neutral files (such as IGES) be used?

11

Mechanical and Thermomechanical Concerns

Introduction

An important aspect of electronics packaging is developing an understanding of the stresses that its components undergo and their relationship to the system's failure and/or reliability. The cause of these stresses may be temperature and its variations, vibration, or physical properties such as weight. It may occur at the board and component level, enclosure levels, and up to the system itself.

Stresses are internal distributed forces, which are caused by external applied loads. Strains are changes in the form under the same loads. Consider a rod of length L and diameter A. One may intuitively recognize that the displacement of the end of this rod depends directly on the magnitude of the applied force—very similar to the force–deflection relationship of a spring–mass system as shown in Figure 11.1.

Now consider what happens inside of this rod in Figure 11.2. The concentrated load is (internally) developed over the area of the cross section. Thus, one may express this distributed force as follows:

$$\sigma = \frac{F}{A}$$

Similarly, a distributed (average) displacement may also be calculated.

$$\varepsilon = \frac{\Delta}{L}$$

It turns out that σ and ε have a relationship similar to a force–deflection curve in a spring–mass system. The slope of this line (E) is called tensile modulus, Young's modulus, or modulus of elasticity.

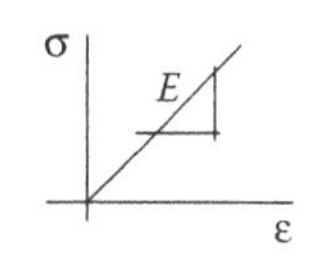

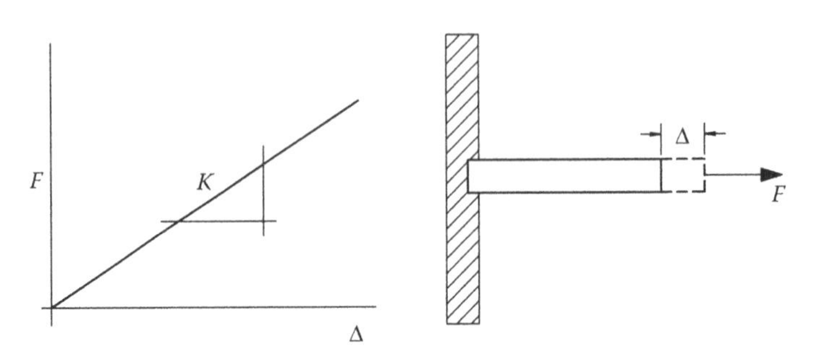

FIGURE 11.1
Force–deflection relationship.

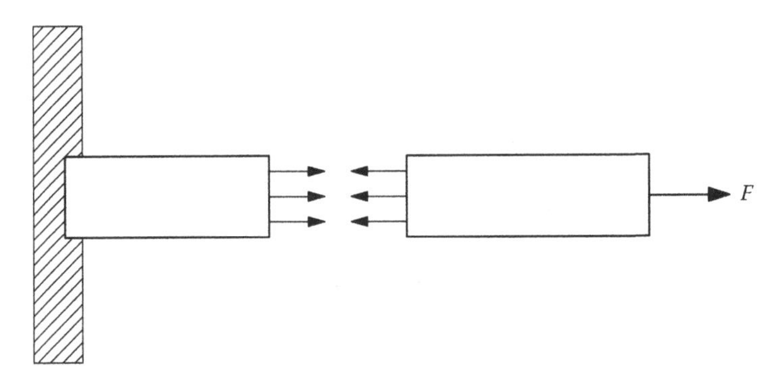

FIGURE 11.2
Internal forces.

Consider another scenario. A block under a shear force will also deflect. In shear, the force–deflection relationships are defined as follows:

$$F = \tau A$$

where A is the area and τ is the shear stress. Furthermore, there is a relationship between the shear stress and shear strain (γ) similar to that of the stress–strain relationship.

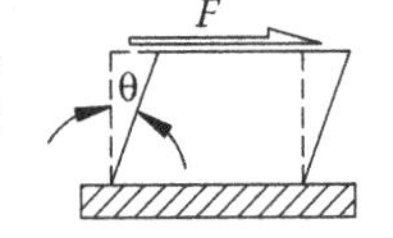

$$\tau = G\gamma$$

where G is shear modulus and γ is shear strain.

In general, both normal and shear stresses develop in solids under a general loading. For example, a cantilever beam under a simple load at the free end exhibits both normal and shear stresses, as shown in Figure 11.3.

Considering that deformations depend on the general state of stress, to calculate stresses or deflections under a general load both constants E and G must be known. In this case, as a minimum the material has to be isotropic, that is, material behavior is independent of the direction of applied loads.

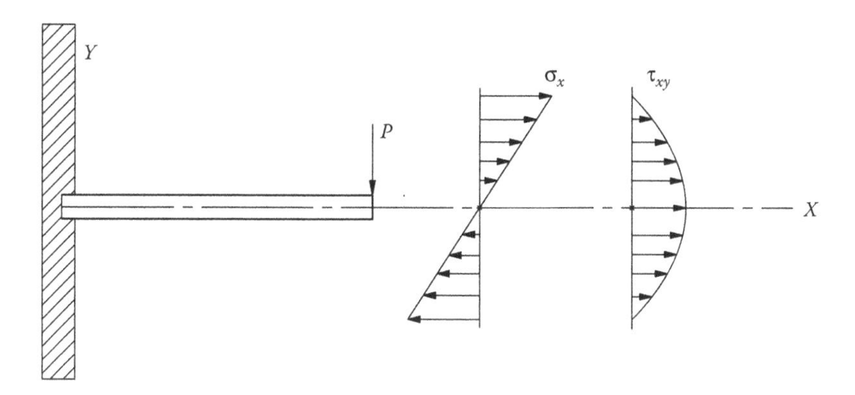

FIGURE 11.3
Normal and shear forces developed in a cantilever beam.

General Stress–Strain Relationship

A small cube of a material under general loading exhibits the stresses shown in Figure 11.4.

The strains are related to stresses through the following matrix relationship

$$\{\varepsilon\} = [\mathbf{S}]\ \{\sigma\}$$

The matrix **[S]** has 36 constants (unknowns); however, it can be shown that these 36 unknowns may be reduced depending on material behavior. They are as follows:

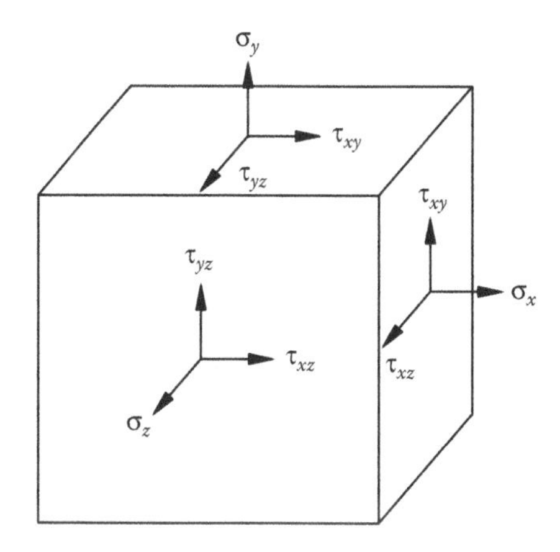

FIGURE 11.4
General state of stress.

Anisotropic Behavior

Materials such as bone exhibit different behavior depending on the loading direction and location. In this case, there are 21 independent constants (unknowns) in the **[S]** matrix. This matrix is symmetric.

$$\begin{Bmatrix} \varepsilon_x \\ \varepsilon_x \\ \varepsilon_x \\ \gamma_{yz} \\ \gamma_{xz} \\ \gamma_{xy} \end{Bmatrix} = \begin{bmatrix} S_{11} & S_{12} & S_{13} & S_{14} & S_{14} & S_{16} \\ & S_{22} & S_{23} & S_{24} & S_{25} & S_{26} \\ & & S_{33} & S_{34} & S_{35} & S_{36} \\ & & & S_{44} & S_{45} & S_{46} \\ & Sym. & & & S_{55} & S_{56} \\ & & & & & S_{66} \end{bmatrix} \begin{Bmatrix} \sigma_x \\ \sigma_y \\ \sigma_z \\ \tau_{yz} \\ \tau_{xz} \\ \tau_{xy} \end{Bmatrix} \tag{11.1}$$

Orthotropic Behavior

Materials such as wood or composite lamina have different properties in two principle directions. These materials may be modeled with nine independent constants (unknowns) in the **[S]** matrix. This matrix is symmetric.

$$\begin{Bmatrix} \varepsilon_x \\ \varepsilon_x \\ \varepsilon_x \\ \gamma_{yz} \\ \gamma_{xz} \\ \gamma_{xy} \end{Bmatrix} = \begin{bmatrix} S_{11} & S_{12} & S_{13} & 0 & 0 & 0 \\ S_{12} & S_{22} & S_{23} & 0 & 0 & 0 \\ S_{13} & S_{23} & S_{33} & 0 & 0 & 0 \\ 0 & 0 & 0 & S_{44} & 0 & 0 \\ 0 & 0 & 0 & 0 & S_{55} & 0 \\ 0 & 0 & 0 & 0 & 0 & S_{66} \end{bmatrix} \begin{Bmatrix} \sigma_x \\ \sigma_y \\ \sigma_z \\ \tau_{yz} \\ \tau_{xz} \\ \tau_{xy} \end{Bmatrix} \tag{11.2}$$

Isotropic Behavior

Most engineering material properties are independent of direction. These materials may be modeled with only two independent constants (unknowns) in the **[S]** matrix. This matrix is symmetric.

$$\begin{Bmatrix} \varepsilon_x \\ \varepsilon_x \\ \varepsilon_x \\ \gamma_{yz} \\ \gamma_{xz} \\ \gamma_{xy} \end{Bmatrix} = \begin{bmatrix} \frac{1}{E} & -\frac{\nu}{E} & -\frac{\nu}{E} & 0 & 0 & 0 \\ -\frac{\nu}{E} & \frac{1}{E} & -\frac{\nu}{E} & 0 & 0 & 0 \\ -\frac{\nu}{E} & -\frac{\nu}{E} & \frac{1}{E} & 0 & 0 & 0 \\ 0 & 0 & 0 & G & 0 & 0 \\ 0 & 0 & 0 & 0 & G & 0 \\ 0 & 0 & 0 & 0 & 0 & G \end{bmatrix} \begin{Bmatrix} \sigma_x \\ \sigma_y \\ \sigma_z \\ \tau_{yz} \\ \tau_{xz} \\ \tau_{xy} \end{Bmatrix} \tag{11.3}$$

Note that this relationship requires the knowledge of the following three constants: E, G, and ν, where E is the modulus of elasticity, G is the shear modulus, and ν is Poisson's ratio defined as the ratio of the transverse strain in the j direction to strain in the i direction when the stress is in the i direction. All three variables may be determined experimentally; however, the following relationship exists among them:

$$G = \frac{E}{2(1+\nu)} \text{ and } -1 < \nu < \frac{1}{2}$$

Material Behavior and the Stress–Strain Curve

The choice of material is an important factor in the overall system design from the point of view of both budget and structural integrity. As it was pointed out, different materials behave differently, particularly if their behavior depends on the direction of application of the load as in anisotropic or orthotropic materials.

If various specimens from different isotropic materials are loaded to complete fracture, three different behaviors are observed:

1. Brittle materials—The stress–strain curve is completely linear as depicted in Figure 11.5.
2. Ductile material with strain hardening—The stress–strain curve is nonlinear, but the initial portion of the curve is linear as shown in Figure 11.6.
3. Ductile material without strain hardening—The stress–strain curve is nonlinear, but the initial portion of the curve may or may not be linear (Figure 11.7).

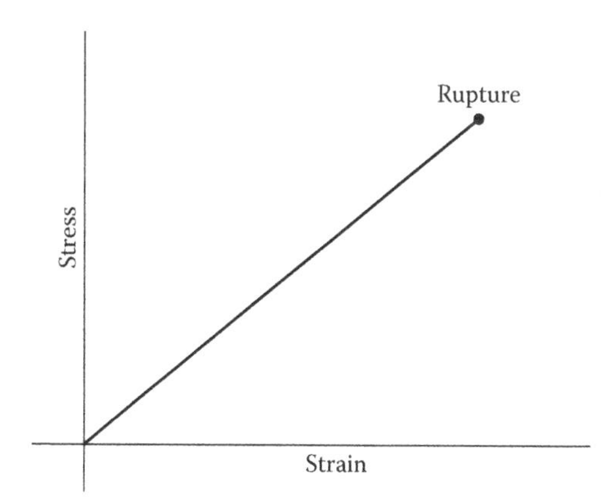

FIGURE 11.5
The behavior of brittle materials.

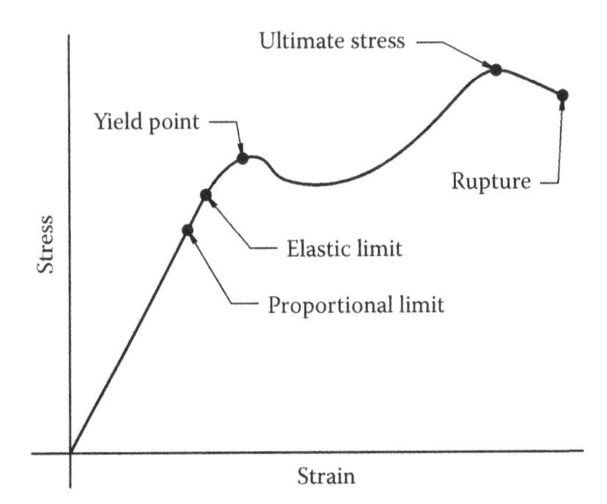

FIGURE 11.6
Ductile material with strain hardening behavior.

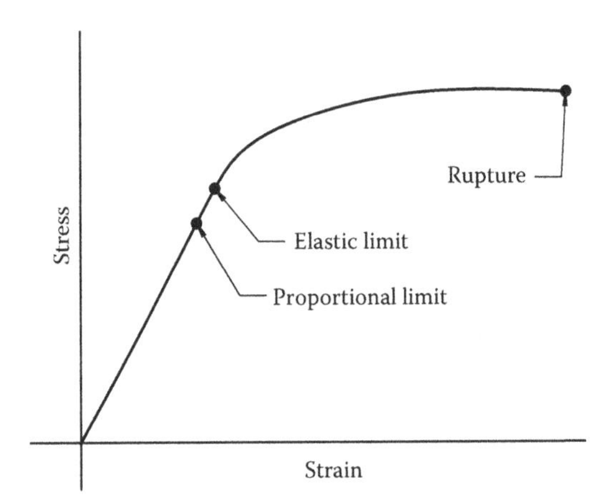

FIGURE 11.7
Ductile material without strain hardening behavior.

Examples of these types of materials in electronics packaging are as follows: Most ceramics and certain types of printed circuit boards may be considered as brittle. Most metallic cases and chassis are ductile but exhibit strain hardening. Solder as well as most plastics are ductile but do not have any strain hardening.

When dealing with ductile materials, there are several points of interest on the strain–stress curve. Knowledge of these points enables us to verify whether a system has failed or has the potential of failure. These are defined as follows:

1. Proportional limit: The point on the stress–strain curve where the curve begins to deviate from a straight line.
2. Elastic limit: Maximum stress to which a specimen may be subjected when, upon removal of the load, no permanent deformation is caused.

3. Yield point: A point on the curve where there is a sudden increase in strain without a corresponding increase in stress.
4. Yield strength: The maximum stress that can be applied without permanent deformation of the test specimen. This is the value for which there is an elastic limit or at 2% deformation.
5. Ultimate strength: Also called tensile strength, it is the maximum stress value obtained on a stress–strain curve.

Determining Deformations under Application of General Loads

The general equations of elasticity are formulated first by balancing the external forces with the internal stresses and obtaining the differential equations of equilibrium. The general form of these equations is as follows:

$$\frac{\partial \sigma_x}{\partial x} + \frac{\partial \tau_{xy}}{\partial y} + \frac{\partial \tau_{xz}}{\partial z} + X = 0$$

$$\frac{\partial \sigma_y}{\partial x} + \frac{\partial \tau_{xy}}{\partial y} + \frac{\partial \tau_{yz}}{\partial z} + Y = 0$$

$$\frac{\partial \sigma_z}{\partial x} + \frac{\partial \tau_{xz}}{\partial y} + \frac{\partial \tau_{yz}}{\partial z} + Z = 0$$

The solution to these equations must satisfy not only the boundary conditions but also what is called the conditions of compatibility. These conditions are a set of six differential equations between various components of strain (Timoshenko and Goodier 1970). Once the stresses and strains are calculated, the overall deformations may be determined.

Thermal Strains and Stresses

As electronics equipment is operated, the internal temperature rises to a steady-state value. Once the equipment is shut down, the temperature is lowered to that of the environment.

There is also true temperature transience as electronics equipment is operated. Suppose that the electronics equipment is used for number crunching. As the CPU is engaged in this activity, its power consumption increases and there is a corresponding increase in temperature. As this activity is reduced for I/O activities, or once the calculations are completed, the power consumption is reduced, thereby reducing the temperature levels.

Materials generally expand (or shrink) as the temperature increases (or decreases). As components begin to expand at different rates, they "push" against each other, leading to what is generally known as thermal stresses. Thermomechanical analysis, thus, involves the impact of temperature change on material behavior and its internal state of stress and strains.

In electronics packaging design, one must be aware of the thermal load variations that the components may undergo and the impact they may have on the overall system. In general, this impact is seen primarily at chip, component, and board levels.

Thermal Strains and Deflections

In a uniform temperature field, the behavior of a uniform component depends on its geometric configuration. In general, three conditions may exist:

1. *No restrictions*: The component is free to deform. While there are deformations, no significant stress state is formed.
2. *Constraints*: The component may be constrained minimally. In this case, only deflections and deformations take place, but again no significant stress state is formed.
3. *Properly constrained*: As the material expands, there is not enough "room to move." Therefore, significant stresses may develop.

However, if the temperature field is not uniform and a temperature gradient exists, then various regions of the component expand (or shrink) differently from their neighboring regions, caused by the nonuniform temperature distribution. This may cause severe stresses leading to a failure or fracture. An example of this phenomenon is that when hot liquid is poured in a cold glass, the glass shatters.

There is yet a third condition. In the last two cases, it was assumed that the material is uniform. Many components in electronics equipment are nonuniform, each segment having a different rate of thermal expansion. Furthermore, many engineering plastics used today exhibit a dependency on the magnitude of temperature. This combined with a nonuniform temperature distribution can potentially develop into severe stress gradients.

Basic Equation

In a linear static problem, total deformation is a summation of deformations caused by mechanical loads and deformations caused by thermal loads. In terms of strain, this may be expressed as

$$\varepsilon_{\text{Total}} = \varepsilon_{\text{Mechanical}} + \varepsilon_{\text{Thermal}} \tag{11.4}$$

where $\varepsilon_{\text{Thremal}} = \alpha \Delta T$, α is the coefficient of linear thermal expansion, and ΔT is the temperature change from a "stress-free" state. This equation looks deceptively simple. A more realistic expression is to modify Equation 11.3 based on Equation 11.4 to obtain the following relationship:

$$\begin{Bmatrix} \varepsilon_x \\ \varepsilon_x \\ \varepsilon_x \\ \gamma_{yz} \\ \gamma_{xz} \\ \gamma_{xy} \end{Bmatrix} = \begin{bmatrix} \frac{1}{E} & \frac{-\nu}{E} & \frac{-\nu}{E} & 0 & 0 & 0 \\ \frac{-\nu}{E} & \frac{1}{E} & \frac{-\nu}{E} & 0 & 0 & 0 \\ \frac{-\nu}{E} & \frac{-\nu}{E} & \frac{1}{E} & 0 & 0 & 0 \\ 0 & 0 & 0 & G & 0 & 0 \\ 0 & 0 & 0 & 0 & G & 0 \\ 0 & 0 & 0 & 0 & 0 & G \end{bmatrix} \begin{Bmatrix} \sigma_x \\ \sigma_y \\ \sigma_z \\ \tau_{yz} \\ \tau_{xz} \\ \tau_{xy} \end{Bmatrix} + \begin{Bmatrix} \alpha \Delta T \\ \alpha \Delta T \\ \alpha \Delta T \\ 0 \\ 0 \\ 0 \end{Bmatrix} \tag{11.5}$$

There are very few closed-form simple equations that can solve realistic problems in the electronics packaging field and, in general, finite element analysis must be employed for solving practical problems. One such circumstance where a closed-form solution may be obtained is the procedure for calculating thermal stresses and strains in plates (Jones 1975, Hall 1993). In this formulation, aside from the assumptions of linear elasticity, it is assumed that the plate thickness is much less than its other dimensions. Furthermore, the impact of plate edges has been ignored. Hall (1993) has shown that should the plate bend in one direction (i.e., a cylinder), Equation (11.5) will be reduced to

$$\varepsilon_x = \left(1 - \nu^2\right) \frac{\sigma_x}{E} + \alpha \Delta T$$

and

$$\sigma_y = \nu \sigma_x$$

For an axisymmetric case, again Hall (1993) has shown that Equation 11.5 will become

$$\varepsilon_x = \left(1 - \nu\right) \frac{\sigma_x}{E} + \alpha \Delta T$$

and

$$\sigma_y = \sigma_x = \sigma$$

Recall that to determine the level of deflection and to solve for plate stresses and strains caused by any load, the first step is to balance forces and moments.

The procedure is no different here; however, Hall's simplification enables us to develop a relationship between temperature change and deformations without the need for solving any differential equations (Hall 1993). A further treatment of this subject in any detail is beyond the scope of this book, and the interested reader is referred to *Thermal Stress and Strain in Microelectronics Packaging* (Lau 1993).

Earlier it was pointed out that the impact of temperature gradient is primarily at chip, component, and board levels. Now let us examine each of these areas briefly.

Die Attachments

Residual thermal stresses are introduced in the cooling step of the bonding process because of thermal expansion and mismatch among the die, the bonding material, and the package. These stresses may cause the die to crack.

In this regard, voids existing in the bonding layers lead to stress concentration areas. These voids are generated by a variety of things including trapped gas, liquid, or other impurities. Other stress concentration areas may be caused by layer separation as a result of improper bonding, fatigue, creep, or rupture. In addition to stress concentration, voids may also increase the chip-operating temperature and cause hot spots.

Methods of improving die-attach quality include bonding in a pure environment, application of pressure for good contact, and back-grinding and wafer thinning of GaAs devices

Integrated Circuit Devices

As integrated circuit (IC) configurations are generally an encapsulation of a stack of silicon and lead frame, thermal expansions and contractions could potentially have detrimental impacts on the reliability of these devices. One such impact may be residual bow of the package that has been induced during the manufacturing processes. Suhir (1993) studied two types of plastic packages, namely, thin elongated packages with large chips known as thin small-outline packages, and high-lead-count large square packages with relatively small chips, known as plastic quad flat packages. He concluded that the residual bow is primarily influenced by the coefficient of thermal expansion (CTE)—as opposed to Young's modulus. Furthermore, he suggested that the molding compound must have a much higher expansion rate as opposed to the silicon chip and the lead frame. Thus, by having a thin chip and lead frame compared to the overall package thickness, and by placing them in the midplane of the package, the bow may be reduced significantly or be eliminated completely.

Printed Circuit Board Warpage

The sources of thermal stress may be numerated as follows:

1. Mismatch of global thermal coefficient (TCE) between major components
2. Local TCE mismatch between subcomponents such as lead and solder
3. Lead stiffness
4. Thermal gradients in the system (i.e., nonuniform temperature distributions)
5. Unbalanced component layup
6. Nonplanar boards

The following are a few simple rules to minimize warpage of printed wire board (PWB).

1. Geometry symmetry—Asymmetric layups create layer warpage. Use symmetric configurations about the board midplane
2. CTE match—Consider materials that have similar CTEs
3. Thickness tolerance—PWB warpage is very sensitive to layer thickness tolerance; control it rigorously

Considering that the printed circuit board (PCB) warpage is caused by CTE mismatch, it has been proposed (Peak et al. 1997) that the warpage may be modeled if the board is assumed to behave as a bimetallic board. In this case the maximum deflection is given by

$$\delta = \frac{\alpha_b L^2 \Delta T}{t}$$

where L is the diagonal length of the PCB, t is the thickness of the PCB, and ΔT is the temperature rise. Note that for an accurate estimation of warping, α_b is termed "specific coefficient of thermal bending" and must be calculated correctly. It is based on the relative volume of each material as well as their various physical properties. For more information on α_b see Yeh et al. (1993) and Daniel et al. (1990).

Some Tips for Avoiding Temperature-Related Failures

Studying the impact of temperature on a piece of equipment can be as crude as blowing hot air on various segments of a working PCB using an air gun. However, for a more detailed study:

1. Consider the effects of temperature on properties. Do material properties vary greatly with temperature?
2. Conduct a heat transfer analysis to develop a better understanding of temperature variations
3. Where a range of properties is given, use both ends of the spectrum
4. Use the temperature field to assess a need for a thermal stress analysis

Simplifications (or Making Engineering Assumptions)

We can now review a set of simplifying assumptions that may be applicable in about 85% of all stress analysis problems. The first set of such assumptions is based on the idea that materials behave linearly and proportional limit is not reached. The second set of assumptions is based on reducing a three-dimensional geometry into a one- or two-dimensional case.

Linear Elasticity

If loads are such that the stresses within the material never reach the proportional limit and the following conditions are met, it is said that the system follows the theory of linear elasticity or strength of materials. The theory of linear elasticity dictates that

- Deformations, strains, and rotations remain small
- Stiffness through the model does not change
- Boundary conditions remain the same, for example, loading direction with deformation does not change
- Material remains in the linear elastic range

This is a potential pitfall for a great deal of designers in using finite element analysis packages somewhat as a black box. Once, any of these assumptions are violated, the solutions obtained must be questioned and other options exercised.

Geometric Simplifications

The general equations of elasticity are three-dimensional and complicated. However, under certain conditions, it is possible to reduce the three-dimensional stress–strain relationship to two dimensions, thus simplifying the solution procedure greatly. These are

- Beam, plate, and shell theories
- Plane stress
- Plane strain
- Axisymmetric

Beams, plates, and shells are areas that have been studied for a long time and a well-established set of formulae for calculating their deflections, stresses, and natural frequencies under a variety of loading and boundary conditions exists. For more information on these topics see Baumeister et al. (1979).

In general, a structure is under conditions of plane stress if the stresses that develop along one of the three orthogonal axes are so small compared to the rest of the stresses that they can be assumed to be zero (Figure 11.8). This generally occurs if the thickness of the problem is much smaller than other dimensions of the problem and all applied loads are in the plane of the problem, then $\sigma_z = 0$ and $\tau_{yz} = \tau_{xz} = 0$.

In contrast to plane stress, if the thickness of the structure is much larger than other dimensions of the problem and all loads are applied uniformly in that dimension, then conditions of plane strain exist. An example of a plane strain problem is a cylinder under uniform pressure (Figure 11.9). Another example would be the strip foundation of a building. All cross sections along the x-axis of the cylinder and the strip foundation that are away from the boundaries are under the same loading conditions and, therefore, experience the same deformation. As a result, there will be an out-of-plane stress. In general a structure is under the plane strain condition if the strains that develop along one of the three orthogonal axes, say, the x-axis are zero. The longitudinal stress along the x-axis is not zero and can be evaluated in terms of σ_z and σ_y, and the shear stress along the x-axis is zero.

If the geometry is a body of revolution and the loading and boundary conditions are also symmetric around the same axis, then it is possible to invoke the axisymmetric assumption and reduce a three-dimensional problem into two dimensions (Figure 11.10).

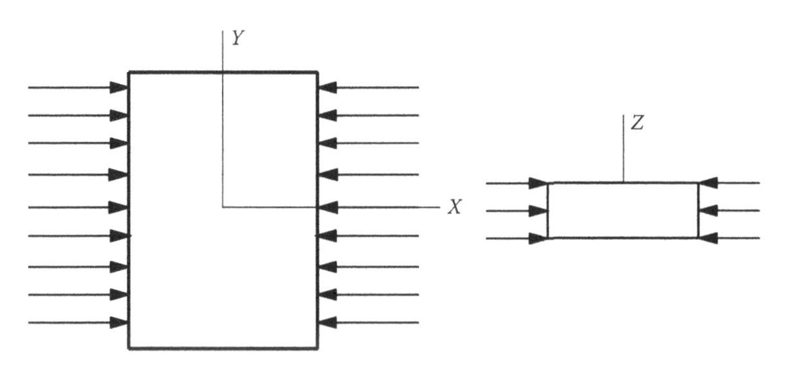

FIGURE 11.8
Plane stress condition.

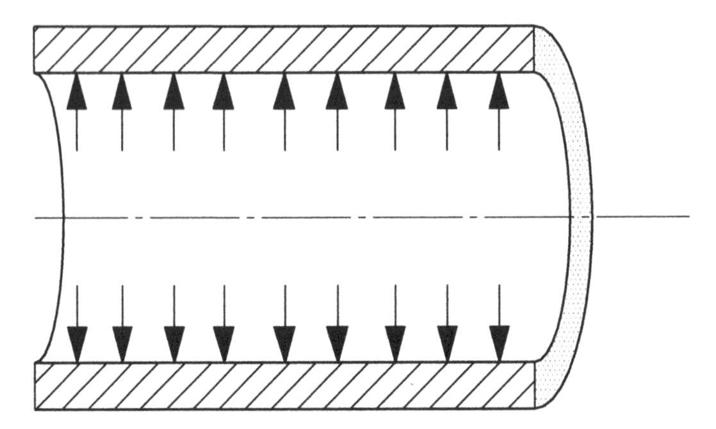

FIGURE 11.9
A cylinder under uniform pressure; plane strain conditions.

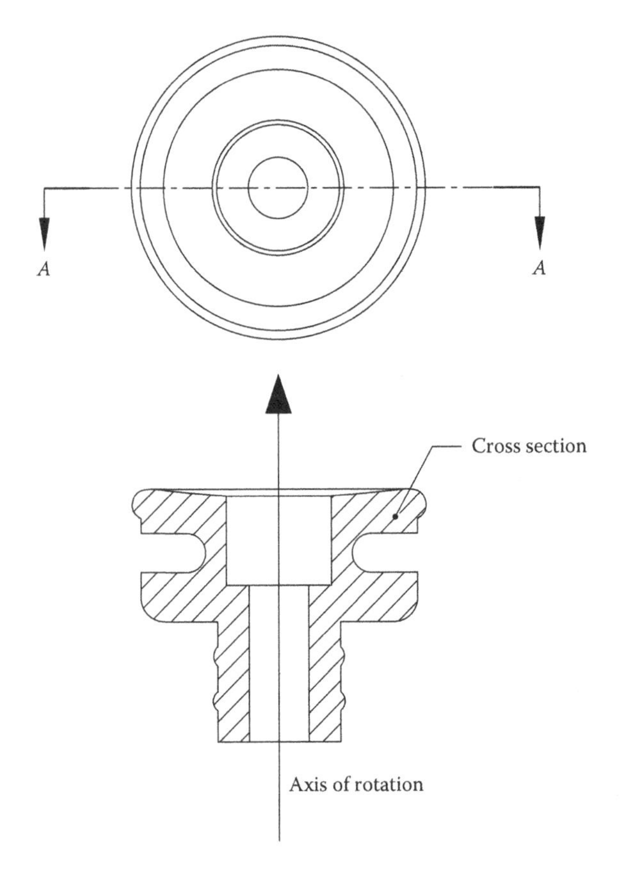

FIGURE 11.10
Axisymmetric conditions.

Stress Concentration

In a certain class of problems, the stress field is uniform over the domain, with the exception of isolated regions where it becomes complicated. An example of this problem is a simple cantilever beam loaded with a uniform load along its axis but with a notch near its end. Most engineers recognize this notch as a stress concentration point and expect that it would generate higher stress levels. A notch is a common stress concentration point. Other common causes of stress concentration are as follows:

- Abrupt changes in section geometry such as the bottom of a tooth on gear
- Pressure at the point of application of the external forces
- Discontinuities in the material itself, such as nonmetallic inclusions in steel
- Initial stresses in a member that result from over-straining and cold working
- Cracks that exist in members caused by part handling or manufacturing

In the theory of strength of materials, a stress factor is provided depending on the geometry of the point, but further discussions are beyond the scope of this book, and the reader is referred to other sources such as Timoshenko and Goodier (1970), Boresi et al. (1978), and MacGregor et al. (1979). It may be added, however, that in practice, a stress concentration factor for circular holes in plates is approximately given by the following relationship (Boresi et al. 1978):

$$K = \frac{3\kappa - 1}{\kappa + 0.3}$$

where κ is the ratio of width of the strip to the diameter of the hole.

Example

A 1-lb printed circuit board assembly (PCBA) is mounted on a plastic tray as shown in Figure 11.11. We need to design this tray so that it can withstand a base vibration of 5 Gs with a transmissibility of 12. The PCBA's center of gravity is located at the center of the line connecting the two top standoffs.

Acrylonitrile Butadiene Styrene is selected as the molding resin. It has a Young's modulus of 160 ksi and a yield strength of 2685 psi and an ultimate strength of 4000 psi.

Assume that dynamic forces caused by vibration are transferred to the plastic housing in the vertical direction. The maximum forces acting on

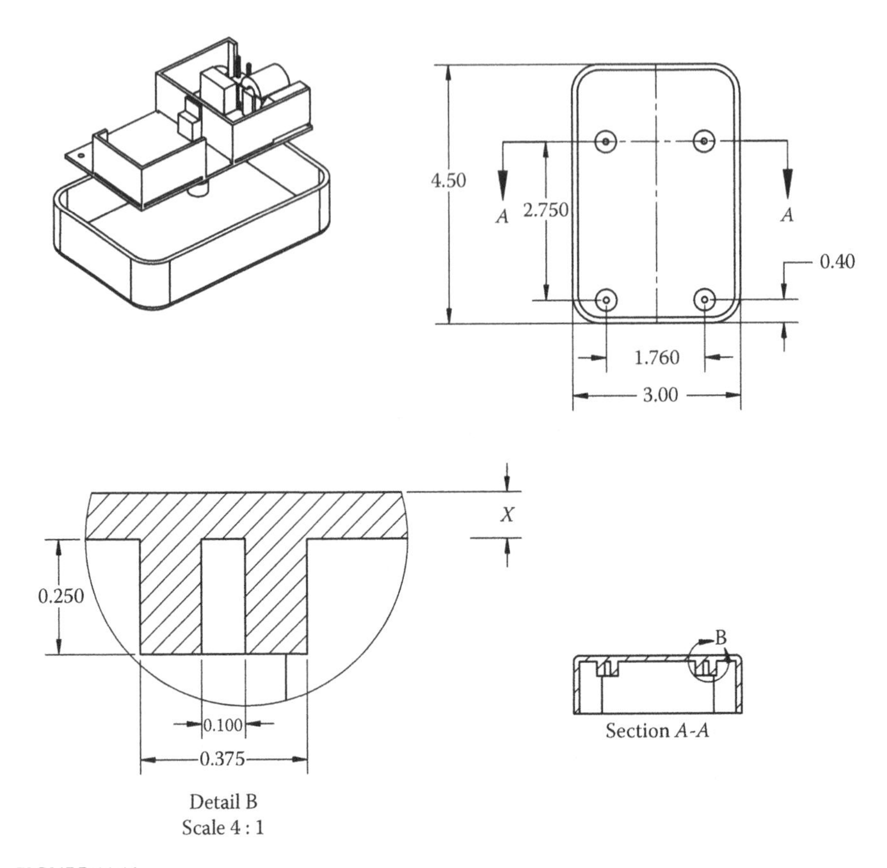

FIGURE 11.11
A plastic enclosure housing a 1-lb PCB.

only one standoff is the product of the proper portion of the weight (here 50%), the applied acceleration, and the unit's transmissibility.

$$F_{dynamic} = \left(\frac{1}{2}\right)(1\text{lbs})(5G)(12) = 30\text{lbs} \text{ exerted on each of the two top standoffs.}$$

Thickness Calculation Based Solely on Shear

One approach to calculating the wall thickness is to consider how the standoff may shear in a direction normal to the plate (effectively leaving a hole in its wake). This assumption would necessitate the fact that the back plate does not deform and is essentially rigid:

$$\text{Effective area} = \pi(0.375)(x)$$

$$\text{Shear stress} = \frac{30\text{lbs}}{\pi(0.375)\ X} \times \text{stress concentration factor} \le 2685$$

The stress concentration factor is caused by the sharp corner of the back plate and the standoff and may be assumed to be near three for sharp transition areas (Boresi et al. 1978):

$$x \geq \frac{30\,\text{lbs}}{\pi(0.375)\,2685} \times 3 = 0.028\,\text{in}$$

This wall thickness would theoretically be sufficient if the standoff were to move perpendicular to the back plate. However, there is another component that needs to be included; as the PCBA moves back and forth because of vibration, it causes deformation of the back plate. This deformation causes both shear and normal stresses. The stress field may have a detrimental effect on the ability of the enclosure to survive the dynamic loads.

Thickness Calculation Based Solely on Back Plate Deflection

The more realistic approach would be to calculate the wall thickness based on the induced stressed caused by bending and deflection of the back plate (MacGregor et al. 1979):

$$S_m = \frac{kP}{x^2}$$

where S_m is the maximum stress, P is the applied load, and x is the thickness. k is a geometric factor that depends on the plate's aspect ratio. Here k is 0.454 for an aspect ratio of 1.5. Rewrite this equation to get

$$x = \sqrt{\frac{kP}{S_m}} = \sqrt{\frac{0.454(30+30)}{2685}} = 0.1\,\text{in}$$

Clearly, the stresses caused by back-plate bending are more severe than the previous case and the wall thickness must be based on 0.1 in thickness. In general, we would need to specify a factor of safety and use a slightly larger value. However, in using the plate deflection formula, we have assumed that the side walls of the enclosure are perfectly rigid, whereas in reality the side walls do in fact deform and take some of the vibration energy. Furthermore, we have assumed that the only two posts (standoffs) are carrying the entire load—and we have ignored the role of the PCBA in distributing the weight. All in all, the calculated wall thickness is a conservative value. Having said this, it is noteworthy to consider that a finite element analysis conducted on this part reveals that the Von Mises stresses under the dynamic loads are 2777 psi if we apply two 30-lb loads on each post and would be about 1300 psi if we model the PCBA as well.

12

Acoustics

Introduction

I recall my first encounter with noise in electronics equipment as an engineer when I was leading the design and development of an electronics system used for education of children with disabilities. In the first iteration of the design, a combination of an internal fan and a series of louvers was utilized. The first prototypes were astonishingly quiet—which we only noticed in comparison with the noticeably noisier subsequent designs. The first design's drawback was that the hot spot temperature rises were exceeding critical components' limits. In an effort to meet production deadlines, we replaced the internal fan with an external fan, removed louvers, and developed a design that met the thermal specifications. However, one could hear a humming sound while operating the unit. A noise that is only too familiar to anyone who operates a personal computer, and is just about taken for granted.

Since I had noticed the humming sound in the new design, one day, I decided to quantify my suspicions by measuring the sound levels. Using a handheld device, I began to measure the sound levels at several locations near the two systems. To my astonishment, I realized that although I could clearly hear one system much louder than the other, the handheld device was registering very similar sound levels. After a few iterations in taking measurements, I finally decided that I knew nothing about this field and that I need to ask for help.

This chapter is a synopsis of what I have learned about sound, noise, their management, and impact on electronic equipment. It is meant to bring an appreciation of issues and challenges involved, and an awareness that the design team should have of acoustics and noise. It is by no means an exhaustive review. For an in-depth study of noise and acoustics, *Industrial Noise Control and Acoustics* (Barron 2003), *Designing for Product Sound Quality* (Lyon 2000), and the 2013 *ASHRAE Handbook* (Chapter 8) are recommended.

Noise, Sound, and Their Difference

Let us begin by developing some basic understanding of this field. When pressure waves propagate through a material, sound is said to have been generated. Noise is generally defined as any sound which is unpleasant or harmful. It is the subjective perception of the heard sound which would render it as noise. While some have divided sound into noise and music, others have proposed the concept of sound effects and how it may be manipulated in design—particularly in architectural designs (Hellström 2001).

Regardless of one's definition, it is well accepted that there is a real danger of permanent hearing loss if one is subjected to "high-enough" sound levels. This reality has prompted many governments to establish regulations to protect the welfare of people in industrial and/or public environments. Aside from hearing loss, emotional/psychological stress and fatigue have also been associated with noisy environments.

Ironically, it is not always government regulations advocating for lower sound levels. Many a time, consumers' response to a product may be based on how quiet a product may be. A system or a product with moving parts does not generally *just* happen to be quiet. It must be designed quiet.

Electronics equipments have penetrated just about any aspect of modern life from the obvious to the obscure. As electronics products become more sophisticated, they produce more heat. Removal of this heat requires use of active cooling technologies such as fans. These technologies along with storage devices such as CD ROMs, and hard drives, are the primary sources of noise in consumer electronics equipment. In medical electronics, such as infusion pumps or dialysis equipment, pneumatics and moving parts such as valves may be additional sources of noise.

To design a quiet electronics system, first, we must develop an understanding of what is loud and what is quiet. In other words, we need to know measurement standards and the basic terminology used in acoustics. Next, we need to learn the principles of noise control. Finally, we need to develop an awareness of the acoustic issues specific to electronics packaging.

Basic Definitions

Three basic terms and their associated definitions are provided as background information:

Sound (or acoustic) power: This metric is a measure of emitted sound energy at its source, and is the total radiated sound energy per unit time (Talty 1988).

Sound intensity: This metric is defined as sound power per unit area.

Sound (or acoustic) pressure: This metric, perceived by the human ear and measured, is defined as the difference between atmospheric pressure and the pressure level in the vibrating medium (e.g., air). This metric is not related to frequency or wavelength (Talty 1988).

Governing Equations

Sound is perceived when pressure disturbances propagate. Mathematically, it is described as a wave and may be represented as

$$\frac{\delta^2 P}{\delta t^2} = C^2 \nabla^2 P \tag{12.1}$$

where P is pressure, t is time, and C is speed of sound in the medium. Clearly, the speed of sound depends on the medium properties. For an ideal gas,

$$C = \sqrt{\gamma RT}$$

where $\gamma = c_p/c_v$ is the ratio of specific heats, R is specific gas constant, and T is absolute temperature.

The solution of this equation requires a knowledge of both the initial and boundary conditions. In other words, the same sound generating source can have different pressure distribution fields if it is operated outdoors versus inside a small (or large) room; with curtains and drapes or with reverberating walls. Hence, a product that is perceived as quiet in one environment may be deemed as annoying in another.

One way of comparing different pressure fields is to calculate the "sound pressure levels" (SPLs) associated with that field. The solution to Equation (12.1) provides the root mean square pressure (P_{RMS}) needed for this calculation.

$$\text{SPL} = 20\log\left\{\frac{P_{\text{RMS}}}{P_{\text{ref}}}\right\} \tag{12.2}$$

Typically,

$$P_{\text{ref}} = 20 \times 10^{-6}\ \frac{N}{m^2}$$

Similarly, it may be shown that sound intensity (I) is related to the root mean square pressure as shown below (Talty 1988).

$$I = \frac{P_{\text{RMS}}^2}{\rho C}$$

where, ρ is the density of the medium. Considering that sound intensity is defined as sound power (W) per unit area, for a spherical wave, the relationship between power and pressure may be written as

$$I = \frac{W}{4\pi r^2} = \frac{P_{RMS}^2}{\rho C}$$

where r is the distance from the source. This equation may be manipulated to produce

$$W = \frac{4\pi r^2 P_{RMS}^2}{\rho C}$$

Alternatively, should the sound power of a source be known, the root mean square of pressure at a given distance r from the source may be calculated:

$$P_{RMS} = \sqrt{\frac{\rho C W}{4\pi r^2}}$$

For the sake of completeness, another term, "sound power level", is also defined:

$$L_W = 10\log\left\{\frac{W}{W_{ref}}\right\} \tag{12.3}$$

Typically,

$$W_{ref} = 1\times10^{-12}\text{ W}$$

Measurement and Standards

Recall my initial experience with the noise measurement. Because I lacked a fundamental understanding of sound and its behavior in different environments, I had placed my trust solely in the sound meter that I had used. If fact, when I think back on this incident, I do not recall if it was even calibrated.

In addition to making sure that the equipment is properly calibrated, Barron (2003) recommends identifying the "electrical noise floor" of the equipment either through the unit's documentation or by replacing the microphones with an equivalent electrical impedance. Electrical noise floor is signal noise generated by the equipment which will impact the readings.

In addition to qualifying the test equipment, it is important that a detailed test plan be developed. International standards such as ISO 3744, ISO 2450, or ISO 7779 may be referenced for test setup, proper placement of the unit under test (UUT) relative to the test chamber (or the environment), and the placement of microphone (or microphones) relative to UUT. Two other elements of this test plan should be the required (or needed) accuracy and well as acceptable levels of uncertainty.

Barron (2003) further recommends that prior to developing the test plan, one should develop an understanding of the specific nuances of the UUT as well as its use environment or the test room/facility. In short, the test designer should listen to the unit in operation and make note of any anomalies. These observations will enable the test engineer to understand if the produced noise is uniform or directional, continuous or not, and to identify any special testing requirements.

Another factor worth considering is this: measuring SPLs in an environment with a sound source is akin to measuring temperature of an environment with a space heater. For the same levels of power, temperature distribution varies depending on the environment, that is, is the room large or small; are there any open windows or not? Because of the influence of these various factors, temperature measurements alone may not be directly related to the heat radiating from a space heater.

Similarly, measuring SPLs of a sound-radiating source may vary significantly depending on the environmental conditions, and hence, the readings cannot be directly related to how strong the sound source may be. For this reason, sound-level measurements are often made in anechoic or reverberant chambers. By having the exact environmental conditions under control, SPLs may then be related to sound intensity levels associated with equipment under test.

Alternatively, the sound intensity levels may be directly measured without the need of any special environmental conditions. It is similar to measuring the radiating heat of a space heater directly. A sound intensity meter measures not only the magnitude of the intensity but also its direction. This is extremely helpful in identifying the equipment with the highest noise level from a bank of noisy equipment.

Now, this identifies the root of my measurement error that I described in the introduction of this chapter. By using a sound meter, I was actually measuring the SPLs present in the room that was the result of the combined sound sources in the room. There is now little wonder that I could not get different readings. For my purposes, I should have used a sound intensity meter.

Acoustics as a Design Priority

It is not uncommon that mitigating the noise generated by an electronics product be a design afterthought. Often, a vague requirement is written to the effect that "the system shall be quiet." From a product development point of view, good designs begin with proper requirements. By having a measurable, and verifiable, noise requirement, the design team—including marketing—have given priority to the end product's acoustics. The acoustics requirements would then flow down to various subsystems which would impact design specifications.

Once the detailed design begins in earnest, the design engineer should be aware of three factors that impact noise generated within an equipment. These may be enumerated as follows:

1. Inefficient fans
2. Aeroacoustics
3. Resonance or near resonance conditions

Inefficient Fans

In Chapter 5, fan performance curves were introduced and three distinct segments were identified. These were block-off, stall, and free flow regions. A fan performs optimally when it operates to the right of the stall region where the pressure drop is low. Based on Equation 12.2 the lower the pressure, the lower the sound pressure level. As we learned in Chapter 5, the operating point of a fan is dictated by the unit's impedance curve. Although the design engineer does not often have much control over the impedance curve, there is some flexibility in selecting the fan, and its associated curve.

Having said this, we need to note that manufacturer's published fan curves are developed in an environment where the flow into the fan frame is fully developed and uniform (Sloan 1985). Deviation from this ideal flow will impact the fan curve and ultimately, not only the noise levels but also the operating point of the fan. Thus, placement of the fan in the enclosure and its spatial relationship to other components in the device impacts the flow pattern and is a crucial factor in noise generation. The best design would be one where the airflow patterns to the fan are fully developed. This may not always be possible in practice.

If the fan noise becomes excessive, other technologies such as valve-less or fan-less cooling devices (Mahalingam et al. 2007; Staley et al. 2008) may need to be employed.

Aeroacoustics

Another factor that contributes to a "noisy" fan is the speed of blade rotation; the higher the rotation per minute (RPM), the noisier the fan. So, the general remedy is to design-in a bigger fan if possible, and to slow the fan down or use it intermittently using a control algorithm.

The phenomenon that fan noise is correlated to its rotational speed is associated with the blade geometry and aeroacoustics. Aeroacoustics is the study of the sound generated as aerodynamic forces interact with a solid surface.

Developing fan blade geometry for noise reduction is a well-studied field, and new fan blade designs for noise reduction are proposed (Bianchi et al. 2012). In addition to the fan blade, aeroacoustics noise manifests itself when air is blown across sharp edges or corners. This happens often, when either printed circuit boards or fan guards are placed too close to the fan.

Resonance

The third cause of noise in an electronics device may be attributed to resonance of some of the installed components. In many products, resonance or near resonance conditions may be caused by aerodynamic loads such as air flow in a duct, across an edge, or on a shielding plate. Other sources of resonance may be moving parts such as an internal motor or a pump. Often, these vibration-producing components are mounted on vibration isolators, but over time, common synthetic rubber isolator lose their elasticity and transmit their vibration to the unit's enclosure.

As was suggested earlier, often design teams think of noise and its reduction as an afterthought. A best practice approach is to study competitive devices for their noise levels and then by specifying a measurable and verifiable noise requirement, the development team can give priority to this aspect of product design as well.

13

Mechanical Failures and Reliability

Introduction

Reliability concerns those characteristics that are dependent on time: stability, failure, mean time to failure/repair, and so on. It may also be defined as the probability of a product or device performing adequately for the period of time intended under the operating conditions encountered.

However, if harm or injury occurs because of poor reliability, the manufacturer or distributor may have a responsibility to compensate for these losses and/or injuries. The general liability law holds that those who introduce a defective product into the market are generally liable for the product and any harm or injury caused.

There is another definition, one that is to one's disadvantage. This is not a legal definition, but in many circumstances, it may mean loss of the customer's good faith and business. So, it may be said that if a product has low reliability, the distributor and the manufacture are still liable—even if no harm or injury has occurred. As Taguchi and Clausing (1990) have explained:

> When a product fails, you must replace it or fix it. In either case, you must track it, transport it, and apologize for it. Losses will be much greater than the costs of manufacture, and none of this expense will necessarily recoup the loss to your reputation.

Reliability engineering is an extensive field concerned with managing failures in equipment and systems (MIL-HDBK-338B 1998), but there are, in general, three aspects of reliability that concern electromechanical design engineers, namely, mechanical, electrical, and chemical issues. Ironically, thermal concerns impact one or all of them and do not act independently.

In this and the next two chapters, we will talk about each to some extent. Additionally, we need to remember that other factors impacting reliability include electromagnetic or radio frequency interference. Furthermore, shelf life, chemical reactions, solderability, moisture, aging, etc., have an important influence on reliability as well. In addition, the reliability of the software controlling the device should not be ignored. A discussion of these other factors, however, is beyond the scope of this book.

Failure Modes

Since reliability is, in a way, a study of failure, it is important to first define failure. Failure is defined as the inability of a system to meet its design objectives. It may be that it was poorly designed and never met its objectives, or that it initially met its design objectives but after some time it failed. In these two scenarios, clearly, there have been certain overlooked factors.

Assume that you have designed a sensor to be operated inside a freezer; however, it was operated at room temperature and it gave erroneous readings. Would you consider this a system failure or an operator error?

A different scenario: Your colleague designed an entertainment radio for a Boeing 777. It passed all the specified tests but the support brackets would permanently bend only after a few hundred hours of flight, resulting in damaged cables. Once the cables were replaced, the unit would function as expected. Would you consider this a failure?

Now consider a third scenario: Another colleague designed a system and, during testing, ignored signs of potentially high-temperature components. Furthermore, test conditions did not replicate the operating environment. As a result, the unit experienced a thermal run away, and heat buildup caused the plastic enclosure to melt. What are your thoughts now?

Clearly, many factors lead to system failures; some could be caused by lack of proper design and verification, whereas others could be attributed to improper usage and still others may be a result of the designer's lack of insight into potential factors affecting the system. Failure modes and effects analysis (FMEA) helps the design engineer and his or her team to clearly identify sources of failure and indicate whether the solutions may be sought in mechanical, electrical, or other causes.

There are four causes of mechanical failures. These are as follows:

1. Failures by elastic deflection—These failures are caused by elastic deformation of a member in the system. Once the load causing the deformation is removed, the system functions normally once again. An example of this type of failure is resonant vibration of relays. Another example is linear buckling of a support structure.
2. Failures by extensive yielding—These failures are caused by application of excessive loads where the material exhibits a ductile behavior. Generally, the applied loads and the associated stress factors create stress fields that are beyond the proportional limit and in the neighborhood of the yield point. In these scenarios, the structure is permanently deformed and it does not recover its original shape once the loads are removed. This is generally a concern with metallic structures such as chassis and racks.

3. Failures by fracture—These failures are caused by application of excessive loads where the material exhibits a brittle behavior, or in ductile materials where stresses have surpassed the ultimate value.
4. Progressive failures—These failures are the most serious because initially the system passes most, if not all, test regiments and yet after some time in the field, it begins to fail. Creep and fatigue belong to this category.

Thus, in designing a part or an assembly, the design engineer must determine possible modes of failure and then establish suitable criteria that accurately predict various failures. In the next segment, these failures and their criteria are examined.

Failures by Elastic Deflection

Failure by elastic deflection means that deflections are too large to be acceptable; however, stresses are below the proportional limit and as such are not important as a design criterion. Quite frequently these types of undesirable deflections are observed in vibration environments. When the amplitude of vibrating components is large enough that parts collide, the system has failed. In electronics systems, this failure may be catastrophic if components are permanently damaged or temporarily disabled as in the case of resonance of mechanical relays. Other examples involve structures such as beams or shells, which may buckle under compressive loads. Under these circumstances, the structure regains its original shape once the loads are removed.

The failure criterion for failures caused by extensive elastic deflection is set based on the knowledge of the maximum allowed deflections—particularly between elastic components such as printed circuit boards or cold cathode tubes. These and similar structures easily deform as a result of vibration or other inertial forces and could possibly collide causing what is known as chatter. The design engineer should calculate the maximum allowable deformation values and take precautions to prevent the part movements from exceeding these values.

A criterion that is often neglected is buckling. A simple method to identify potential problem areas is to identify components that are under compressive loads and then to ensure that these components have not exceeded their buckling load-carrying capacities. A classic example to demonstrate this failure is column buckling. A simple slender column would be able to carry a load so long as the load is below P_{max}.

$$P_{max} = \frac{\pi^2 EI}{L^2} \tag{13.1}$$

where E is Young's modulus of elasticity, I is second moment of inertia, and L is the column's length. Once the applied load surpasses this value, the column would buckle. Thus, if the equipment rack is made of long slender members (such as trusses), it would be prudent to calculate the loads in each member and apply Equation (13.1) to each member that has a compressive load.

Another area where buckling failures may play a role is in the design of flexible buttons using thermoplastic elastomers (TPE). Often, a design requirement dictates that a certain activation force should not be exceeded for these buttons. A general configuration for these TPE buttons is shown in Figure 13.1 and involves a relatively heavy plug in the center surrounded by a narrow web. Again, depending on the web's thickness and the distance to travel, it is possible to create a condition that would cause the button to buckle and either tear the web rendering the button useless, or push the plug off its track and away from the switch underneath.

Generally speaking, in failures involving elastic deflection, a stress analysis is not needed; however, deflections and/or buckling loads need to be calculated. Determining buckling loads may require the use of finite element analysis software.

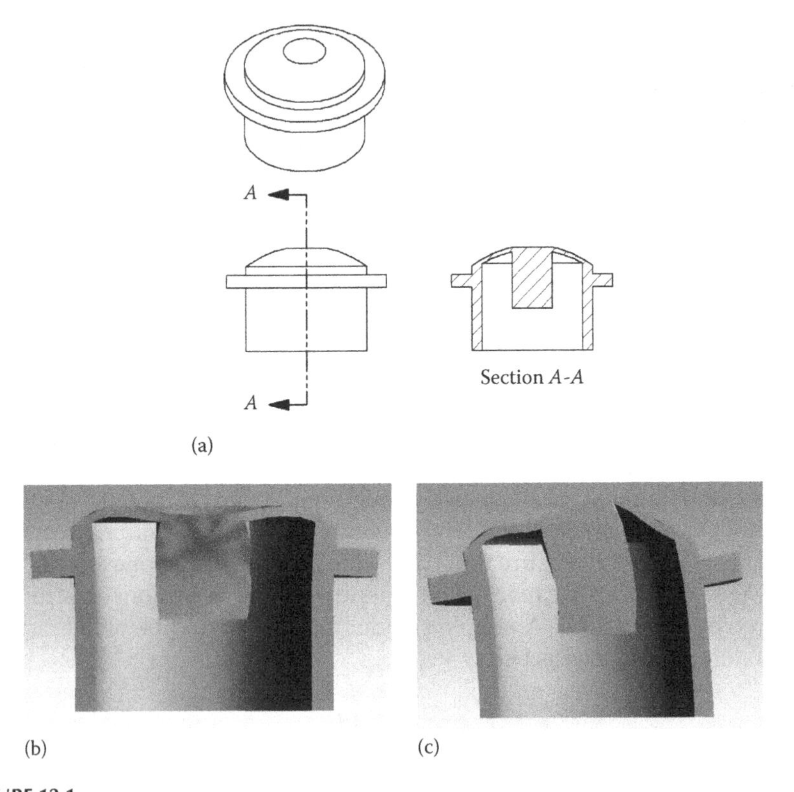

FIGURE 13.1
A push button made with TPE is susceptible to bucking. (a) Initial geometry, (b) Deformation with loads below bucking level, and (c) Deformation with loads above bucking level.

Failures by Extensive Yielding

For a ductile material, the initial portion of the stress–strain curve may be linear; however, as the loads cause stresses to surpass proportional limit and eventually the yield point, permanent deformations take place in the system. Under such circumstances, the system may have deformed to the point where it might have lost its load-carrying capacity and the equipment may collapse. These are most significant with regard to simple structural members such as axially loaded members, beams, columns, torsion members, or possibly thin plates subject to in-plane forces. Here, the stress field and its distribution play an important role as a design criterion. The design engineer should conduct a stress analysis and then identify whether the conditions for material yielding exist or not. Ironically, the criteria used to identify yielding depend on the material to a large extent. These are enumerated below.

1. Maximum shear stress (Tresca's criterion) is a good criterion for ductile materials. The part fails when maximum shear stress has reached yielding.
2. Maximum strain (St. Venant's criterion) is also a good criterion for ductile materials. The part fails when maximum strain has reached yielding. It gives slightly more reliable results than maximum principal stresses.
3. Von Mises failure criterion is based on strain energy density of distortion being equal to energy of distortion at yield.
4. Octahedral shear stress is the same approach as the strain energy but the formulation is based on shear energy. Generally speaking, Tresca and Von Mises give the best results.

Failures by Fracture

There are two types of failure by fractures: sudden with no evidence of plastic flow and fracture of cracked or flawed members. Often—though not always—this occurs when the material behavior is brittle. The best criterion used to identify failure by facture is maximum principal stress (Rankine's criterion). Again, the design engineer should conduct a stress analysis and evaluate the stress field and its distribution. The part fails when maximum principle stress has reached yielding (or rupture).

In the case of cracked or flawed members, the crack tip acts as the point for stress concentration, thus allowing local stress levels to go beyond the yield values and into the plastic region. If the material has a ductile behavior, it is possible for the part to carry the load because the strain field around the crack will be in the plastic region and the stresses remain below ultimate. In a brittle material—or if the material has a brittle behavior around the crack tip—the plastic region would not form and the crack would grow larger and propagate to the point where a catastrophic failure would take place.

Progressive Failures

There are two other phenomena that must not escape the design engineer. One is creep and the other is fatigue. Both have been studied extensively as separate causes of failure and when the two are combined as well as when they couple with corrosion (Pecht 1991; Lau 1993; Bisbee et al. 2007; Holdsworth et al. 2007; Sabour and Bhat 2007; Wereszczak et al. 2007).

Briefly, creep is the mechanical action by which a strain field changes without a change in the stress field. This action is generally accelerated at temperatures near the melting temperature of the material. Under a proper environment, once a stress–strain field is developed, a lower stress would be needed to maintain the same strain. Thus, strains begin to grow for a given constant stress. This starts a very unstable trend whereby the rate of change in strain increases, or at best is a constant. Eventually, micro-cracks begin to form within the material, which eventually combine and grow to form into macroscopic fractures and ultimately catastrophic failures.

For metals such as solder, the creep activation temperature could be near the operating temperature of the equipment. This problem becomes particularly acute when the equipment undergoes thermal cycling and the creep phenomenon couples with fatigue and creates a condition known as creep ratcheting which leads to the failure of solder joints (Lau 1993). For a more in-depth review of creep, see Appendix F.

Fatigue is the process of progressive fracture. It happens when stresses cycle from a low value to a high value and back. It becomes an acute problem in vibration because the stresses not only go from low to high but also change sign. In other words, in vibration problems, stresses in vibrating parts go from compressive to tensile creating very adverse conditions in the material at microscopic levels. Under these conditions, microscopic cracks are formed. With continued changes in the stress field, these cracks grow and eventually form macroscopic fractures which then lead to sudden fractures.

Although conducting an analysis for creep requires in-depth knowledge and understanding of the creep strain-rate relationships and modeling, it is generally possible to conduct a fatigue analysis using standard techniques for stress analysis—particularly finite element analysis. Once the stress field is determined, the following may be used as measures for fatigue calculations:

1. A good criterion to evaluate stresses for fatigue is Von Mises failure criterion. This approach is based on strain energy density of distortion being equal to energy of distortion at yield.
2. Another criterion is octahedral shear stress which is the same approach as the strain energy but the formulation is based on shear energy.

Once the stress value is evaluated based on one of these criteria, the S–N diagram (see Appendix G) for that material may be consulted to learn whether the endurance limit (if it exists) has been exceeded or not. Stress values below the endurance limit indicate that the component would survive indefinitely otherwise; its life expectancy should be estimated. For a more in-depth review of fatigue, see Appendix G.

Life Expectancy

When we speak of life expectancy, it is implied that we are concerned with progressive failures because the other modes of mechanical failure occur rather rapidly. In this regard, theoretically, it seems to be a simple matter to calculate the life of a product if we have the data that give us its life under various conditions. For instance, say, using the S–N curve in Figure G.2 (Appendix G), the material would survive to 1×10^9 cycles of stress reversals if the values are about 82.0 ksi. If the frequency of stress reversals (i.e., vibration) is 100 Hz, it would take 10,000,000 (=1,000,000,000/100) s or about 2778 h for the part to reach its failure point. At a frequency of 1000 Hz, its life is reduced to 1,000,000 s or about 277.8 h. Similarly, in case of creep, if we have the strain-rate relationship for a given material, we can easily calculate how long it would take before maximum sustainable strains have been reached.

In practice, however, calculating equipment life is not simple. On the one hand, there are factors such as creep, corrosion, and fatigue that interact, and on the other hand, operating conditions vary, and, this variation needs to be taken into account.

First, let us consider the pure fatigue case and then consider pure creep. Then, we will discuss a simple model where the combination may be taken into account.

Life Expectancy for Pure Fatigue Conditions

Miner's index is used to predict the lifetime of a member of a system that experiences different levels of stress and the corresponding frequencies. This equation is particularly useful in vibration environments:

$$D_f = \sum_i \frac{n_i}{N_i} \tag{13.2}$$

N_i is the cycles to failure at stress level σ_i and n_i is the actual cycles of vibration at that stress level. Clearly (and theoretically), at end of life $D_f = 1$. However, many have argued to set D_f to values less than unity depending on the environmental and working conditions.

Example

Suppose that a vibratory member made of 300-M alloy undergoes a stress level of 82,000 psi at a frequency of 60 Hz. Considering that the S–N curve (Figure G.2) gives the number of cycles it can endure at this stress level to be 1×10^9 cycles, determine the time to failure.

Frequency is defined as number of cycles per unit of time (usually seconds). Thus, to calculate the number of cycles for a given frequency and time, one may multiply frequency and time to obtain the number of cycles:

$$D_f = \sum_i \frac{n_i}{N_i}, i = 1$$

$$D_f = \frac{n}{N} = \frac{\text{number of cycles}}{N} = \frac{\text{time} \times \text{frequency}}{N} = \frac{tf}{N}$$

$$1 = \frac{t \times 60}{1{,}000{,}000{,}000}$$

$$t = 16{,}666{,}666.67\,\text{s or } 4629.63\,\text{h}$$

Does this mean that the part will fail after 4629.63 h? The correct answer is that if the used S–N curve provided data for a 50% probability of failure, then it is expected that by $t = 4629.63$, 50% of the samples have failed. A different way of expressing this number is to say

$$L_{50} = 4629.63\,\text{h}$$

This annotation is often referred to L_x life, where the subscript x refers to the percentage of failed samples.

Life Expectancy for Random Vibration Conditions

In Section "Random Vibration" in Chapter 7, it was pointed out that random excitations cause all the natural frequencies (and mode shapes) of a continuous system to be present. However, the notable response of the system takes place at what is termed as "apparent" frequency, which is mainly influenced by the first natural frequency and contributions from other higher ones. In a random vibration environment, the number of positive zero crossings that occur per unit of time significantly influences the apparent resonant frequency. The number of positive zero crossings refers to the number of times that the stress (or displacement) crosses the zero line with a positive slope. Based on this definition, one apparent cycle is between two positive zero crossings (Sloan 1985; Steinberg 1988).

The lifetime of a piece of equipment under random vibration may be estimated using the following relationship (Sloan 1985):

$$T = \frac{B}{4f\sigma^b\left[0.2718(2/3)^b + 0.0428\right]}\,\text{Sec} \tag{13.3}$$

where f is the response frequency, σ is 3S stress. B and b are constants and depend on the material used:

$$\text{For G10 glass epoxy: } b = 11.36, B = 7.253 \times 10^{55} \text{psi/cycle}$$

$$\text{For aluminum: } b = 9.10, B = 2.307 \times 10^{47} \text{psi/cycle}$$

This equation is based on a skewed Gaussian distribution and incorporates the number of zero crossings. The drawback of this formulation is the relatively limited information on constants B and b for a large class of materials.

For a single-degree-of-freedom system, only one frequency is possible; thus, the apparent frequency and the resonant frequency are identical.

Example

Suppose that in a random vibration environment the 1S stress values for the 300-M alloy has been calculated to be 60 ksi. What would be the expected life of the part should the system be approximated as a single degree of freedom with a frequency of vibration of 60 Hz?

Appendix G suggests that a relationship may be developed between the stress values and the number of cycles to failure. For the 300-M alloy, the following relationship has been proposed (MIL-HDBK-5J 2003):

$$\log N = 14.8 - 5.2 \log\left[\frac{\sigma_{max}}{(1-R)^{0.38}} - 94.2\right]$$

where R is the stress ratio, which is the ratio of minimum stress to maximum stress in a fatigue cycle and σ_{max} is in ksi. For a fully reversible load, $R = -1$.

For $R = -1$, we have

At 1S: $\sigma_{max} = 60\text{ksi}, N_1 \approx \infty$,

Figure G.2 indicates that endurance limit is about 80ksi,

$n_1 = (\text{frequency}) \times (68\% \text{ of times at this stress level})$

At 2S: $\sigma_{max} = 120\text{ksi}, N_2 = 3.0 \times 10^5$,

$n_2 = (\text{frequency}) \times (27\% \text{ of times at this stress level})$

At 3S: $\sigma_{max} = 180\text{ksi}, N_3 = 4175$,

$n_3 = (\text{frequency}) \times (4\% \text{ of times at this stress level})$

or

$$N_1 \approx \infty,\ n_1 = (60)(0.68t)$$

$$N_2 = 3.0 \times 10^5,\ n_2 = (60)(0.27t)$$

$$N_3 = 4175,\ n_3 = (60)(0.04t)$$

Now apply Miner's rule:

$$\frac{n_1}{N_1} + \frac{n_2}{N_2} + \frac{n_3}{N_3} = 1$$

$$\frac{60 \times 0.68t}{\infty} + \frac{60 \times 0.27t}{3 \times 10^5} + \frac{60 \times 0.04t}{4175} = 1$$

$$t = 1590 \text{ s}$$

That is, 1590 s or 26.5 min. If a number of these systems were being tested in a laboratory, this number means that about 50% of the samples would have failed within nearly half an hour (or L_{50} = 26.5 min).

Life Expectancy for Pure Creep Conditions

For pure creep damage, Robinson's index may be used to predict the lifetime of a member of a system that experiences different levels of temperature under a static load.

$$D_c = \sum_i \frac{t_i}{t_{r_i}}$$

where t_i is the hold time spent at T_i. t_{r_i} is the rupture time at the same temperature.

This equation, while useful, does not serve much of the purpose in electronics packaging. On the one hand, the premise of Robinson's equation is that the part is under a static load. On the other hand, the impact of creep becomes significant when local temperatures approach the melting point of metals and some plastics and the glass transition point of other plastics. In electronics, only solder and some encapsulants have melt or glass transition temperatures low enough that creep may play an important role in their behavior.

Life Expectancy for Creep–Fatigue Interactions

In a static sense, the only stresses found in electronics packages are residual stresses in solder or encapsulants. Relaxation of these stresses because of creep does not cause any damage and yet tends to reduce stresses in soldered leads in certain instances involving low frequency vibrations (Steinberg 1991). Thermal cycling, however, creates a more severe environment for encapsulants and particularly for solder. This is because that in addition to creep, fatigue begins to play a role because of induced cyclic stresses caused by thermal coefficient mismatch between various components and materials on the board.

A significant amount of research has been conducted to develop a reliable model for predicting solder life in a variety of environments. A very popular model is the Coffin–Manson or modified Coffin–Manson relationship. These models are constructed based on empirical data and as such are accurate only for the specific materials and/or conditions that they were based (Tasooji et al. 2007). A basic form of the Coffin–Manson equation is as follows.

$$N_f = \frac{A}{(\Delta T)^B} \quad (13.4)$$

In this equation, N_f is the number of temperature cycles to failure, ΔT is the temperature range of thermal cycling, and A and B are dependent constants. Although A depends mainly on material type and behavior, it has been shown that B has a range of one (1) to three (3) for solders (Jeon et al. 2005), with a typical value of two (2). Considering the empirical nature of this relationship, a number of improvements have been proposed in order to improve the applicability and accuracy of the life predictions. These improvements are collectively referred to as *modified Coffin–Manson* equations and are generally accurate within the criteria and conditions for which they were developed (Tasooji et al. 2007).

One such modification has been proposed by Ross and Wen (1993) and has the following ramification:

$$N_1^{\beta_1} = N_2^{\beta_2}$$

$$\beta = 0.442 + 6.0 \times 10^{-4}\, T_m - 0.0174 \ln\left(1 + \frac{360}{\tau}\right)$$

where T_m is the mean solder temperature (°C), and τ is the half-cycle dwell time (min), and the subscripts 1 and 2 refer to two different dwell times when all other conditions are kept the same. In other words, if the number of cycles to failure at a given temperature and dwell time is known, it may be possible to calculate the number of cycles to failure at a different temperature and a different dwell time.

Example 1

Suppose that a system fails after 1500 cycles at a temperature of 80°C and a dwell time of 30 min. What would the number of cycles to failure be if the dwell time is reduced to 15 and 5 min at the same temperature?

$$\beta = 0.422 + 6.0 \times 10^{-4} T_m - 0.0174 \ln\left(1 + \frac{360}{\tau}\right)$$

$$\beta_1 = 0.422 + 6.0 \times 10^{-4} (80) - 0.0174 \ln\left(1 + \frac{360}{30}\right)$$

$$\beta_1 = 0.445$$

$$\beta_2 = 0.422 + 6.0 \times 10^{-4} (80) - 0.0174 \ln\left(1 + \frac{360}{15}\right)$$

$$\beta_2 = 0.434$$

$$\beta_3 = 0.422 + 6.0 \times 10^{-4} (80) - 0.0174 \ln\left(1 + \frac{360}{5}\right)$$

$$\beta_3 = 0.415$$

Now that βs are calculated, we can calculate the number of cycles to failure.

$$N_1^{\beta_1} = N_2^{\beta_2}$$

or

$$N_2 = N_1^{\beta_1/\beta_2}$$

$$N_2 = 1500^{(0.445/0.434)}$$

$$N_2 = 1817\left(\text{life at 15 min dwell time}\right)$$

$$N_3 = 1500^{(0.445/0.415)}$$

$$N_3 = 2545\left(\text{life at 5 min dwell time}\right)$$

These results indicate that the life of the product increases dramatically when it is subjected to high temperatures for lower dwell times.

Example 2

Suppose that a system is tested in a chamber that varies the temperature between –10°C and 80°C. The dwell time in each chamber is 30 min. The unit is operational during the testing and it is known that its critical components have a temperature rise of 45°C above environment. Fifty percent of the units tested fail after 825 cycles.

We need to determine the life of this product if it is being operated at room temperature for nearly 2 h per day, 200 days a year.

$$\beta = 0.422 + 6.0\times10^{-4}T_m - 0.0174\ln\left(1+\frac{360}{\tau}\right)\lim_{x\to\infty}$$

$$\beta_1 = 0.422 + 6.0\times10^{-4}\left(80+45\right) - 0.0174\ln\left(1+\frac{360}{30}\right)$$

$$\beta_1 = 0.472$$

Room temperature is assumed to be 25°C.

$$\beta_2 = 0.422 + 6.0\times10^{-4}\left(25+45\right) - 0.0174\ln\left(1+\frac{360}{120}\right)$$

$$\beta_2 = 0.460$$

Now that βs are calculated, we can calculate the number of cycles to failure.

$$N_1^{\beta_1} = N_2^{\beta_2}$$

or

$$N_2 = N_1^{\beta_1/\beta_2}$$

$$N_2 = 825^{(0.472/0.460)}$$

$$N_2 = 990$$

Thus the units will survive 990 cycles. Since each cycle is one day, the data indicate that 50% of the units will fail after (990/200) = 4.95 years or $L_{50} = 4.95$.

Design Life, Reliability, and Failure Rate

In the previous section, we developed formulas for calculating life expectancy of mechanical components undergoing either fatigue or creep. It was also mentioned that there may be interaction with other factor from the environmental conditions impacting component's life. Furthermore, it was pointed out that the calculated expected life should not be viewed in an absolute sense—that it is a probabilistic value. In the examples that were presented earlier, what we calculated as expected life meant that 50% of the samples had failed within the calculated time period. Had we used fatigue or creep data based on, say, 10% failure, we could have calculated a 10% life expectancy.

Often, life expectancy is equated with design life of a component. Design life is defined as the length of time during which the component is expected to operate within its specifications. However, this definition appears to suggest that failures do not happen at all—although that is not the case at all. For this reason, "operating life expectancy", or "operating life", has been used in place of design life. Furthermore, operating life expectancy has been defined as the length of time that a component operates within its "expected failure" rate.

The question is how component failure rates may be calculated based on their life expectancy values. Chapter 16 will demonstrate how this information at a component level may be used to calculate the overall failure rate of the entire unit. To make this calculation, the concept of reliability should be introduced. Reliability is considered to be a function of time and failure rate. In other words

$$R = F(\lambda, t)$$

where R is defined as reliability of a component or a part, λ is defined as the component's failure rate, and t is time.

Example

A component under cycling fatigue load exhibits a L_{50} of 60 months (in other words, 50% of samples fail in 60 months) with a standard deviation of 20 months. What is the expected monthly failure rate of a population of these components?

A generic definition of failure rate is

$$\text{Failure rate} = \frac{\text{number of failed units}}{\text{time}}$$

$$\lambda = \frac{50\%}{60} = 0.833\%$$

The average monthly failure rate is 0.833% of the total population.

Another approach is to use the exponential form of the reliability equation expressed as

$$R = e^{-\lambda t}$$

One may argue that the exponential form of the reliability equation may not be valid for most mechanical failures. In this regard, there are two points worth mentioning. First, strictly speaking, this equation holds true if mechanical-wear failures are smaller than failures occurring by chance (Bazovsky 2004). Second, the point of this exercise is to illustrate the relationship between reliability and failure rate.

When 50% of samples fail, one may suggest that reliability is at 0.5. Thus, we get the following relationship

$$0.5 = e^{-\lambda(60)}$$

$$\ln 0.5 = -60\lambda$$

$$\lambda = -\frac{\ln 0.5}{60} = 0.0115$$

Failure rate using this equation is 0.0115 or 1.15% per month.

The reader may be puzzled why there are two different answers to this problem. It should be noted that this example provides a back of the envelop method of calculating the average monthly failure rate; as such depending on the approach taken (and its associated assumptions) different numerical results may be obtained. For this particular example, a more precise formula is to use Weibull distribution discussed in Chapter 16.

Chapter 16 will provide more details and other possible ways of studying reliability. For now, the purpose of this section is to bring the reader's attention to the fact that ultimately, any life calculations should be tied to the concept of reliability, and from a practical point of view, to an expected failure rate.

14

Electrical Failures and Reliability

Introduction

Although the focus of this book is on the mechanical aspects of the design of electromechanical products, for the sake of completeness, a review of electrical failure modes is provided here. For a more in-depth study of failure modes and mechanisms in electronics packages refer to Viswanadham and Singh (1988).

In addition to mechanical failures, electronic components and assemblies fail for a variety of other reasons. We, as engineers, program and/or QA managers, should be aware of these causes and failures and take precautions to either eliminate them or mitigate their impact. This will result in developing more robust designs and products, and furthermore, understanding how our product fails will form a basis for creating maintenance and repair scheduling. To do this, let us first review the failure modes and mechanisms associated with electronic components.

Failure Modes and Mechanism

In the previous chapter, it was pointed out that reliability was in a way a study of failure. Thus, it was important to first define failure. It is also important to define failure modes. In the previous chapter, failure was defined as the inability of a system (including its components) to meet its design objectives—or the specified requirements—under specified operating conditions.

As indicated before, many factors would lead to failures such as designer's lack of knowledge of potential factors having an adverse impact, poor design verification activities, or even poor requirements. Failure modes and effects analysis (FMEA) should be employed to identify sources of failure and indicate potential mitigation activities.

Failure Modes

In the most general sense, printed circuit board assemblies (PCBAs) and electronics (sub)systems exhibit three modes of failures. These are open circuits, short circuits, and intermittent.

Open Circuit Failure Mode

When an undesired discontinuity forms in an electrical circuit, it is said that the circuit is open. This is caused often by solder cracks due to a variety of thermomechanical stresses. Earlier, it was pointed out that the printed circuit boards may warp due to temperature differences both in-plane and through its thickness. The excessive warping or thermal cycling contributes to solder failures leading to developing an open circuit.

It should also be pointed out that open circuits may also develop at the chip level due to solid state diffusion leading to degradation and wear out which are often aggravated as temperature or electrical stresses increase.

Another contributor to developing this failure mode is chemical attack (such as corrosion) which will be treated in Chapter 15.

Short Circuit Failure Mode

When an undesired connection forms between two electrical circuits or between a circuit and ground, it is said that the circuit is shorted. Major contributors to short circuits are based on chemical attacks such as dendrites and whiskers. Chemical attacks and their impact on failure and reliability are studied in Chapter 15.

However, another contributor to development of shorts is contaminants left behind in the manufacturing process combined with humidity and high temperatures (Viswanadham and Singh 1988).

Intermittent Failure Mode

An intermittent failure is a microscopic open (or a short) circuit which is also under other environmental influences. For instance, a crack in solder may be exasperated under thermal loads leading to an open circuit failure; however, as the board is removed from its environmental conditions and cools down, the gap closes, and it would work as intended.

In this example temperature was the contributing factor. Vibration is another: if the PCBA are not properly secured, they can develop chatter under vibration and cause short circuits. Chatter due to vibration has also been observed in mechanical relays.

Thus, in designing a part or an assembly, the design engineer must first determine possible failure modes and the associated failure mechanisms then either modify the design or provide a suitable criterion to mitigate it.

Failure Mechanism

Failure mechanism in electronics packages has been well studied over the years. Failure mechanisms may be attributed to the following factors:

1. Design inefficiencies
2. Fabrication and production issues
3. Stresses such as thermal, electrical, and mechanical and environmental factors such as humidity

A treatment of electronic design guideline and best practices is beyond the scope of this work along with an in-depth study of physics of electronics failures due to temperature and/or other stressors. Viswanadham and Singh (1988) have provided an in-depth discussion of failure mechanism with emphasis on fabrication and production issues.

The next segment reviews approaches on estimating life expectancy of electronics equipment under temperature or other stress factors.

Life Expectancy of Electronics Assemblies

From the previous discussion, it was clear that the life of electronic assemblies depends on temperature and electrical, mechanical, and other environment factors such as moisture, dust, even radiation. As many as 20 different life expectancy models have been proposed. For a summary of these equations see Viswanadham and Singh (1988). Herein, more commonly known models are reviewed. These are Arrhenius, Black, and Eyring models.

The Arrhenius Model

Arrhenius's model is based on empirical data and is generally used to describe thermally activated mechanisms (Condra 2001). This model is a rate equation which describes temperature dependency as follows:

$$\text{Rate} = C\,\text{Exp}\left(-\frac{E_a}{K_b T}\right)$$

where:

C is a constant
E_a is the activation energy
K_b is the Boltzmann constant (8.617×10^{-5} eV/K)
T is temperatures in absolute value

This relationship clearly does not provide a life expectancy period. The implication of this equation is that at a given temperature the rate of a given mechanism multiplied by the time required for completion of the mechanism is constant. In other words

$$\text{Rate}_{@T_2} \times \text{Time to failure}_{@T_2} = \text{Rate}_{@T_1} \times \text{Time to failure}_{@T_1} = \text{Constant}$$

where T_1 and T_2 are two different temperature values.

$$t_{@T_2}\,\text{Exp}\left(-\frac{E_a}{K_b T_2}\right) = t_{@T_1}\,\text{Exp}\left(-\frac{E_a}{K_b T_1}\right)$$

$$t_{@T_2} = t_{@T_1}\,\text{Exp}\left(\frac{E_a}{K_b}\left(\frac{1}{T_2} - \frac{1}{T_1}\right)\right) \tag{14.1}$$

or

$$\text{AF} = \frac{t_{@T_2}}{t_{@T_1}} = \text{Exp}\left(\frac{E_a}{K_b}\left(\frac{1}{T_2} - \frac{1}{T_1}\right)\right) \tag{14.2}$$

Therefore, by running an experiment at two different temperatures, failure times t_1 and t_2 may be measured leading to a determination of E_a for a particular device:

$$E_a = \frac{T_2 T_1}{T_1 - T_2}\quad K_b \ln \text{AF} \tag{14.3}$$

Then, the life expectancy of that device at a particular temperature may be calculated using Equation 14.1 or 14.2.

Example

A new electronics product has been developed. In order to determine its expected life, two different experiments were conducted. In the first, a sample of units was placed in a 100°C oven and operated. The average failure time of this population was 148 h. A second set was tested at 75°C; and their average failure time was 876 h. What should the expected life be, if these units are operated at 40°C?

From Equation 14.2, we have

$$\text{AF} = \frac{t_{@T_2}}{t_{@T_1}} = \frac{876}{148} = 5.919$$

And from Equation 14.3, we can obtain the activation energy.

$$E_a = \frac{T_2 T_1}{T_1 - T_2} K_b \ln \mathrm{AF} = \frac{(75+273)(100+273)}{100-75}\left(8.617\times10^{-5}\right)\ln(5.919)$$

$$E_a = 0.796\,\mathrm{eV}$$

From Equation 14.1,

$$t_{@40} = t_{@100}\,\mathrm{Exp}\left(\frac{E_a}{K_b}\left(\frac{1}{T_2} - \frac{1}{T_1}\right)\right)$$

$$= 148\,\mathrm{Exp}\left(\frac{0.796}{8.617\times10^{-5}}\left(\frac{1}{(40+273)} - \frac{1}{(100+273)}\right)\right)$$

$$t_{@40} = 17017\,\mathrm{h}$$

Hence, this device will have an average life expectancy of nearly 17,000 h (or nearly two years) at 100% duty cycle.

The Eyring Model

While Arrhenius equation includes only temperature as the stressor, a similar equation may be developed which includes a second stressor in addition to temperature. The general form of this equation is called the Eyring model (Viswanadham and Singh 1988, Condra 2001).

$$t_f = \frac{A}{B} S^{-n}\,\mathrm{Exp}\left(\frac{E_a}{K_b T}\right)$$

where:

- t_f is the expected life
- A and B are the constants
- S is the applied stress—(such as percentage relative humidity or electrical current)
- n is a parameter related to the stress
- E_a is the activation energy
- K_b is the Boltzmann constant
- T is temperatures in absolute value

Black's Equation for Electrical Stress Effects

Other factors such as electrical stress (current, voltage, and power) may have adverse effects on the device life as well. A component is generally rated for a specific voltage, current, power, and so on. Reliability studies have shown that once this rating is exceeded, the failure rates increase.

Black applied Eyring's model to relate electrical stress to the component's life as follows (Viswanadham and Singh 1988, Condra 2001):

$$t_f = \frac{a_m b_m}{A} J^{-2} \mathrm{Exp}\left(\frac{E_a}{K_b T}\right)$$

where:

t_f is the expected life
a_m and b_m are the thickness and width of metallization (conductor), respectively
A is a geometry and material-related constant
J is the current density
E_a is the activation energy
K_b is the Boltzmann constant
T is temperatures in absolute value

Design Life, Reliability, and Failure Rate

In Chapter 13, the concept of design life, reliability, and failure rate was discussed and a generic equation for reliability was provided. In that chapter, data based on physics of failure (such as fatigue and creep) were used to estimate life expectancy. These data ultimately included information about reliability, that is, if we used a fatigue curve based on a 50% failed samples, it was understood that the calculated life expectancy time t was associated with a 50% accumulated failure leading to a 50% reliability.

In general, reliability equations describing mechanical parts are complicated. In contrast, reliability equations for electronics equipment may have the following simple form:

$$R = e^{-\lambda t}$$

where R is defined as reliability of a component or a part, λ is defined as the component's failure rate, and t is time. Hence, this equation provides a simple relationship between life, failure rate, and reliability. This represents one equation and three variables. If λ is to be calculated both R and t should be known. One may be tempted to use one of the aforementioned models to estimate time; however, it soon becomes evident that unlike their mechanical counterparts, none of these models provide an associated reliability value. As a result, the relationship between life expectancy and reliability may be more complicated and require more experimental data.

Example

Consider the previous example where the life expectancy of the new electronics product was calculated. Now, this added information is available: the provided data were representative of failure of 10% of each sample set. In other words, 90% of the samples were still operational beyond the reported times. In this case, what is the reliability of the product and what should be the expected failure rate?

Since only 10% of the sample sets had failed, it stands to reason that the reliability of the product is at 90% for the given times; hence, one could calculate the failure rate as follows.

$$0.9 = e^{-\lambda(17017)}$$

$$\lambda = 6.19 \times 10^{-6}/h$$

Reliability equations will be visited once again in Chapter 16 where the concept of reliability and failure rate will be given a more detailed attention.

15

Chemical Attack Failures and Reliability Concerns

Introduction

In Chapters 13 and 14, we considered two important influences on reliability, namely, mechanical and electrical factors. A third and, to some extent, less considered influence is chemical and electrochemical factors, which may be divided into two broad categories: electrochemical attacks and migration. Migration, in turn, may be divided into three areas: whiskers, dendritic growth, and diffusion. Although metallic migration has been known for decades, it has received more attention recently as the industry is moving toward lead-free soldering.

Electrochemical Attacks

The most common chemical attack in electronics packaging is of an electrochemical nature and is generally known as corrosion. It occurs in the interconnects where the design engineer has not been careful in specifying the right material for the contacts, or in equipment that is operated near saltwater bodies, or in the support structure where the enclosure is bolted. It is also observed in equipments where moisture condensates and forms small pools of water.

Although emphasis has traditionally been placed on corrosion of components such as the CPU, as shown in Figure 15.1, it should be pointed out that the support structure of the electronics is also susceptible to corrosion, which could lead to catastrophic failures. An example is shown in Figure 15.2 where the fan bracket is corroded and could eventually fall apart. In the same system, there is evidence of corrosion under the screw as well as at the spot weld location (Figure 15.3).

If we consider corrosion as a type of an environmental attack on metals, plastics, too, are attacked, but the mechanism of failure is different. Some

FIGURE 15.1
Corrosion of a CPU.

FIGURE 15.2
Corrosion of a fan bracket.

plastics are sensitive to oxygen, and others are damaged by UV light, and still another class of plastics outgas corrosive vapors that may be harmful to neighboring elements.

Corrosion

Corrosion is a natural two-step process whereby a metal loses one or more electrons in an "oxidation" step resulting in free electrons and metallic ions. The free electrons are conducted away to another site where they combine

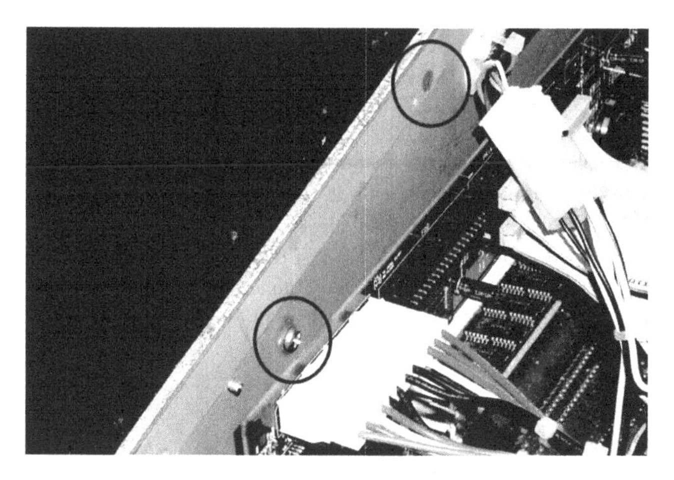

FIGURE 15.3
Corrosion under the screw and at spot weld location.

with another material in contact with the original metal in a "reduction" step. This second material may be either a nonmetallic element or another metallic ion. The oxidation site where metallic atoms lose electrons is called "anode" and the reduction site where electrons are transferred is called "cathode." These sites can be located close to each other on the metal's surface, or far apart. There is an oxidation electric potential (or energy level) associated with each step, and this potential level, electromotive force (EMF), depends greatly on the material. Referring to Table 15.1, the EMF for gold is +0.15 and for beryllium is −1.70. If the electrical potential level at the reduction level is much higher than that of the oxidation, corrosion takes place rapidly and the anode may be consumed very quickly (Davis 2003).

The electron flow from anode to cathode may even take place in elements present outside of the base metal of interest. For example, in a buried pipe scenario, the soil may become the recipient of electrons relative to the pipe metal and if a medium (such as water) is present to provide and transport ions, the pipe can corrode.

For corrosion to take place, four factors must be present at the same time, as shown in Figure 15.4. These are:

1. There must be an anode.
2. There must be a cathode.
3. There must be an electrolyte.
4. There must be an electrical connection.

Anode/cathode pairs, known as "corrosion cells," come in a variety of forms. The ones most relevant to electronics packaging are galvanic cells, stress cells, and fretting corrosion.

TABLE 15.1

Galvanic Potential of Various Metals

Group	Metal	EMF(V)
1	Gold: solid and plated; gold–platinum alloys	+0.15
2	Rhodium, graphite	+0.05
3	Silver: solid or plated; high silver alloys	0.00
4	Nickel: solid or plated; high nickel–copper alloys, titanium alloys	−0.15
5	Copper: solid or plated; low brasses or bronzes; silver solder; German silver, high copper–nickel alloys; nickel–chromium alloys, 300 Series Stainless Steels	−0.20
6	Commercial yellow brasses and bronzes	−0.25
7	High brasses and bronzes	−0.30
8	18% chromium-type corrosion-resistant steels	−0.35
9	Chromium plated; tin plated; 12% chromium-type corrosion-resistant steels	−0.45
10	Tin plate; tin–lead solder	−0.50
11	Lead: solid or plated; high lead alloys	−0.55
12	Aluminum, wrought alloys of the 2000 Series	−0.60
13	Iron: wrought, gray, or malleable; plain carbon and low alloy steels	−0.70
14	Aluminum, wrought alloys other than 2000 Series aluminum, cast alloys of the silicon type	−0.75
15	Aluminum, cast alloys other than silicon type, cadmium: plated and chromate	−0.80
16	Hot-dip zinc plate; galvanized steel	−1.05
17	Zinc, wrought; zinc-base die-casting alloys; zinc plated	−1.10
18	Magnesium and magnesium-base alloys, cast or wrought	−1.60
19	Beryllium	−1.70

Source: MIL-HDBK-1250A (1995).

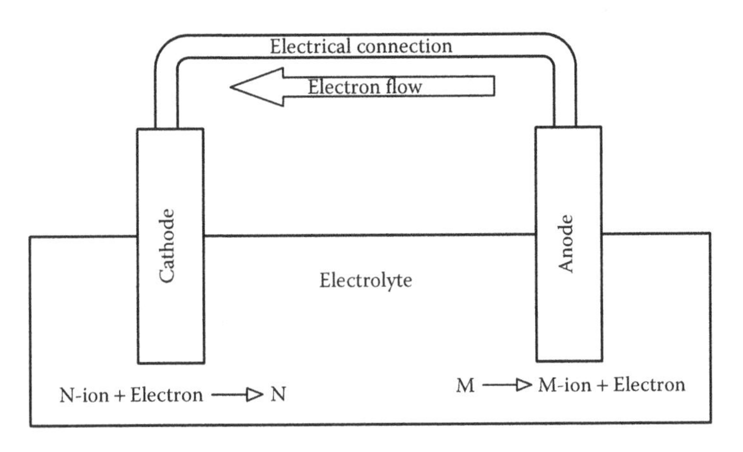

FIGURE 15.4
A typical corrosion cell: M represents the anodic metal, whereas N represents the material in or near the cathode such as another metal or available hydrogen in the electrolyte.

Galvanic Cells

Galvanic cells are formed when one of the following two criteria is satisfied: two dissimilar metals exist in close proximity or the metal is a multiphase alloy in the presence of an electrolyte.

Dissimilar Metals

Earlier, it was pointed out that there is an electric potential associated with both oxidation and reduction steps. If the oxidation potential of one metal is higher than the adjacent metal, then the metal with the higher (negative) potential acts as an anode and the other becomes the cathode. Examples of dissimilar metals are iron and zinc, or nickel and gold. Generally, metals with a lower potential are placed near the top of an electromechanical chart, and the ones with higher potential are placed near the bottom of the chart (MIL-STD-889 1976; MIL-HDBK-1250A 1995). It should be pointed out that the potential value and hence the metal's placement is influenced greatly by the nature of the electrolyte used. If the electrolyte is seawater, then gold and platinum are at the top of the chart. The term noble is used to refer to the elements on the top of the electrochemical chart (Table 15.1). Noble materials are cathodes and survive the corrosion process, and less noble metals will be the anodes and corrode.

General Corrosion of Multiphase Alloy

When metallic alloys are made, metallic and nonmetallic elements are dissolved in a base metal. There are two ways that solutes combine with the base metal or solvent.

1. *Substitutional solid solution*—The alloy's atoms are similar to the solvent and they replace the parent metal's atom in the lattice.
2. *Interstitial solid solution*—The alloy's atoms are smaller and they fit between the atoms in the parent's metal lattice.

Often, neither of the two solutions can completely dissolve all the added atoms, resulting in a mixed atomic grouping. In other words, different crystalline structures occur within the same alloy; thus, multiple phases are present. Examples of these multiple phase alloys are stainless steel, cast iron, and aluminum alloys. The individual phases of alloys have different electrode potentials, which can result in one phase acting as an anode, another as the cathode, and the entire alloy may be subject to corrosion.

Stress Cells

Another cell formation that could potentially induce corrosion is called a stress cell. The basis of this cell is founded on the principle that stored strain energy may be a source of electropotential difference. Thus, if the metallic body is under localized stresses, electrons tend to flow out of the stressed region, thus creating an anodic condition. Three configurations may be cited as examples of this type of corrosion cell.

Grain Boundaries

On a micro level, if atoms along grain boundaries are displaced so the crystal array loses its regularity, they experience deformations from their natural state, causing stored strain energy. This energy translates into an electrode potential, leading to anodic conditions along grain boundaries, which leads to selective corrosion along these boundaries.

An example of this type of corrosion may be found in welding of stainless steels. Welding heat causes carbon to precipitate out at the boundary in the form of chromium carbide, thus creating localized strains leading to the development of corrosion cells. The spot weld shown in Figure 15.3 is an example of this type of corrosion.

Cold Working

Similar to grain boundaries, areas subjected to cold working in a metal contain a higher concentration of dislocations, and as a result will be anodic to non-cold-worked regions. Thus, cold-worked sections of a metal will corrode faster. For example, nails that are bent will often corrode at the bend, or at their head where they were worked by the hammer.

Stress Concentration Regions

On a macro level, areas of stress concentration will also contain metal atoms at higher strain energy states. As a result, stress concentration areas will be anodic to low-stress regions and can corrode selectively. An example of this is as follows: bolts under load are subject to more corrosion than similar bolts that are unloaded. The corrosion taking place under the sheet metal screw as shown in Figure 15.3 is an example of this type of corrosion.

Fretting Corrosion

Technically, fretting corrosion is not a different type of corrosion cell, but as it relates mainly to corrosion in between two joining surfaces, it is given its own category. It is the main cause of failure in otherwise well-designed electrical connector contacts. Fretting corrosion is caused by the relative micromovement of contacts, which may be caused by vibration, change of temperature, and so on. (Antler 1999). Fretting corrosion causes a nonuniform formation of insulating oxides across the contact surface, leading to an increase in contact resistance. In addition, movement of surfaces causes the stable oxidized layer to be "wiped" away, which is replaced by a fresh layer.

Antler (1999) reports that tin and palladium and many of their alloys are particularly sensitive to fretting corrosion; however, contact lubricants tend to form a barrier and prevent corrosion. Antler further makes note of other commonly used metals. For example, he notes that tin-to-gold mating contacts corrode faster than tin-to-tin contacts, whereas if tin is replaced with palladium (i.e., the pair becomes palladium–gold), corrosion becomes rather stable.

Chemical Attacks on Plastics

Chemical attacks do take place on plastics, but modes of failure are generally different than that of metals; however, with the advent of engineered plastics and various fillers such as carbon or steel fibers, galvanic cells may form between a fiber-filled plastic and an adjacent, adjoining metal. The corrosion process may even be exasperated because many resins absorb and retain moisture leading to a process that is referred to as dry corrosion (Goodman 1998).

Aside from the possibility of galvanic cell formation, specifying plastics as a packaging material must take into account the impact of the service environment. If a plastic material is in contact with a corrodent (or solvent), it will either resist it or fail very rapidly. However, there are other environmental factors such as ozone or UV light that may have a detrimental impact on the reliability of a plastic part. The causes and/or mechanisms of this failure are beyond the scope of this book, and references such as Ezrin (1996) and Lustinger (1989) are recommended for further study on plastics failures. For the sake of completeness, the following chemical failure modes in plastics may be noted:

1. Oxygen (and ozone) in the environment attacks the chemical bonds of some plastics, causing damage from discoloration to embrittlement.
2. Moisture in the environment is absorbed by some plastics which may attack the cross-link bonds or in other plastics may cause a change in physical dimensions leading to mechanical failures.
3. Pollutants in the environment such as noxious gases are absorbed and cause either oxidation similar to oxygen and ozone or hydrolysis similar to moisture.
4. Thermal degradation causes the process of de- and re-polymerization.

In short, a plastic material must be properly selected for its environment.

Corrosion Control through Proper Design Techniques

Recall that three conditions must exist simultaneously for corrosion to occur. These are as follows:

1. Two dissimilar (electrochemically) metals must be present.
2. An electrical path must exist between the two metals.
3. An electrolyte must be present.

Corrosion will not occur if any one of these three conditions does not exist. Thus, the most basic corrosion prevention method is to thwart the combination of all three factors. However, if the design circumstances are such that dissimilar metals must come in contact (in the case of fasteners, for example), finishing, plating, and/or special alloys may be used to reduce the chance of corrosion.

MIL-HDBK-1250A (1995) recommends that should dissimilar metals be used, they should be within what is called a "permissible galvanic couple." These are dissimilar metals that have a minimal difference in their EMF voltage and have been shown to have a very slow rate of corrosion. A review of the EMF voltage between members of the permissible galvanic couples reveals that the difference between the EMF of the two metals is generally less than 0.15 V. However, it is generally accepted that a permissible galvanic couple is the most ideal combination of dissimilar metals and is needed if the equipment is to function in harsh environments. If the equipment is to function in benign conditions, then metals with EMF differences as much as 0.25 V may be used (Galvanic Compatibility 2007).

In addition to the aforementioned guidelines, there are a few more precautions that need to be followed to develop an acceptable design from a corrosion point of view. A few of these conditions are enumerated here:

1. *Avoid*
   a. Trapping moisture in hollow sections, crevices, or joints. Enclosures should be designed with proper drain holes, that is, having both proper size and location.
   b. Metallic construction, if plastic housings are acceptable.
   c. Graphite composites in conjunction with metal structures.
   d. Hinge mechanisms susceptible to corrosion.
   e. Materials that are not moisture and/or fungus resistant.
   f. Materials that outgas corrosive vapors.
   g. Sealant materials such as rubber that may be damaged by elemental conditions such as ozone or UV light.
   h. Materials that are prone to excessive moisture absorption.
2. *Mount*:
   a. Printed circuit board (PCB) assemblies vertically with connectors on the vertical side.
   b. Electrical connectors horizontally.
3. *Use conformal coatings with* PCB assemblies.
4. *If possible, maintain temperatures above the dew point to prevent condensation.*

Migration and Electromigration

Migration is the process by which material moves from one area to another area. If migration happens between two adjacent metals such as copper and solder, it is called diffusion. If it happens as a result of internal stresses and in the "absence" of an electric field, it is called whiskers. Finally, if it happens in the presence of an electric field and between "similar" metals, it is called dendritic growth.

Migration and electromigration have traditionally been very important failure mechanisms in microelectronics packaging (Viswanadham and Singh 1988); however, two factors have given this topic relevance in PCB and system package design. The first factor is that as the PCBs are more densely populated, it becomes exceedingly difficult to clean and wash away all the processing chemicals. The presence of the pollutants along with an electric field provides an ideal environment for electromigration and dendritic growth (Krumbein 1987; Bumiller et al. 2004; Kim et al. 2004; Lieberman and Brodsky 2004; Lee et al. 2006; Cyganowski et al. 2007). The second factor is more related to legislature and lead-free environment. Pure tin solder which is a natural replacement for a leaded solder, has a tendency to grow conductive needles (known as whiskers) in an out-plane direction (*z*-axis) (Asrar et al. 2007; Kuhlkamp 2007; Xu et al. 2007; Zhang et al. 2007).

In addition to the formation of whiskers, diffusion may potentially pose yet another problem. Although the intermetallic layer formed between copper and solders is needed to form strong bonds, the thickness of this bond may have a detrimental effect on the ductility of the joint—the thicker the layer, the more brittle the joint. Thus, the choice of solder and the pad material are quite important (Haimovich 1993; White 1993; Zribi et al. 1999; Roubaud et al. 2001; Kuhlkamp 2007; Siewert et al. 2007; Xu et al. 2007; Zhang et al. 2007).

Dendritic growth, as shown in Figure 15.5, consists of leaf-like structures that grow in the presence of an electrolyte and an electric field. They are aggravated as temperatures and voltage levels increase. Eventually, dendrites can cause electrical shorts and failures.

Whiskers as shown in Figure 15.6 are straight needle-like structures with diameters typically up to 3 μm and can grow to over 1 mm in length (Asrar et al. 2007). Their growth has been primarily attributed to the compressive stresses in the tin plating as evidence in a bright tin finish, and as a result bright finishes have been prohibited in microelectronic packages (Xu et al. 2007; Zhang et al. 2007). Whiskers break due to vibration and are a source of debris. They have been known to cause complete failures of several on-orbit commercial satellites (Asrar et al. 2007).

Finally, Figure 15.7 shows a typical copper–solder interface before aging. As may be expected, at the interface there is a formation of a thin intermetallic layer. This layer grows with aging at elevated temperatures, and a variety of compounds may be formed (Asrar et al. 2007). It has been shown that

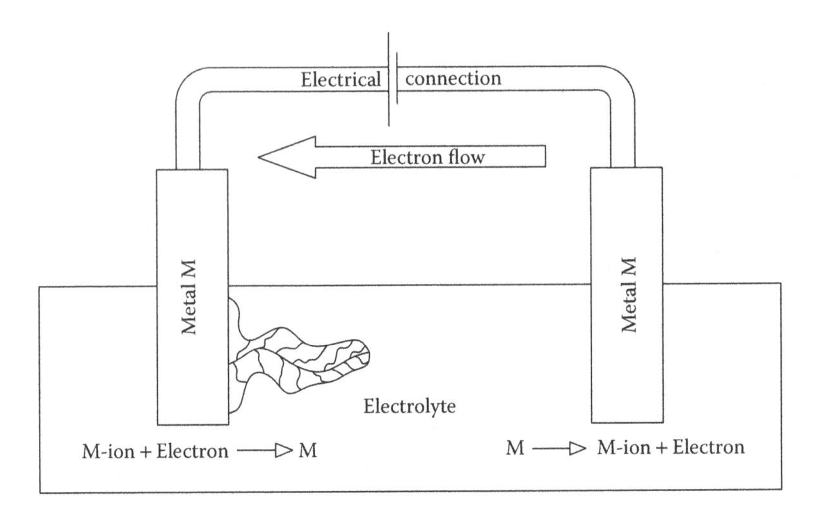

FIGURE 15.5
Dendritic growth.

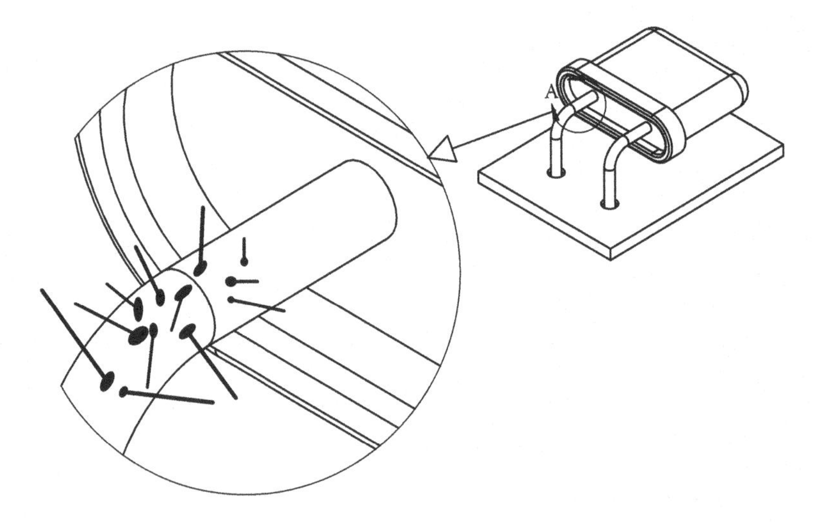

FIGURE 15.6
Whiskers on a crystal oscillator. (Adapted from NASA Goddard Space Flight Center Stockpile, http://www.nepp.nasa.gov/whisker/)

depending on the materials used, and particularly when precious metals are involved, the solder joint embrittlement becomes a serious reliability issue (Zribi et al. 1999; Zhang et al. 2007).

The growth of whiskers is related to the materials and processes used to plate and solder the components on the PCB as well as the environment to which the equipment is subjected. It does not depend on the components used and a variety of components such as diodes, transistors, integrated

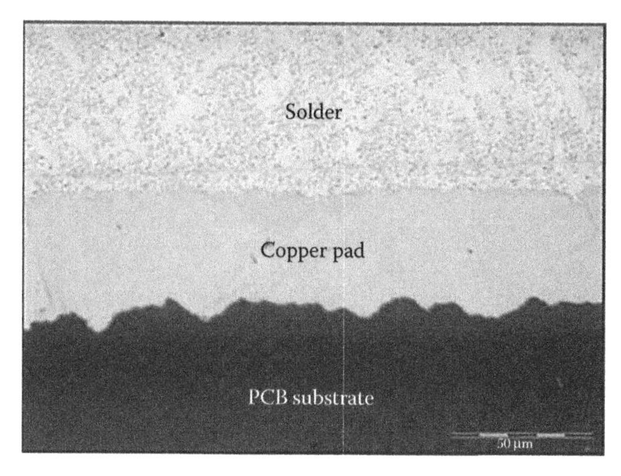

FIGURE 15.7
Copper–solder interface. (Adapted from Siewert et al. *Formation and Growth of Intermetallics at the Interface between Lead-Free Solders and Copper Substrates*, Retrieved on November 23, 2007, from http://www.boulder.nist.gov/div853/Publication%20files/NIST_Apex94_Siewert.pdf.)

circuits, microcircuit leads, and even PCBs have been affected. Many metals have been known to whisker and as such, designers and engineers should be mindful to avoid components plated with materials as well as processes that would facilitate whisker formation.

Although a full discussion of migration is beyond the scope of this book, a review of literature has revealed that there are many potential failure mechanisms that call for more investigation and research. Two good sources of information on this subject are found in NASA's information page (http://nepp.nasa.gov/WHISKER/) and SMTnet – Electronics Industry Web Portal (http://www.smtnet.com).

16

Reliability Models, Predictions, and Calculations

Introduction

Although this book is about the thermal and mechanical aspects of electronics packaging, it is in a way about developing robust and reliable products. In this light, the product development teams need to ask questions such as, "How reliable is my product?" and, "How frequently does the product need to be repaired/maintained?"

Basic reliability calculations are simple but they can be quite sophisticated if statistical theories are employed. Herein, the most basic approaches will be presented in order to enable the product development teams to evaluate the design from a reliability point of view as well.

A number of good resources have been published for a more detailed study of this topic. As an example, *Practical Reliability Engineering*, 5th edition (O'Connor and Kleyner 2012), *AT&T Reliability Manual* (Klinger et al. 1990), *Practical Reliability of Electronic Equipment and Products* (Hnatek 2003), *Reliability Improvement with Design of Experiments* (Condra 2001), and MIL-HDBK-338B (1988) are recommended.

Basic Definitions

Before we begin, we need to become familiarized with some terms. These are as follows:

Probability density function (pdf) is the probability that the equipment is functional for a specific time interval. It is generally denoted as $f(t)$.

Cumulative distributions function (cdf) is the probability of system's first failure at or before time t. It is generally denoted as $F(t)$.

$$0 < F(t) < 1$$

Also

$$F(t) = \int_{-\infty}^{t} f(\tau)\mathrm{d}\tau$$

Survivor function (or Reliability) is the probability of surviving to time t (without failure).

$$S(t) = R(t) = 1 - F(t)$$

Hazard rate [$\lambda(t)$] is the instantaneous rate of failure of a population that have survived to time t. A typical hazard rate curve is shown in Figure 16.1.

There are three regions in this curve:

1. *Infant mortality*—The initially high but rapidly decreasing hazard rate corresponding to inherently defective parts. This portion is also called early mortality or early life.
2. *Steady state*—The constant or slowly changing hazard rate.
3. *Wear out*—This generally occurs in mechanical electronic parts (such as relays) or when component degradation and wear exists. In this stage reliability degrades rapidly with ever-increasing hazard rates.

Hazard rate is commonly expressed in units of FITs. One FIT is equal 10^{-9} per hour. In other words, there is a 10^{-9} probability of failure in the next hour. Military standards are different in that they use "percent failing per 1000 h of operation."

Mean time to failure (*MTTF*) is a term commonly used in reliability calculations and is self-evident. Mathematically, it is the area under the survivor (reliability) function as time goes to infinity.

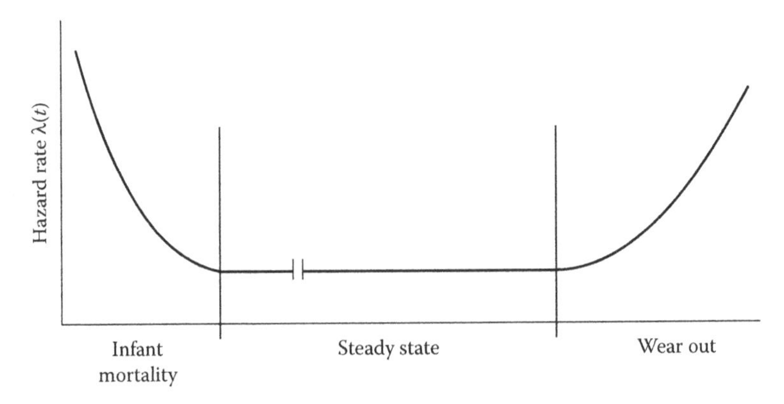

FIGURE 16.1
Hazard rate versus operating time.

$$\text{MTTF} = \int_0^\infty R(t)dt \text{ or MTTF} = \int_0^\infty S(t)dt$$

It may be shown that for a constant hazard rate (λ), we have

$$\text{MTTF} = \frac{1}{\lambda}$$

Mean time between failures (MTBFs) is the average time that a system functions between two consecutive failures. It should be noted that once a system fails, it has to be repaired. Thus, MTBF is the sum of MTTF and the mean time to repair (MTTR).

$$\text{MTBF} = \text{MTTR} + \text{MTTR}$$

Note that these four variables, that is, $F(t), S(t), R(t)$, and $\lambda(t)$ are related and if one is known, the other three may be calculated. Furthermore, time (t) is generally expressed in terms of hours.

Reliability Models

Although there are a number of reliability models, three common models are discussed here; namely, the "exponential distribution" model, the "lognormal distribution" model, and the "Weibull distribution" model, which is widely used particularly for infant mortality and/or wear out calculations.

Exponential Distribution Model

In this distribution model, it is assumed that the hazard rate (λ) is constant and the pdf varies exponentially with time (Hnatek 2003).

$$f(t) = \lambda e^{-\lambda t}$$

Thus the reliability (or survivor function) becomes

$$R(t) = e^{-\lambda t}$$

MTTF is defined as

$$MTTF = \frac{1}{\lambda}$$

Probability of failures is

$$F(t) = 1 - R(t)$$

Example 1

A device has a 250 FIT hazard rate. Determine the MTTF, reliability, and the probability of failures for the first year.

$$\lambda = 250 \times 10^{-9} = 2.5 \times 10^{-7}$$

$$\text{MTTF} = \frac{1}{\lambda} = \frac{1}{2.5 \times 10^{-7}}$$

$$\text{MTTF} = 4{,}000{,}000\,\text{h}$$

$$R(t) = \mathrm{e}^{-\lambda t}$$

$$R(8760) = \mathrm{e}^{-8760(2.5 \times 10^{-7})}$$

$$R = 0.998$$

$$F = 1 - R = 1 - 0.998$$

$$F = 0.002$$

Therefore, for this device the reliability (or survival rate) is 99.8% in the first year and only 0.2% failure.

Example 2

A device has a MTTF of 250,000 h. What is the first year probability of failure?

$$\lambda = \frac{1}{\text{MTTF}} = \frac{1}{250000}$$

$$\lambda = 4 \times 10^{-6} \text{ or } 4000 \text{ FITs}$$

$$R(8760) = \mathrm{e}^{-8760(4 \times 10^{-6})}$$

$$R = 0.965$$

$$F = 1 - R = 1 - 0.965$$

$$F = 0.0344 \text{ or } 3.44\%$$

In other words, a MTTF of 250,000 h represents a 3.44% first year failure!

Lognormal Distribution

Hnatek (2003) provides this definition: "If the natural logarithm of a function is found to be distributed normally, then the function is said to be lognormal." Although Bethea et al. (1991) suggest that there may not be a universally accepted form of the lognormal distribution, a generally accepted form is as follows (Klinger et al. 1990, Bethea et al. 1991, Condra 2001).

$$f(t) = \frac{1}{\sigma t\sqrt{2\pi}} \exp\left[-\left(\frac{1}{2}\right)\left(\frac{\ln t - \mu}{\sigma}\right)^2\right]$$

where σ, called the shape parameter, is the standard deviation of the natural logarithm of the failure time. μ, known as the location parameter, is the mean of the natural logarithm of the failure time.

The cdf becomes

$$F(t) = \frac{1}{\sigma\sqrt{2\pi}} \int_0^t \frac{1}{x} \exp\left[-\left(\frac{1}{2}\right)\left(\frac{\ln x - \mu}{\sigma}\right)^2\right] dx$$

Probability of reliability is

$$R(t) = 1 - F(t)$$

And the hazard rate is

$$\lambda(t) = \frac{f(t)}{R(t)}$$

It should be noted that unlike the exponential distribution, there is no longer a simple inverse relationship between the MTTF and hazard rate λ. Furthermore, conducting "back-of-the-envelop" calculations using the lognormal distribution becomes exceedingly difficult.

Weibull Distribution

A general form of Weibull's pdf is as follows.

$$f(t) = \frac{\beta}{\eta}\left(\frac{t-\gamma}{\eta}\right)^{\beta-1} \exp\left(-\left(\frac{t-\gamma}{\eta}\right)^{\beta}\right)$$

where β is called the shape parameter, η is called the scale parameter or characteristics life—the life at which 63.2% of the population has failed, and γ is called the location or time delay, also known as failure free time (O'Connor and Kleyner 2012).

Based on this formula, the cdf is

$$F(t) = 1 - \exp\left(-\left(\frac{t-\gamma}{\eta}\right)^{\beta}\right)$$

And the reliability becomes

$$R(t) = 1 - F(t) = \exp\left(-\left(\frac{t-\gamma}{\eta}\right)^{\beta}\right)$$

The hazard rate λ is

$$\lambda(t) = \frac{f(t)}{R(t)}$$

$$\lambda(t) = \frac{\beta}{\eta}\left(\frac{t-\gamma}{\eta}\right)^{\beta-1}$$

There are at least two noteworthy points here. First, for $\gamma = 0$, $\eta = 1/\lambda$, and $\beta = 1$, Weibull's distribution reduces to the exponential distribution. The second point is about the versatility of this formula. By adjusting the three parameters of γ, η, and β almost any distribution may be represented:

$$\beta < 1 \text{ gamma}$$

$$\beta = 1 \text{ exponential}$$

$$\beta = 2 \text{ lognormal}$$

$$\beta = 3.5 \text{ approximates normal distribution}$$

For this reason, Weibull distribution is widely used to analyze field or test data. Due to its complexity, this formulation is not suitable for the so-called back-of-the-envelop calculations.

AT&T's Weibull Distribution

Klinger et al. (1990) have created a special version of Weibull distribution to describe the early life behavior (up to 10,000 h). In this formulation, the following assumptions are made:

$$\gamma = 0$$

$$\beta = 1 - \alpha$$

$$\eta^{\beta} = \frac{1-\alpha}{\lambda_{\circ}}$$

Where $0 < \alpha < 1$ is the new shape parameter, and $\lambda_{\circ}$ is an initial (or scale factor) hazard rate. With these assumptions, Weibull's pdf takes the following form.

$$f(t) = \lambda_{\circ} t^{-\alpha} \exp\left(-\frac{\lambda_{\circ} t^{1-\alpha}}{1-\alpha}\right)$$

Based on this model, the reliability, cdf, and the hazard rate become

$$R(t) = \exp\left(-\frac{\lambda_{\circ} t^{1-\alpha}}{1-\alpha}\right)$$

$$F(t) = 1 - R(t) = 1 - \exp\left(-\frac{\lambda_{\circ} t^{1-\alpha}}{1-\alpha}\right)$$

$$\lambda(t) = \lambda_{\circ} t^{-\alpha} \quad \lambda_{\circ} > 0, 0 < \alpha < 1, t > 0$$

It should be noted that $\lambda_{\circ}$ is a scale parameter and is the failure (hazard) rate at 1 h device usage. In addition, the long-term hazard rate is assumed to be achieved by 10,000 h of operation. This model reduces to the exponential distribution for $\alpha = 0$.

Example

A device has a 250 FIT long-term hazard rate, using the AT&T Weibull model determine the reliability of the device in the first 6 months. Assume that $\alpha = 0.75$.

$$\lambda = 250 \times 10^{-9} = 2.5 \times 10^{-7}$$

This is the long-term hazard rate. To recover the initial rate, use the following equation:

$$\lambda(t) = \lambda_{\circ} t^{-\alpha}$$

$$2.5 \times 10^{-7} = \lambda_{\circ} 10000^{-0.75}$$

$$\lambda_{\circ} = 0.25 \times 10^{-4}$$

$$R(t) = \exp\left(-\frac{\lambda_{\circ} t^{1-\alpha}}{1-\alpha}\right)$$

$$R(4380) = \exp\left(-\frac{(0.25 \times 10^{-4}) \times 4380^{(1-0.75)}}{1-0.75}\right)$$

$$R = 0.9992$$

$$F(t) = 1 - R(t)$$

$$F = 0.0008$$

Therefore, for this device the failure rate in the first 6 months is only 0.08%.

Selecting the Right Distribution

There is no simple answer to the question of what is the right distribution to choose. The point to consider is this: the distribution models of which three most common were just introduced should represent the failure behaviors of the product of interest. For instance, if the product of interest is a purely mechanical systems that experiences a steady-state level of vibration, there is a good chance that a lognormal model represents its reliability better than any other distribution. For an electronic equipment with no mechanical components, the exponential model may describe its reliability best.

The dilemma is that most real-life products are a mixture of both electrical and mechanical components that exhibit failures caused by chance and wear. Bazovsky (2004) has demonstrated that the exponential distribution model is suitable for electromechanical systems exhibiting a degree of component wear. The shortcoming of the exponential model is that it cannot predict whether the infant mortality period for a new product has come to an end or whether the wear-out portion of the so-called bathtub curve has begun. The Weibull distribution enables the data analyst to determine whether the product reliability belongs to the infant mortality, steady-state, or wear-out phases but it is more complicated.

From a practical point of view, we need to ask how accurate do we need to be. If either a back-of-the-envelope estimate of reliability or even a more in-depth probabilistic reliability prediction is needed, then, the only choice is exponential distribution. However, as we move into using deterministic reliability predictions involving life data (i.e., both test and field data), then either lognormal or Weibull distributions may be more appropriate. The choices of model should be backed up by goodness-of-fit tests to ensure that the correct model selection has been made.

The Exponential Distribution

The exponential distribution is probably the most widely used distribution. It is the basis of what is called the "parts-count" calculations in MIL-HDBK-217, and Bellcore TR-332 (now Telcordia). There are a number of advantages to the use of this distribution:

1. Ease of use—a single reliability metric; namely, λ.
2. It has a wide applicability and is mathematically traceable.
3. It is additive in the sense that the sum of a number of independent exponentially distributed variables is exponentially distributed.

To use this distribution, the following assumptions should hold:

1. The failure rate is constant—not age dependent.
2. The "infant mortality" region has been eliminated through a post-production environmental stress screening test.
3. There is no or very little redundancy in the device.

This distribution is often used to develop an understanding of the reliability of a product design at its early stages when test data is not readily available.

The Lognormal Distribution

The lognormal distribution is suitable in the reliability analysis of semiconductors (Klinger et al. 1990) and life degradation due to progressive damage such as fatigue or wear of mechanical components (Condra 2001). This distribution is also commonly used in maintainability analysis (a topic beyond the scope of this chapter).

This distribution is better suited for reliability predictions when either field or test data is available.

The Weibull Distribution

The Weibull distribution is widely used in reliability work since it is a general distribution which, by adjustment of the distribution parameters, can be made to model a wide range of life distribution characteristics of different classes of engineered items. It is a powerful tool particularly to identify either infant mortality or wear-out sections of the bathtub curve.

Similar to lognormal distribution, this distribution is better suited for reliability predictions when either field or test data is available.

Device Failure Rate Prediction

To calculate a device (or system) reliability or failure rate, we need to have a knowledge of the failure rates associated with the components and then a knowledge of the system itself. Often, databases (such MIL-HDBK-217F) provide a generic reliability or failure rate for electronics components. Mechanical devices are generally associated with a life expectancy such as L_{10} or L_{50} (alternatively, the B_x nomenclature is used in place of L_x). In Chapters 13 and 14, we learned how to calculate a generic failure rate based on a component's life expectancy.

This generic rates should be adopted to the specific conditions that may exist in the particular device of interest. To do so, we need to develop an understanding of the impact of temperature, electrical, and environmental stresses on the failures rates. Mechanical components are not typically affected by these stressors but electric and electronics components are.

Temperature, Electrical, and Environmental Impacts on Failure Rate

In Chapter 14, we reviewed how temperature and electrical stresses impact the life of electronics components. Now, we can examine how the hazard rates of these components are affected.

Temperature Effects

For reliability calculations, reference temperature is generally set at 40°C and the corresponding hazard rate is denoted by subscript *r* (Klinger et al. 1990). If the operating temperature is different than 40°C, then

$$\lambda_u = A\lambda_r$$

where

$$A = \exp\left[\frac{E_a}{K_b}\left(\frac{1}{T_r} - \frac{1}{T_u}\right)\right]$$

E_a is the activation energy, K_b is the Boltzmann constant, T_r is the reference temperature, and T_u is the use temperature, that is, actual component temperature in the device. Temperatures are in absolute values. Often the reference temperature is 313°C (= 40°C + 273).

Electrical Stress Effects

Other factors such as electrical stress may have adverse effects on the device reliability as well. A component is generally rated either for a specific voltage, current, power, and so on. Reliability studies have shown that once a particular percentage of this rating is exceeded, the hazard rate is increased. In MIL-HDBK-217F (1991), this percentage is assumed to be 25% and is denoted by P_0.

For electrical stress (Klinger et al. 1990), MIL-HDBK-217F (1991) provides the following relationship.

$$\lambda_u = A_E\,\lambda_r$$

where

$$A_E = \exp\left[m\left(P_1 - P_0\right)\right]$$

P_0 and P_1 are percentages of maximum rated electrical stress. P_0 is a reference value and is usually set to 25% for critical applications and 50% for other noncritical applications. P_1 is the percentage of the maximum rated electrical stress. m is provided by MIL-HDBK-217F (Klinger et al. 1990) and other standards such as Bellcore TR-332 (1995). Generally, each component is rated at a particular electrical stress level. At that prescribed value A_E is equal to one (1). Now, should the component be operated at stress levels below the set value its reliability does not change, however, should the operating stress go beyond this limit, its reliability decreases.

Environmental Factors

It stands to reason that an identical product would have a different reliability value if it is used in a benign environment as opposed to a harsh environment. These effects are accounted in a factor generally denoted as π_E. MIL-HDBK-217F (1991) provides a table for π_E which varies between 0.5 for benign environments and 220 for extremely severe conditions.

System Failure Rate

To calculate a system's reliability and its associated hazard rate, we need to have knowledge of the hazard rates associated with the components and then knowledge of the system itself. As it will soon become clear, this approach is also called a "parts-count" method. Because of its simplicity it is frequently used to estimate product reliability at relatively early stages of design.

Component Hazard Rate

The component hazard rate is simply a product of a hazard rate multiplied by various stress factors, namely, electrical, temperature, and quality—another factor undefined as yet. The quality factor takes into account manufacturing issues. Generally, this factor is equal to 3 for commercial items, and 2 or 0.9 for parts that are to be used in military applications.

$$\lambda_{SS_i} = \lambda_{G_i} \pi_{Q_i} \pi_{S_i} \pi_{T_i}$$

λ is the hazard rate, π denotes a stress multiplier, λ_{G_i} denotes the generic value of the hazard rate which is generally provided in handbooks and supplier published data, π_{Q_i} is the quality factor as defined previously, and π_{S_i} is electrical stress factor. Its value is unity for the reference stress value. π_{T_i} is the thermal stress factor. Similarly, its value is unity at a temperature of 40°C.

System Hazard Rate

A system may be composed of a group of subsystems, each made up of many circuit boards. The circuit boards in a subsystem and, indeed, the subsystems within a system may be arranged in series, parallel, or a complex arrangement. A discussion of parallel and complex arrangements is beyond the scope of this work, and the reader is referred to other sources such as Klinger et al. (1990) and MIL-HDBK-338B (1988).

For a series network, the hazard rate of the system may be calculated as follows:

$$\lambda_{SS} = \pi_E \sum_{i=1}^{n} N_i \lambda_{SS_i}$$

where π_E is the environment factor and N_i is the number of each component in the system. This approach provides the strictest sense of reliability calculation in which the failure of any component will flag the failure of the entire system.

To calculate the hazard rate for a system, tabulate the following items for each component:

1. Generic hazard rates
2. Thermal stress factor
3. Electric stress factor
4. Quality factor

The steady-state hazard rate is obtained by multiplying the above numbers. To obtain the failure rate of the entire system, add component hazard rates and multiply by an environmental factor. Values of various stress and environmental factors are available from various sources such as Bellcore (1995) or other standards.

Example

Determine the hazard rate and the MTTF of an electronic medical device. This system is designed to be used only 1 h a day in a benign environment. Next, evaluate the first year repair volume should there be an annual production of 12,500 units. Finally, determine whether these units should be subjected to a post-production environmental stress screening prior to shipment. The bill of materials (BOM) along with generic reliability data are provided in Table 16.1.

Let us assume that in the long term, the failure distribution is exponential and the constant failure (hazard) rate holds. On this basis, the system failure rate may be based on the sum of the component hazard rates.

TABLE 16.1

Failure (Hazard) Rate Calculation Data

Item	Qty	Description	Generic	Quality	Electric	Thermal	Multiply All
1	13	0.1 uF Cap.	1.00×10^{-9}	3	1	0.9	3.51×10^{-9}
2	1	10 uF Cap.	1.00×10^{-9}	3	1	0.9	2.70×10^{-9}
3	2	22 pF Cap.	1.00×10^{-9}	3	1	0.9	5.40×10^{-9}
4	1	Diode	3.00×10^{-9}	3	1	0.9	8.10×10^{-9}
5	1	Net. Res.	5.00×10^{-10}	3	1	0.9	1.35×10^{-9}
6	4	1.0 M Res.	1.00×10^{-8}	3	1	0.9	1.08×10^{-8}
7	1	39 K Res.	1.00×10^{-9}	3	1	0.9	2.70×10^{-9}
8	2	470 K Res.	1.00×10^{-9}	3	1	0.9	5.40×10^{-9}
9	3	Potentiometer	1.70×10^{-7}	3	1	0.9	1.38×10^{-6}
10	2	10 K Res.	1.00×10^{-9}	3	1	0.9	5.40×10^{-9}
11	1	IC	1.00×10^{-5}	3	1	0.9	2.70×10^{-5}
12	1	LCD	3.00×10^{-9}	3	1	0.9	8.10×10^{-9}
13	1	Sensor	2.50×10^{-8}	3	1	0.9	6.75×10^{-8}
14	1	Battery	1.00×10^{-7}	3	1	0.9	2.70×10^{-7}
15	3	Connector	2.00×10^{-10}	3	1	0.9	1.62×10^{-9}
16	3	Switches	1.00×10^{-8}	3	1	0.9	8.10×10^{-8}
						Total	2.89×10^{-5}

$$\lambda_{\text{system}} = \pi_E \sum_{i=1}^{n} N_i \lambda_{SS_i}$$

Based on the data provided in Table 16.1, we have

$$\lambda_{\text{system}} = \pi_E \times 2.89 \times 10^{-5}$$

Since the environment is benign, the environmental factor π_E is 0.5. Thus, the system hazard rate is

$$\lambda_{\text{system}} = 1.45 \times 10^{-5}$$

Base on this rate, the number of failed units in the first year and the associated MTTF is calculated as follows:

$$R(t) = e^{-\lambda t}$$

$$R(8760) = e^{-8760\left(1.45 \times 10^{-5}\right)}$$

$$R = 0.881$$

$$F = 1 - R = 1 - 0.881$$

$$F = 0.119$$

$$MTTF = \frac{1}{\lambda_{\text{system}}} = \frac{1}{1.45 \times 10^{-5}} = 69204\,\text{h}$$

This figure indicates that a 11.9% ($F = 0.119$) first year failure rate may exist, which is equivalent to 1488 (= 0.119 × 12,500 annual production) failed units. This would be true only if the system was operated continuously. However, the duty cycle is only 1 h per day. Thus, a new adjusted MTTF must be found based on the duty cycle. In this particular example, each hour of operation is equivalent to 1 day. Thus, the adjusted MTTF must be calculated by multiplying the calculated MTTF by 24:

$$\text{MTTF}_{\text{adjusted}} = 24 \times \text{MTTF}_{\text{calculated}}$$

$$\text{MTTF}_{\text{adjusted}} = 1660900\,\text{h}$$

$$\lambda_{\text{adjusted}} = \frac{1}{\text{MTTF}_{\text{adjusted}}} = \frac{1}{1660900}$$

$$\lambda_{\text{adjusted}} = 6.02 \times 10^{-7}$$

$$R(t) = \mathrm{e}^{-\lambda t}$$

$$R(8760) = \mathrm{e}^{-8760\left(6.02 \times 10^{-7}\right)}$$

$$R = 0.9948$$

$$F = 1 - R = 1 - 0.9948$$

$$F = 0.0052$$

This indicates that only 0.5% of production or 65 units will likely fail in the first year.

To answer the final question of whether a post-production stress screening should be used, we take the following approach. First, if there are no stress screening in place, we need to take the infant mortality into account. A good model for approximating the infant mortality is the AT&T Weibull equation. Then, based on prior data and experience, we select a shape factor (α). Finally, the scale parameter ($\lambda_{\circ}$) is calculated based on the steady-state failure rate.

Assume that the shape factor $\alpha = 0.75$. AT&T model assumes that the steady-state (or constant hazard rate) begins at 10,000 h. Hence, we have

$$\lambda = \lambda_{\circ} t^{1-\alpha}$$

$$6.02 \times 10^{-7} = \lambda_{\circ} (10{,}000)^{1-0.75}$$

$$\lambda_{o} = 6.037 \times 10^{-4}$$

Now, the number of failed units in one year (8640 h) may be calculated:

$$R(t) = \exp\left(-\frac{\lambda_{o} t^{1-\alpha}}{1-\alpha}\right)$$

$$R(8640) = \exp\left(-\frac{(6.037 \times 10^{-4}) \times 8640^{(1-0.75)}}{1-0.75}\right)$$

$$R = 0.9770$$

$$F(t) = 1 - R(t)$$

$$F = 0.0230$$

The number of first year failures in a year is 288 (=0.0230 × 12,500) units, should they be shipped with the infant failures not screened out. This number is expected to drop to 65 should such a stress screening be put into place. This data should provide the information needed to make a decision on instituting stress screening—which is generally a management decision.

Reliability Testing

In the beginning of this chapter, it was suggested that each design team develop the answers to the following two questions:

1. How reliable is my product?
2. How frequently does the product need to be repaired or maintained?

To answer these two questions, the design team needs to know what parts of the product are likely to fail and, the manner in which they fail. In the field of reliability, it is a commonly held belief that we learn more from product failures than from a product operating successfully. Hence, reliability engineer tends to "break" their products early and often.

Well, maybe not at first. It is possible to develop a best guess estimate for both an expected reliability of the product and a preventive maintenance schedule. These estimates will be based on the reliability block diagram of the product and whether there are redundancies, assumption of the exponential failure distribution, and the use of generic and published failure rates or predicate devices. However, as mentioned, this would only be an estimate.

It is through systematic reliability testing that we can verify whether or not these estimates are acceptable and the assumptions appropriate.

Categorically, disproportionate failures are rooted in one or a combination of the following areas: poor design or design elements, defective materials, weak manufacturing techniques (i.e., excessive variations), and/or poor service procedures. Proper reliability testing exposes these deficiencies.

In general, there are two types of reliability testing; namely, those tests which seek to induce failures; and, those which seek to identify problematic products. Typically, in the design stages of the product development, the reliability engineer tries to gather as much information from testing to failure as possible. The so-called "highly accelerated life testing" (HALT) is used to push components, assemblies, and eventually the finished product well beyond their stated limits and usage. Once the product is launched, and manufacturing becomes the primary activity in the product's life cycle, batches of finished products are subjected to "highly accelerated stress screening" (HASS). The purpose of this test is to separate flawed products from robust units.

Before, we define how to use HALT or HASS, let us develop an understanding of what is involved in these tests.

HALT and HASS

In Chapters 13 through 15, we reviewed both mechanical and electrical failure modes and mechanisms. We learned that in the context of this book, when we speak of reliability failures,* we are primarily concerned with progressive failures whose initiations are not detected; whose end results are loss of performance or a complete device breakdown. This is because of two reasons. First, immediate or sudden failures are easily detected: the unit is not operational even on its first day. Typically, the units are not functional once the user takes the unit out of its packaging, or it has a very short life. The second reason is that the majority of nonprogressive failures are primarily removed either in the design process—if design related, or in manufacturing end-of-line testing—if material/process related.

Hence, the focus on progressive failures is a focus on fatigue and/or creep in a mechanical (thermomechanical) environment; diffusion in an electrical environment; and, corrosion followed by whisker and dendrite growth in a chemical environment.

These mechanisms are generally activated by vibration, heat, electric power, and moisture. Thus, to accelerate latent failures in any product, it would suffice to subject the product to one or a combination of these accelerants; and, then increase their levels. This, in a nutshell, describe the mindset of designing an accelerated life or stress test.

* The field of software reliability is an independent and complex field in and by itself. For simplicity sake, it is ignored here.

The purpose of a HALT is to identify all the potential failure modes of a product in a very short period of time, so that the design team would take one of the two actions: either to improve the design by replacing the failed component with a more robust one, or to develop a service maintenance plan to replace a less-reliable component on a regular basis.

The purpose of HASS is to identify faulty units in manufacturing. This is done in an effort to eliminate the early (or infant) mortality failures. An earlier example demonstrated how effective this screening may be. The applied stresses in HASS are similar to HALT but their levels and duration are much shorter. The reason is that good units should not be damaged during the screening process, while faulty units should become noticeable. These defective units may have faulty components, or have been assembled with sub-par manufacturing process. HASS enables engineers to understand whether excessive manufacturing or process variations exist.

A major difference between HALT and HASS is that in HALT, the product is tested in small quantities; whereas, in HASS, 100% of production quantities are tested. Therefore, larger equipment are needed.

HALT at the Design Stage

At various design stages, different elements may be examined: if information about a component is not available, then, a sample of components may be tested. The same may be said for assemblies and subsystems, and, when a production-equivalent prototype is ready, the entire system may be subjected to HALT.

There is no right or wrong way of conducting a HALT. Having said this, it is important that a test plan be developed first. This plan should include the requirements as well as the design specifications of the unit to be tested. For example, suppose that a power supply is to be tested. It is not sufficient to say that the power supply provides two (2) amperes of a DC current at 12 V. It should state that the power supply provides up to two (2) amperes of DC current at a voltage of 12 V using an input voltage of 90 to 135 V at 60 Hz at a temperature range of –10°C to +55°C.

The test plan should also specify what stressors will be employed and the number of samples. The most common stressors are heat and vibration; though moisture and electric power may also be used. Other tests may introduce salt mists among others. It is important that each stressor be introduced individually first and then once its impacts are known, then, they can be combined to study their interactions.

Typically, the units under test are subjected to room (ambient or laboratory about 25°C) temperatures and humilities (30%–40% RH). Then, while their operation is being monitored, their temperature is lowered at increments of 10°C and soaked at each increment for about 10–15 min. Once the temperature has stabilized, the temperature is reduced by another 10°C and units are soaked while under observation. This temperature reduction cycle is continued until the performance of units under test begin to degrade. The conditions when the expected outputs have drifted are recorded. The test continues until the units fail completely and no longer recover their function. Again, these final conditions are also recorded.

From this test, two data points are identified: the low temperature at which units begin to malfunction and the lowest temperature at which units fail completely. It should be noted that these values may be within or outside of the products specification. It is the relationship between the specification and the observed limits which is indicative of the unit's robustness and reliability.

Once the cold temperature test is completed, a similar test is conducted whereby temperatures increase—usually in 10°C increments—until both malfunction and total loss of function is observed. Again, the upper temperature bounds are identified and compared with the specifications.

The next test is to shake the units. Accelerometers are placed on units. Depending on the test plan, and the shake table available, samples are subjected to either random vibrations or sinusoidal sweeps with increasing amplitude. If random vibration is available, the starting point is about one (1) or two (2) g_{rms} increasing with one (1) or two (2) g_{rms} increments. As the input vibrations propagate through the unit, they may grow due to transmissibility values present in the system. Hence, a one (1) g_{rms} input may translate to a 15 g_{rms} experienced by components in the unit. It is not uncommon to raise the input load to values as high has 50 or 60 g_{rms}. Similar to the temperature test, the vibration test is carried out until both malfunction an complete breakdown is observed and recorded.

The test plan might indicate that once each stress has been applied separately, the test should be repeated for a simultaneous application of temperature and vibration. Or, a temperature cycling from hot to cold and reverse should be done. In addition, other possible stressor such as electric power, moisture, or salt fog may be necessary to expose other potential failures.

Once failures have been obtained, units are investigated to identify root causes of failures. On the basis of this information, the design team would entertain whether a redesign is warranted.

A word of caution is warranted here. A test engineer may easily induce a product failure. The question is whether the induced failure is realistic and has it been observed in either previous designs or competitive devices. Care must be taken to avoid inducing unrealistic failures.

For a more detailed guideline on temperature and vibration steps and durations, refer to *HALT, HASS and HASA Explained* (McLean 2009).

HASS at the Manufacturing Stage

Once HALT is completed, the upper and lower bounds of the operating and destruction limits are identified. Also, the interaction of various stressors on the unit will be known. As a results of this knowledge, the design team has made a conscious decision on how to make the product more robust. By the time HASS is being considered, the design stage of product development has been completed and the manufacturing phase has begun.

From this point on, the product reliability is vulnerable to excessive manufacturing variations as well as material defects that may not be identified at incoming inspection. To ensure against these vulnerabilities, HASS is implemented. The stresses and their combinations will be the same as the ones identified to be effective at the HALT stage; however, their durations and levels are reduced in order to identify variations and undesired changes. McLean (2009) recommends that the temperature limits of HASS be set at 20% of the operating limits of HALT, and the vibration limits be at 50% of HALT destruct limits. All other stressors such as electric power should be set at unit specifications.

Reliability Predictions Based on Test Data

Earlier, we focused on HALT or HASS. It is important now to emphasize while both a HALT and HASS are tools within the reliability tool kit, they do not provide us with the information needed to make predictions. The primary reason is that the induced failures are the results of accumulated stresses at various levels. Therefore a single Arrhenius or fatigue equation does not apply. For this reason, life and/or reliability predictions should be based on other accelerated life tests (ALTs). In the simple ALTs, generally two or more levels of one stress (e.g., temperature) is applied as follows. One stress is applied at a constant level until failure is induced. The time of failure and the level of stress is recorded. Then, the same stress at a different level is applied until failure is obtained. Again, the time of failure and this different level of stress is recorded. By fitting the data into an appropriate model, life expectancy at nominal conditions may be calculated. Operating a device in an elevated-temperature oven to failure is an example of this test.

One of the challenges in conducting accurate reliability calculations based on test data is that no two people use the same device the same way. Different people use their products differently; some use their product with care and delicacy, and some are just "rough." Consider, for instance, a day in life of an automobile. A car model may be driven by a teenager in a mountainous area—driving in high altitudes—with a daily average of 100 miles of travel. The same car model may be used by someone at sea level conditions in flat lands with a daily average driving of only 10 or 15 miles a day. Clearly, the stresses experienced by these two cars will impact their longer functionality—and their reliability—differently. Would you design a product based on the benign or the most stressful conditions?

Having said this, it is nevertheless possible to develop a typical day-in-the-life of a given product, and identify the experienced stresses. On the basis of these stress levels and the results of any ALTs, reliability metrics may be developed. Chapters 13 and 14 have provided examples of how a product's life under nominal conditions may be calculated provided that samples of the same product are tested under accelerated conditions.

There are other life testing scenarios under which a number of samples are tested either for a certain length of time or certain number of failures. While it is tempting to provide the reader with formulas to calculate reliability metrics (MTBF, for instance) for these scenarios, doing so without a detailed review of the theories in this area may lead to confusion and potentially erroneous calculations. Unfortunately, developing a theoretical foundation for this purpose is beyond the scope of this book. The reader interested in this topic is encouraged to read *Reliability Improvement with Design of Experiments* (Condra 2001); *Accelerated Testing: Statistical Models, Test Plans, and Data Analysis* (Nelson 2004); or even MIL-HDBK-338B.

17

Design Considerations in an Avionics Electronic Package

Introduction

In this chapter, the mechanics and thermal design procedure and calculations (based on both analytical means and finite element methods) of an airborne cabin telecommunications unit (CTU) are discussed along with the relevant Federal Airworthiness Regulation (FAR) (2008), as well as Aeronautical Radio, Inc. (ARINC) standards (1991). Reliability and failure rates of components and their effects on the performance of the system are also studied. In designing avionics electronics packages, four areas of engineering need to come together and work concurrently as a team. These are electronic, electromechanical packaging, thermal, and vibration. Oftentimes, thermal and vibration analyses are done only as an afterthought through experimental means to ensure Federal Aviation Authority (FAA), FAR and/or ARINC standards are met.

Equipment design is influenced by its operational environment and its impact on susceptible components. For example, if a system is operating in sunlight, it can experience as much as 350 W of heat loading per square foot of exposure. It may also be that the same system is experiencing random vibrations caused by a variety of sources, adding to the severity of the stress field, especially at critical frequencies. The same considerations must be made for electromagnetic interference (EMI) and other environmental issues. In short, a system cannot be designed adequately by ignoring the operational environment. In the avionics industry, regulatory agencies and standards organizations such as the FAR (2008) and ARINC (1991) impose rules, and, along with the Radio Technical Commission for Aeronautics (RTCA) (1989), have developed sets of test procedures to enable design engineers to test and verify their designs and prevent any safety-threatening failures. Much too often, however, thermal, mechanical, and reliability evaluations have been left up to these tests and as the last item on the design agenda with no analysis taking place. This practice relies heavily on the designer's previous experience and lack of any substantial changes to new systems. Otherwise, time- and budget-consuming design cycles must be undertaken before the product may be released to the market.

There is another drawback to testing without analysis. These tests only reveal whether a system fails or not. They do not give any substantial information in regards to mean-time-between-failure (MTBF) reliability, mean-time-to-repair (MTTR), and so forth. These issues are important when one considers business issues such as scheduling repairs, product warranties, or consumer satisfaction with the product.

Herein, analysis of a hypothetical commercial aviation electronics package is discussed. The particular system is assumed to be an ARINC 746 (1991) CTU.

Design Parameters

The following is a set of design parameters developed based on various regulations and standards. These areas cover operational characteristics and electrical and mechanical design. It should be noted that the design of electronics, although not discussed here, is within the scope of operation with its constraints set forth within the electrical design parameters. Packaging aspects are within the realm of mechanical design parameters.

Operational Characteristics

The model CTU is similar to a public branch exchange (PBX) computer. It generates commands for the establishment and termination of air/ground, air/SATCOM, and seat/seat communications. Furthermore, the CTU has interfaces to the aircraft's networks and to the radio telephone control computer. It is also linked to embedded processors in the radio, allowing users to obtain an air-to-ground modem link.

Reference Documents

To gain FAA certification and industry acceptance for this device, compliance with the following reference documents is mandatory:

- ARINC 746—Cabin Communication System (CCS)
- ARINC 600—Air Transport Avionics Equipment Interfaces
- RCTA/DO-160C—Environmental Conditions and Test Procedures for Airborne Equipment
- FAR Part 25—Federal Aeronautics Regulations

Electrical Design Specifications

The specifications of the electrical design must be as follows:

1. The AC input is to be protected using a single-pole circuit breaker.
2. The DC outputs of the power supply are to be protected using fuses. The DC output return is not permitted to float aboveground potential. The high-voltage DC side of the power supply is to be tied to a power interrupt capacitor. No batteries are to be used for power interrupts unless the battery meets FAR Part 25, or a special containment is used.
3. The CTU must incorporate an internal chassis/safety ground independent from the return line per ARINC 600 and 746 techniques for grounding and bonding.
4. The CTU must incorporate a single-point ground as described in ARINC 600. All printed circuit boards (PCBs) should have an internal ground plane. This ground plane must be tied to ground at a single point. PCB standoff locations are not to be tied to chassis ground, thereby defeating single-point grounding.
5. In case of loss of power, the power supply and capacitor combination shall be capable of maintaining the output load for 2 s, a time in excess of RTCA/DO-160C and ARINC 600 specifications. The loss of power is defined as output drops of 3% below nominal values.
6. The CTU must comply with RTCA/DO-160C for EMI/RFI. Experience has shown that all wires are to be shielded and/or twisted. All inputs/outputs should either be isolated or filtered. ARINC 600 connectors should be used.

Mechanical Design Specifications

The specifications of the mechanical design must be as follows:

1. The unit shall weigh less than 20 lb.
2. ARINC 600 standards dictate a size limit of 6 MCU,* 7.50 in (width) × 12.76 in (length) × 7.64 in (height) (W×L×H).
3. It is imperative that the CTU does not create any fire or smoke under any circumstances, which is considered to be a hazard to flight operations. It is strongly recommended that all wires conform to

* MCU is defined as follows: "The Modular Concept Unit (MCU) is the basic building block module for use in commercial airplane avionics system design. This specification provides for the standard interfaces between the MCU-type Line Replaceable Units (LRUs) and the electrical wiring, environmental control systems, and supporting structures."

MIL-W-22759, and MIL-W16878E and cables to MIL-C-27500. It is against FAA rules to use any cables jacketed with polyvinyl chloride (PVC). All PCBs are to be FR4-G10 or better.

4. The packaging sheet metal must be aluminum. No fractures or tears are permitted on formed sheet metal parts. The assembly must retain its mass during sudden deceleration—12 *g*s per FAR Part 25 and RTCA/DO-160C crash safety regulations.
5. All screws, washers, and nuts must conform to military standards. All fasteners must have a self-locking device, such as nylon or deformed threads. All power entry modules (PEM) style inserts must be self-locking style and only installed after irradiating and plating.
6. The temperature range is from –15°C to 55°C and altitude of 0 to 15,000 ft. No temperature derating is permitted.
7. The CTU will be tray mounted. Air will be moved through the tray metering plate and enter the CTU from the bottom and exit from the top. The bottom and top surfaces will have several hundred 4 mm diameter holes. Every effort must be made to eliminate other air leakage points. ARINC 600 dictates that an easily removable internal air filter be installed that filters particles down to 4 microns. Airflow impedance is critical because vital aircraft systems must not starve for ample airflow. ARINC 600 dictates an air weight flow of 220 kg/h/kW at 40°C and 1013.25 mbar. No exterior side surface is to exceed an average of 60°C and no single point may exceed 65°C at an inlet temperature of 55°C (ground) and an altitude of 15,000 ft.

Electrical and Thermal Parameters

The following parameters in the electrical system have been specified:

1. The power supply in this unit must have 130 W of output with a minimum power factor of 0.90 and a minimum efficiency of 80%.
2. The nominal input voltage of 115 volts AC at 1.6 amps and will have an input of 184 VA. This gives a system thermal dissipation of between 165.6 and 184 W.

Analysis

The scope of this chapter does not allow an electrical and/or electronic analysis of the CTU. The mechanical analysis focuses on the following three areas per ARINC 600 requirements:

1. Thermal analysis is performed to ensure that average surface temperatures do not exceed 60°C and that no localized temperature exceeds 65°C.
2. A static analysis will be performed to investigate a 12 g per axis load per FAR Part 25 regulations and RTCA/DO-160C procedures. This analysis will ensure that the ultimate stresses are not exceeded and is designed to ensure that the unit will remain intact. Functionality of the unit after testing is not an issue.
3. The last analysis is to test for random vibration per RTCA/DO-160C requirements.

Reliability and MTBF calculations are yet another aspect of analysis that will be dealt with briefly. Depending on the field application of the unit, either MIL-HDBK-217F (1991) or Bellcore Standard (1995) TR-332, Issue 5, may be used. In this application, Bellcore Standard TR-332, Issue 5, will be used to derive an estimate of the lifetime of the electronics. Other methods may be employed to calculate the mechanical (fatigue) lifetime of the packaging itself. Figure 17.1 depicts a potential design of this system.

Thermal Analysis

The CTU has the following components and maximum potential heat dissipation distributions:

1. PCBA 1—uP Board, 26 W
2. PCBA 2—Memory board, 26 W
3. PCBA 3—Network board, 19.5 W
4. PCBA 4—Network board, 19.5 W
5. PCBA 5—Power supply, 54 W
6. PCBA 6—E1 network board, 19.5 W, quantity two

This totals 184 W. ARINC 600 requires that no more than 220 kg/kW/h flow be used. The task, therefore, is to find out whether the system is within ARINC's airflow budget.

$$220\frac{\text{Kg}}{\text{kW h}}\times 184\frac{\text{W}}{1}\times\frac{1\,\text{kW}}{1000\,\text{W}}=40.48\frac{\text{Kg}}{\text{h}}$$

In other words, our system may not use more than $40.48\,\text{Kg}/\text{h}$ of available cooling air. To determine whether this is a sufficient amount, recall Equation 5.9:

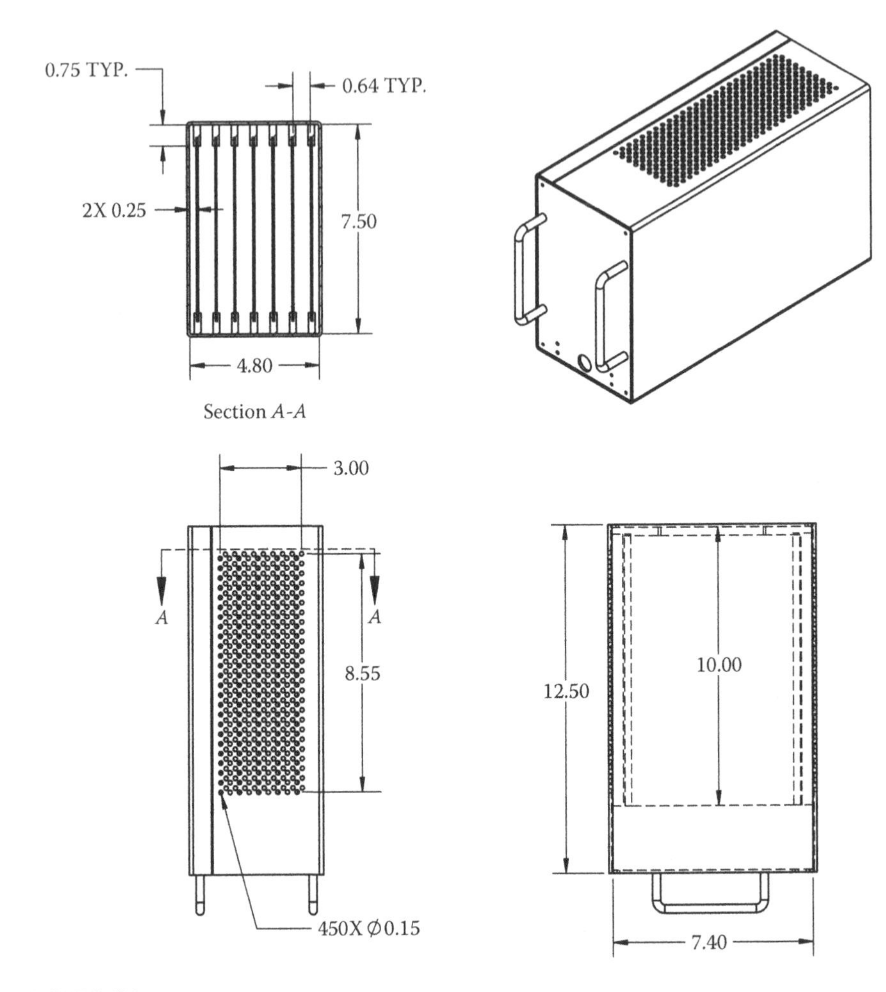

FIGURE 17.1
A potential design of a cabin telecommunication unit (CTU).

$$\dot{m} = \frac{Q}{C_p \Delta T}$$

where $\dot{m}$ is mass flow rate. This equation may be rewritten by substituting the value of $C_p = 0.2789$ for air and maintaining units of W for Q and Celsius for temperature:

$$\dot{m} = \frac{3.5856\ Q}{\Delta T} \frac{\text{Kg}}{\text{h}} \tag{17.1}$$

Rewrite this equation and solve for ΔT:

$$\Delta T = \frac{3.5856\,Q}{\dot{m}}$$

$$\Delta T = \frac{3.5856 \times 184}{40.48} = 16.3°\text{C}$$

In other words, with the available flow, our coolant temperature rise will be 16.3°C. ARINC requires a temperature rise of not more than 10°C. Note that ARINC does not concern itself with the component temperature—just the maximum temperature rise on the unit. Considering that the provided airflow is not sufficient to keep the temperature rise to 10°C, a warning flag has been raised.

The first reaction may be to anticipate changing the electronics design. Since this analysis is taking place after the design cycle is completed, any changes in the electronics will be extremely expensive. An alternative is to study the standards further and investigate any other possible solutions.

ARINC procedures were developed for the standard avionics rack system implemented in the electronics bay of the airplane. The rack system provides coolant air to all electronic equipment including the vital systems intended for survival of the flight. However, it is possible to develop a private rack that delivers coolant air independent of the standard system. The source of coolant air for this "private rack" may be the cabin air circulating through the system. This is a way to eliminate any regulatory or airframe manufacturer restrictions on the availability of coolant airflow.

Based on Equation 17.1, the auxiliary source must supply a 65.88 kg/h flow rate for a 10°C inlet to outlet temperature rise. The steps to calculate the required flow rate as well as hot spot temperatures were outlined in Chapter 5. There, it was determined that the maximum component temperature will be 69°C with a flow rate of 31.7 cubic feet per minute (CFM). To provide this flow, a 400 Hz fan will be used, which will deliver a minimum of 36 CFM.

Load Carrying and Vibration Analysis

To study the characteristics of the system under various mechanical loads, a finite element model using 4446 shell elements was developed. A static analysis was conducted where 12 g loads were applied along the three main directions. Von Mises stresses were calculated as shown in Table 17.1. Based on this analysis, the system will be able to withstand the required 12 g static loading.

Furthermore, the system should be subjected to random vibration excitations in three axes with an input level of 0.1 g^2/hz for 10 to 2000 Hz range, per RTCA/DO-160C procedures. The next step is to determine the deflections of the PCBAs as well as the chassis and the induced stresses. In Table 7.1, the exact frequency of the PCBA was calculated. Here, we use the frequency

TABLE 17.1

Calculated Stresses and the Affected Areas

Direction	Maximum Stress (psi)	Affected Area
X	4716	Edge guides
Y	6512	Welds and joints
Z	2252	Back panels and mounting screws

associated with clamped sides (141.58 Hz). Separately, the first mode of vibration of the chassis was determined to have a frequency of 420 Hz. It is important to realize that although 420 Hz is greater than 2 × 141.58 Hz (in other words, the design follows the frequency rule as discussed in Chapter 7), the resonance vibration of each board will contribute energy to the transmissibility of the chassis at 141.58 Hz frequency and vice versa. The transmissibility curve of each PCBA will not only receive additional energy at the 141.58 Hz peak, but we should also expect a second peak at a frequency of 420 Hz.

We begin by first calculating the response of the chassis to the random excitation.

$$PSD = 0.1g^2 / \text{hz}$$

$$f_{\text{Chassis}} = 420$$

$$T_m^{\text{Uncoupled}} = 0.5\sqrt{420} = 10.25$$

$$f_{\text{PCBA}} = 141.58$$

$$R = \frac{f_{\text{Forcing}}}{f_{\text{Resonant}}} = \frac{f_{\text{PCBA}}}{f_{\text{Chassis}}} = \frac{141.58}{420} = 0.337$$

$$T_m^{\text{Contribution}} = \frac{1}{1-R^2} = \frac{1}{1-0.337^2} = 1.128 \text{ at } 141.58 \text{ Hz}$$

$$T_m^{\text{Coupled}} = 7 \times \frac{1}{1-R^2} = 7 \times 1.128 = 7.9 \text{ at } 141.58 \text{ Hz}$$

This indicates that although one PCBA may not have much of an impact on the chassis, seven PCBAs all having the same resonant frequency could create a second peak on the chassis transmissibility curve. Thus, the response acceleration should be calculated as follows:

$$g_{\text{RMS}} = \sqrt{\frac{\pi}{2} \times 0.1 \times 141.58 \times 7.9 + \frac{\pi}{2} \times 0.1 \times 420 \times 10.25} = 29.18$$

$$X_{\mathrm{RMS}} = \frac{9.78 \times 29.18}{420^2} = 0.0016\,\text{in}$$

Now, we can focus on the PCBA response. Considering that all PCBs are identical, only one board will be analyzed. Similar to the chassis, we expect to have energy contribution to PCB transmissibility at its uncoupled resonance frequency as well as a second peak at 420 frequency.

$$PSD = 0.1\,\text{g}^2/\text{hz}$$

$$f_{\mathrm{PCBA}} = 141.58$$

$$T_m^{\text{Uncoupled}} = \sqrt{141.58} = 11.90$$

$$f_{\text{Chassis}} = 420$$

$$R = \frac{f_{\text{Forcing}}}{f_{\text{Resonant}}} = \frac{f_{\text{Chassis}}}{f_{\text{PCBA}}} = \frac{420}{141.58} = 2.966$$

$$T_m^{\text{Contribution}} = \frac{1}{1 - R^2} = \frac{1}{1 - (2.966)^2} = -0.128 \text{ at } 420\ \text{Hz}$$

The negative sign only implies that the PCBA and the chassis are out of phase. Now, we can calculate the two peaks on the PCBA transmissibility curve.

$$T_m^{\text{Coupled},1} = 11.9 \times 1.128 = 13.42 \text{ at } 141.58\ \text{Hz}$$

$$T_m^{\text{Coupled},2} = 10.25 \times 0.128 = 1.31 \text{ at } 420\ \text{Hz}$$

$$g_{\mathrm{RMS}} = \sqrt{\frac{\pi}{2} \times 0.1 \times 141.58 \times 13.42 + \frac{\pi}{2} \times 0.1 \times 420 \times 1.31} = 19.63$$

$$X_{\mathrm{RMS}} = \frac{9.78 \times 19.63}{(141.58)^2} = 0.0096\,\text{in}$$

The overall 3S random stresses generated in the CTU were 9457 psi. The stresses generated by random vibrations may be used to estimate a mechanical lifetime for the PCBAs and the box.

The lifetime may be estimated using Equation 13.3:

$$T = \frac{B}{4 f \sigma^b \left[0.2718 \left(\frac{2}{3}\right)^b + 0.0428\right]} \text{ Sec} \tag{17.2}$$

where f is the response frequency. B and b depend on the material used:

$$\text{For G10 glass epoxy: } b = 11.36, B = 7.253 \times 10^{55}\,\text{psi/cycle}$$

$$\text{For aluminum: } b = 9.10, B = 2.307 \times 10^{47}\,\text{psi/cycle}$$

$$T = \frac{2.307 \times 10^{47}}{4 \times 141.58 \times 9457^{9.10}\left[0.2718\left(\frac{2}{3}\right)^{9.10} + 0.0428\right]} = 5.436 \times 10^{9}\,\text{Sec}$$

This is equivalent to more than 172 years. Based on this formula, this unit will not suffer fatigue failure. Per RTCA/DO-160C standards, any laboratory testing does not exceed more than two hours. This length of time is only sufficient to screen for systems to fail in their infancy and unfortunately may not be indicative of field behavior and reliability.

Reliability and MTBF Calculations

The reliability calculations for this system were based upon Bellcore TR-332 (1995), Issue 5, rather than MIL-HDBK-217F (1991). The reason for this is that the system is defined as telephone equipment rather than avionics. TR-332 also allows for greater use of commercial grade parts and assigns high quality factors to military-standard components. The following equation was used to generate the system failure rate:

$$\lambda = \lambda_G \pi_Q \pi_S \pi_T \pi_E \tag{17.3}$$

where:

λ = Failure rate (FITs) or failure per 10^9 h
λ_G = Parts generic failure rate or failures per 10^9 h
π_Q = Quality factor
π_S = Electrical stress factor
π_T = Temperature stress factor
π_E = Environmental factor (a factor of 6 was used)

First Year Electronics Failures

It is not practical to provide a bill of material (BOM) for this CTU here. However, for a given BOM, a table similar to Table 13.1 may be constructed. The failure rate for this system has been calculated as:

$$\lambda = 692 \text{ FIT's and may be rewritten as } \lambda \cong 69 \times 10^{-9}\,\text{h}$$

The AT&T Reliability Manual (Klinger et al. 1990) defines the system survival rate (or reliability) of the first 12 months (8760 h) per the following equation:

$$S_t = e^{-\lambda t} = 0.994$$

In other words, 99.4% of the units will be in service after one year without repairs.

Acknowledgment

This segment was completed with the help and input of late Mr. Robert E Walter, President of Airtronic Services, Inc.

Appendix A: Description of Finite Element Model

The following is a brief description of the finite element model used in Chapter 3. Figure A.1 depicts the finite element mesh comprising 3426 tetrahedral elements. This mesh was created using an automatic mesh generation option of a commercial finite element analysis software package. Specified conductivity was 90 BTU/(hr ft °F). The software package had the option of specifying an interface resistance, which was set to 781.25 BTU/(hr ft^2 °F). The back wall was given a constant temperature of 100°F and the heat source was specified at 12 W. Figure A.2 shows the temperature distribution in this system.

Earlier, it was mentioned that the interface condition was modeled in this simulation. Figure A.3 provides a close-up of the temperature distribution in the bracket near the wall. This shows how heat flow spreads in the vertical direction and subsequently toward the sink (i.e., the constant temperature in a wall).

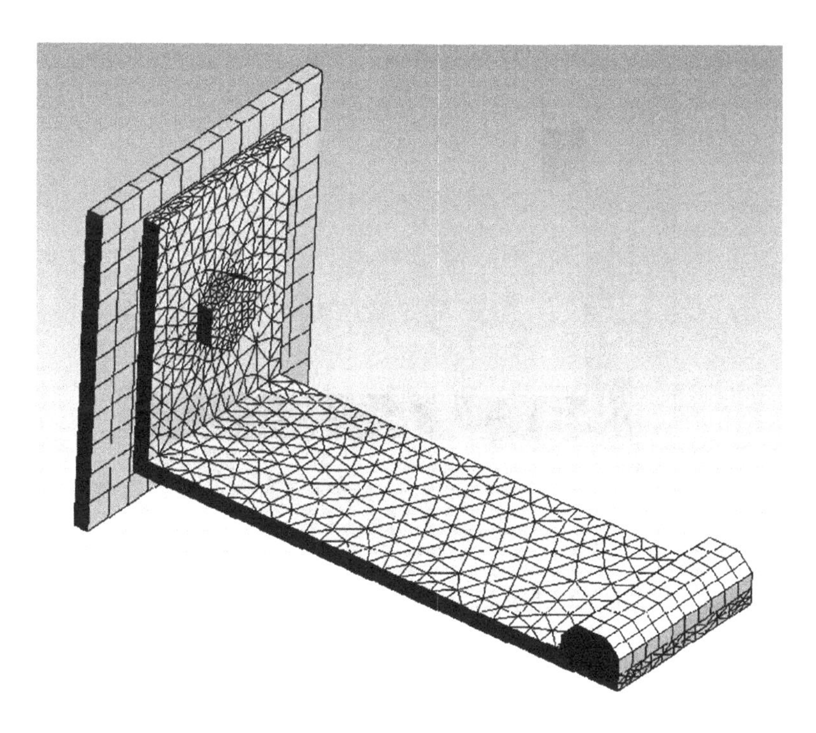

FIGURE A.1
Finite element mesh of the bracket.

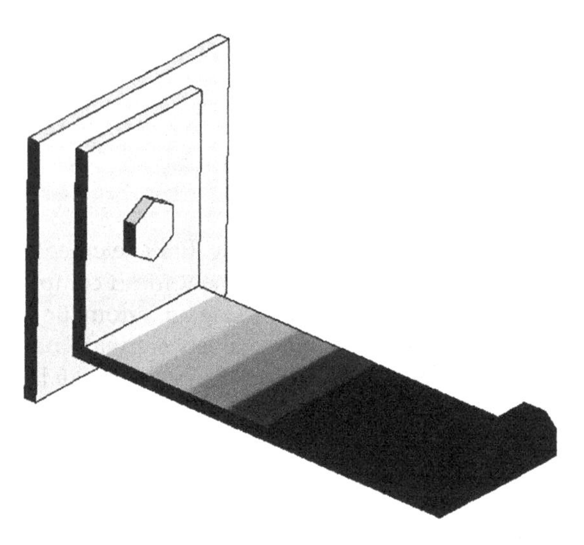

FIGURE A.2
Temperature distribution in the bracket—Max 161.88°F.

FIGURE A.3
Temperature distribution in the bracket near the bend.

Appendix B: Standard Atmosphere

A series of experiments has provided evidence in the variation of temperature as a function of height. These data indicate that as the height increases and up to about 11,000 m (about 36,000 ft), temperature drops linearly. From this height to a height of about 25,000 m (about 82,000 ft), temperature is constant. Ironically, it begins to increase linearly for heights above 25 km. Once we define a temperature variation, both pressure and density may be calculated based on the defined temperature (Anderson 1978; Glenn Research Center 2007).

SI Unit

For the SI units, the following equations are defined for temperature variation with height:

$$T = 15.04 - 0.00649h \quad \text{for} \quad h < 11{,}000\,\text{m}$$

$$T = -56.46 \quad \text{for} \quad 11{,}000\,\text{m} < h < 25{,}000\,\text{m}$$

$$T = -131.21 - 0.00299h \quad \text{for} \quad h > 25{,}000\,\text{m}$$

The pressure equation may be developed as follows:

$$p = 101.29\left[\frac{T + 273.1}{288.08}\right]^{5.256} \quad \text{for} \quad h < 11{,}000\,\text{m}$$

$$p = 22.65\exp(1.73 - 0.000157h) \quad \text{for} \quad 11{,}000\,\text{m} < h < 25{,}000\,\text{m}$$

$$p = 2.488\left[\frac{T + 273.1}{216.6}\right]^{-11.388} \quad \text{for} \quad h > 25{,}000\,\text{m}$$

In these equations, h is the altitude in meters, T is in °C, and p is in kPa. Note that to calculate density, the equation of state $(p = \rho RT)$ should be employed. The result is as follows:

$$\rho = \frac{p}{0.2869(T + 273.1)}$$

The unit of density is kg/m^3.

American (English) Unit

For the American units, the following equations are defined for temperature variation with height:

$$T = 59 - 0.00356h \quad \text{for} \quad h < 36{,}000\,\text{ft}$$

$$T = -70 \quad \text{for} \quad 36{,}000\,\text{ft} < h < 82{,}000\,\text{ft}$$

$$T = -205.05 - 0.00164h \quad \text{for} \quad h > 82{,}000\,\text{ft}$$

The pressure equation may be developed as follows:

$$p = 2116\left[\frac{T + 459.7}{518.6}\right]^{5.256} \quad \text{for} \quad h < 36{,}000\,\text{ft}$$

$$p = 473.1\ \exp(1.73 - 0.000048h) \quad \text{for} \quad 36{,}000\,\text{ft} < h < 82{,}000\,\text{ft}$$

$$p = 51.97\left[\frac{T + 459.7}{389.98}\right]^{-11.388} \quad \text{for} \quad h > 82{,}000\,\text{ft}$$

In these equations, h is in ft, T is in °F, and p is in lb/ft^2. Note that as before, to calculate the density, the equation of state should be employed. The result is as follows:

$$\rho = \frac{p}{1718(T + 459.7)}$$

The unit of density is slugs/ft^3.

Appendix C: Transient Flow Emperical Factor

Predicting flow behavior in the transition region (i.e., $2000 < Re < 10{,}000$) is a difficult task. To calculate the heat transfer coefficient in the transient region, the *Thermal Network Modeling Handbook* (K&K Associates 1999–2000) suggests the following relationship:

$$h_c = JC_pG\left(\frac{C_p\mu}{K}\right)^{-(2/3)}$$

where

$$J = C_1G\left(\frac{\mu_b}{\mu_s}\right)^{0.14}$$

For a definition of terms, refer to Chapter 5. The only variable referred to in this appendix that has not been defined is C_1, which is an empirical factor. The *Thermal Network Modeling Handbook* (K&K Associates 1999–2000) provides a graph for evaluating C_1 for a variety of tube and duct configurations. A surface fitted to these data is provided in Figure C.1. The analysis was performed on the LAB Fit (Silva 2007). LAB Fit is a Windows-based software program developed for the treatment and analysis of experimental data.

The fitted surface may be expressed as follows.

$$C_1 = A\left(\frac{L}{D_h}\right)^{(B+(C/Re))} + D\ln Re$$

where:

$$A = 0.0193071470406$$

$$B = 0.0149681526261$$

$$C = -0.1427130777051\times 10^{3}$$

$$D = -0.164787464170\times 10^{-2}$$

In this formula, L is the flow length, D_h is the hydraulic diameter of the flow cross section, and Re is the Reynolds number. The surface was fitted with $R^2 = 0.988$ and $\chi^2 = 44$.

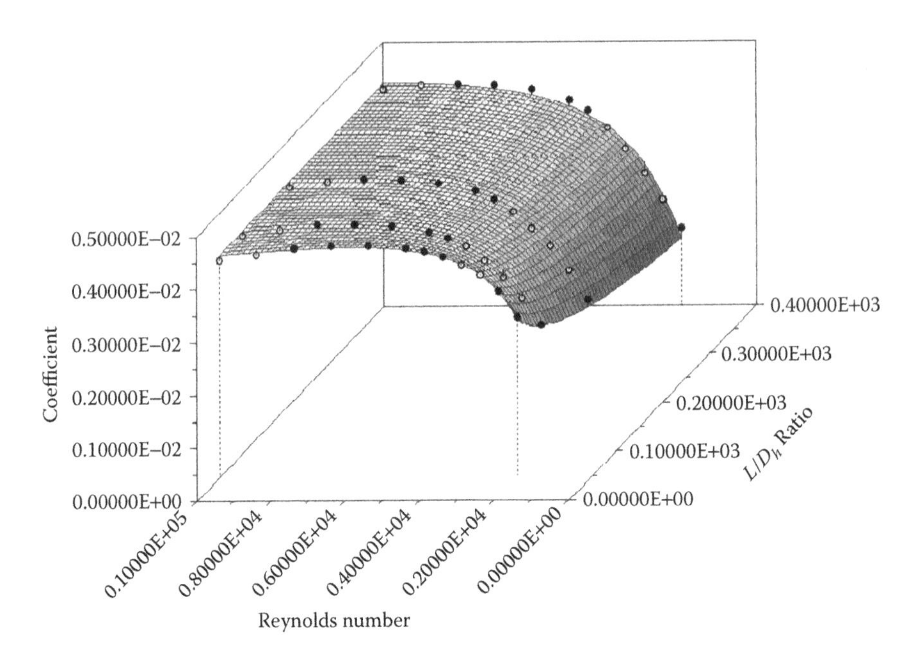

FIGURE C.1
A fitted surface providing values for C_1.

Appendix D: Impact of Convection on Spread Angle

In Chapter 3, we discussed that as heat is conducted in a body, it spreads and as a result, the surface area needed for calculating thermal resistance enlarges. This approach is quite valid for layered structures where heat flows from one layer to another and the direction of the heat flux is perpendicular to the layers. Now, imagine that the last layer is a heat sink and is subjected to convective flow. The spread angle is no longer a function of heat-sink conductivity and would depend on the heat transfer coefficient.

To develop this relationship, we need to make certain simplifying assumptions: first, there is only one point of heat source; second, the heat sink is infinite in both directions; and finally, the temperature throughout the heat sink is constant. Based on these assumptions, we can develop a differential element as shown in Figure D.1 and balance the heat equation.

In Figure D.1, dQ_1 and dQ_2 are the conductive heat flux entering and leaving this element. dQ_3 is the convective heat leaving the same element.

$$dQ_1 = -Kt\,d\theta\frac{dT}{dx}$$

$$dQ_2 = -Kt\,d\theta\frac{d}{dx}(T + dT)$$

$$dQ_3 = \bar{h}\,d\theta\,dx(T - T_{\text{Bulk}})$$

Recall that a $\bar{h}$ is the average heat transfer coefficient over the entire surface. Now, we can write the heat balance equation for this element:

$$dQ_1 + dQ_2 + dQ_3 = 0$$

When we substitute for dQ_1, dQ_2, and dQ_3, we obtain the following differential equation:

$$Kt\frac{d^2T}{dx^2} = \bar{h}(T - T_{\text{Bulk}})$$

To solve this equation, say $\mathfrak{I} = T - T_{\text{Bulk}}$ and rewrite this equation as

$$Kt\frac{d^2\mathfrak{I}}{dx^2} = \bar{h}\ \mathfrak{I}$$

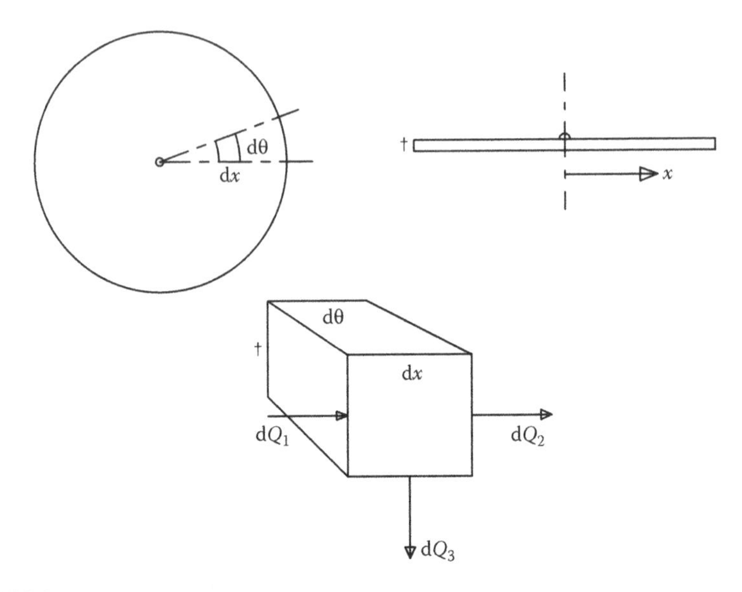

FIGURE D.1
Differential element needed for heat balance.

The solution to this equation is as follows:

$$\Im = \Im_{\circ} \exp\left(-\sqrt{\frac{\bar{h}}{Kt}}x\right)$$

or

$$\frac{\Im}{\Im_{\circ}} = \exp\left(-\sqrt{B_i}\,\frac{x}{t}\right)$$

where B_i is Biot number, defined as follows:

$$B_i = \frac{\bar{h}t}{K}$$

Alternatively, we can write this equation in terms of bulk temperature

$$\frac{T - T_{\text{Bulk}}}{T_S - T_{\text{Bulk}}} = \exp\left(-\sqrt{B_i}\,\frac{x}{t}\right) \tag{D.1}$$

where T_S is the surface temperature at $x/t = 0$ (immediately below the heat source). Figure D.2 compares this ratio for aluminum and a typical polymer for the same value of heat transfer coefficient and thickness (t). Clearly, a high-conductivity heat sink provides a much larger spread area.

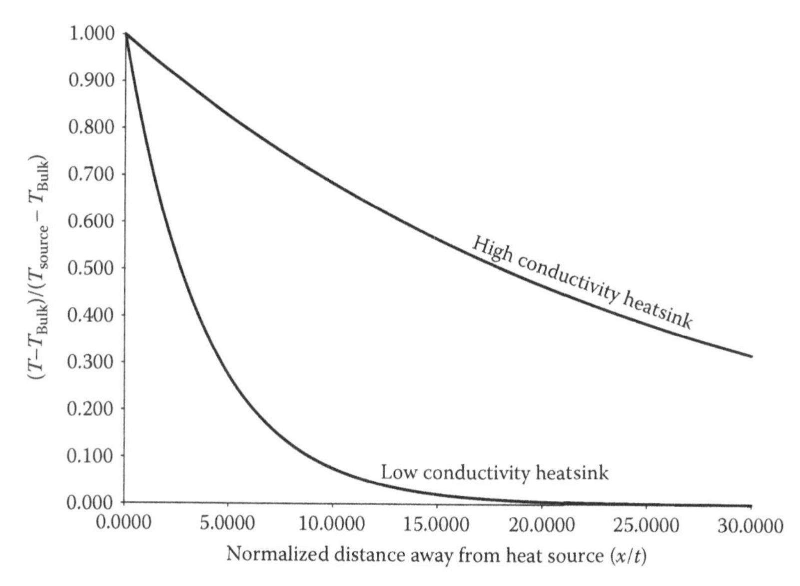

FIGURE D.2
Heat spread in an aluminum heat-sink versus a polymer heat sink.

In calculating surface temperature of a heat sink subjected to convective heat transfer, it is common to use the entire surface area. In electronics packaging, this is generally a very good assumption because the heat sink is generally an aluminum alloy. For instance, in the example solved for Figure D.2, it would take a radius-to-thickness ratio of nearly 200 before the aluminum sink temperature equals that of the ambient (or bulk) temperature, that is, $T - T_{\text{Bulk}} = 0$. In contrast, a low-conductivity heat sink has a radius-to-thickness ratio of nearly 30 before ambient temperatures are reached (Figure D.2). Clearly, in this case, less heat is removed and thus inner components experience higher temperatures.

Another point worth mentioning is this: When we use the area of the entire heat sink, we calculate a uniform heat-sink temperature, whereas in reality there is a temperature gradient as we move away from the heat source. It may be possible to develop an estimate of the hot spot temperature using Equation D.1, and the uniform heat-sink temperature, but the details of this calculation are beyond the scope of this book.

Appendix E: Heat Transfer Cofficients Models—Narrow Gap versus Wide Gap

In Chapter 5, we considered two models for calculating the heat transfer coefficient, namely, one model based on calculating the Colburn factor, and the other based on fluid flow rate. The advantage of using the Colburn factor is that the various flow regimes (i.e., laminar or turbulent) can be taken into account. The implicit assumption for these two models is that the channel is wide enough to accomodate the heat from the walls and does not impact the fluid bulk temperature. Furthermore, static pressure is needed to supply a certain level of flow rate changes to accommodate any change in the channel (or gap) height.

It stands as a reason to investigate the conditions under which these two assumptions are violated. Bejan et al. (1996) suggest that should the channel be too narrow and the fluid bulk temperature be affected, the following relationship may be employed:

$$h = \rho \dot{V} C_p \tag{E.1}$$

Equation 5.15 provided the relationship between pressure drop and flow rate as described in

$$\sigma\ \Delta P = \ \Re\left(\rho\ \dot{V}\right)^2 \text{ inches of water} \left(''\text{H}_2\text{O}\right)$$

where loss coefficient (or factor) is

$$\Re = 0.226 \frac{N}{A^2} \frac{''\text{H}_2\text{O}}{\left(\text{lb/min}\right)^2}$$

Assume that the channel has a width of Y (in) and a height of b (in). We can rewrite the expression for pressure drop as follows:

$$\sigma\ \Delta P = \ 0.226 \frac{N}{Y^2}\ \rho^2 \left(\frac{\dot{V}}{b}\right)^2 \tag{E.2}$$

With the above equation, it would be possible to vary the gap height (b) and calculate the pressure loss for a constant flow rate, or calculate the flow rate for a given pressure loss.

Figure E.1 provides the relationship between heat transfer coefficient (h) and gap height (b) for a constant flow rate (based on Equations 5.22, 5.23, and E.1). Figure E.2 shows the corresponding relationship for the pressure

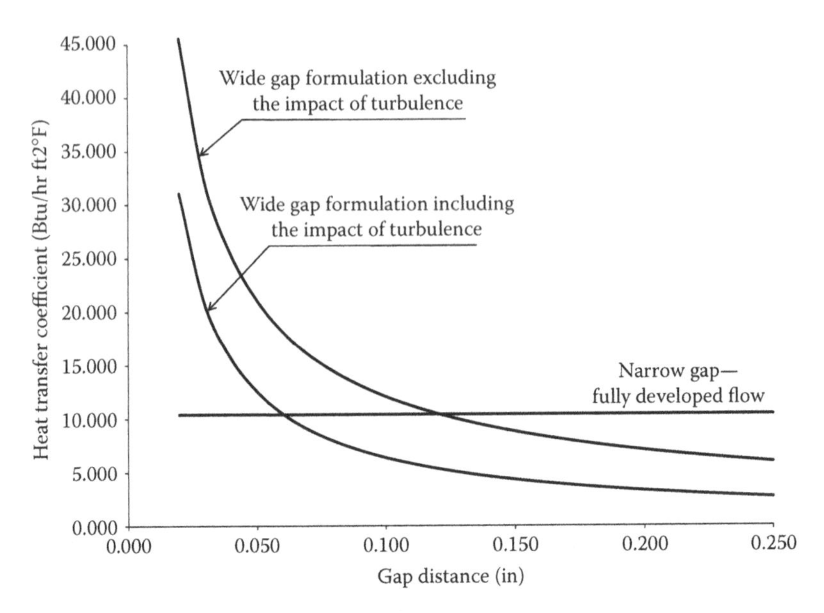

FIGURE E.1
Heat transfer coefficient as a function of gap height for a constant flow rate.

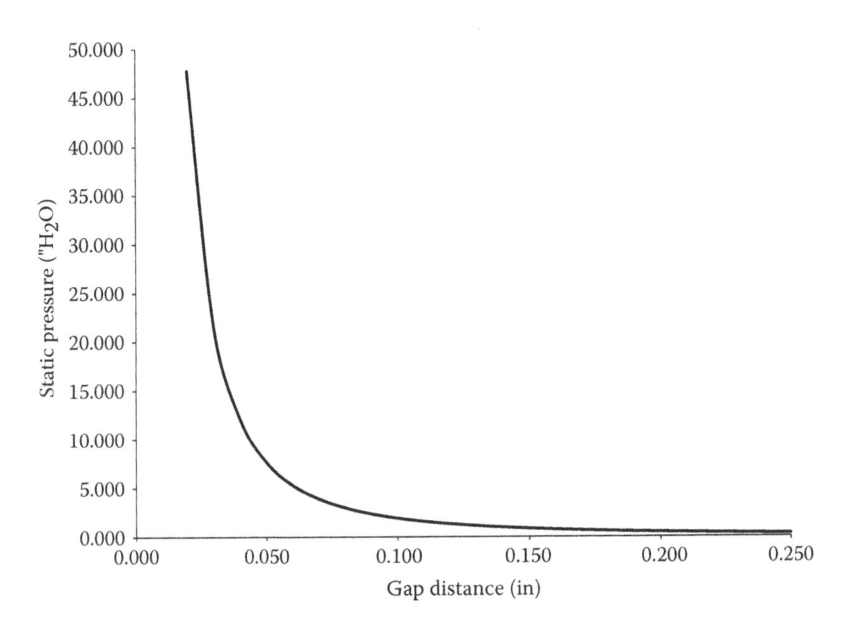

FIGURE E.2
Pressure loss as a function of gap height for a constant flow rate.

loss (based on Equation E.2). Conversely, Figure E.3 provides the same relationship as in between heat transfer coefficient (h) and gap height (b) for a constant pressure loss value (again, based on Equations 5.22, 5.23, and E.1).

One may argue that the optimum value of the heat transfer coefficient will be near the area where the two models coincide. For Figure E.1, this would correspond to a gap height of about 0.06 in (or 0.125 in, depending on the model used), and for Figure E.3, the optimum gap height would be 0.25 in (or 0.2 in, again depending on the model used).

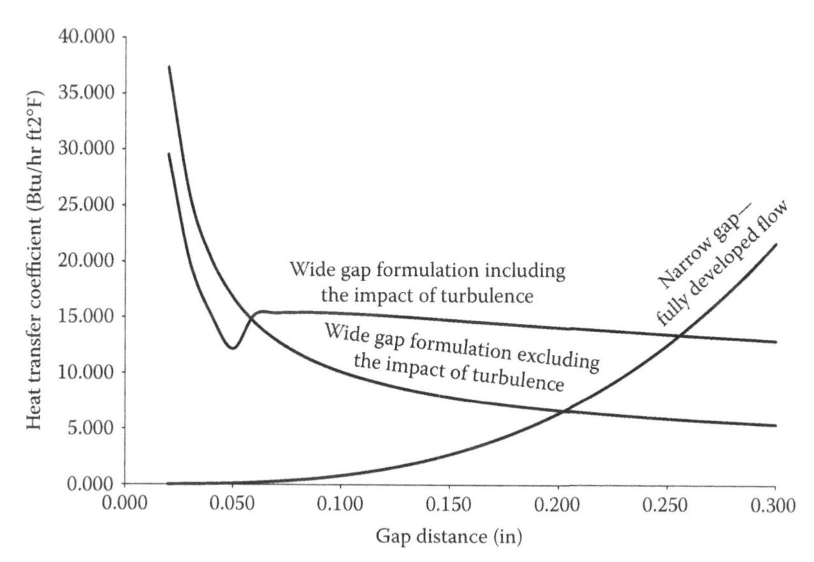

FIGURE E.3
Heat transfer coefficient as a function of gap height for a constant pressure loss.

Appendix F: Creep

When a material is subjected to external loads, the internal structure (i.e., grain lattice, crystals, or molecules, in the case of many plastics) has to move in order to be aligned in a formation that would withstand the external loads. This new alignment may be called a state of stress, considering that there would be a tendency for the material to revert back to its natural state once the loads are removed.

If enough energy (generally in the form of heat) is provided, this internal structure tends to accept the new formation as a natural state, and thus the state of stress is either lessened or relieved altogether. This phenomenon is called creep and is defined by Lau (1993) as a mathematical model to represent the behavior of rate-sensitive elastoplastic materials at elevated temperatures. Clearly, this behavior depends on the material's properties and temperature levels, along with residence time at those levels and applied loads.

In and of itself, creep may not be considered a failure mechanism; however, it may lead to catastrophic failures. If creep is allowed to continue, deformations are enlarged and, eventually, strains become so large that the structure loses its load-carrying capacity and fails either in a buckling mode or a stress rupture.

Nearly all materials exhibit creep near their melting points. Although this proximity to the melting point varies from material to material, in general, once temperature value surpasses 50% of the melting point, creep and strain rates associated with creep become important factors that may not be ignored. In electronics packaging, solders and plastics are particularly at risk because of their relatively low melting temperatures and operating temperatures that may exceed the 50% threshold (Pecht 1991; Lau 1993; Lau et al. 1998).

Stages of Creep

The process of creep development is not uniform or constant. Figure F.1 depicts a typical creep curve with the three stages of creep identified. In the first regime or stage, strain rates are extremely high, but the rates are quickly lowered as time goes on and approach a steady-state value, which marks the onset of the steady-state regime. Finally, when enough time has gone by where there has been sufficient change in the microstructure of the material, strain rates once again increase rapidly—generally exponentially—until there is a fracture. It should be pointed out that the steady-state regime is the best understood stage.

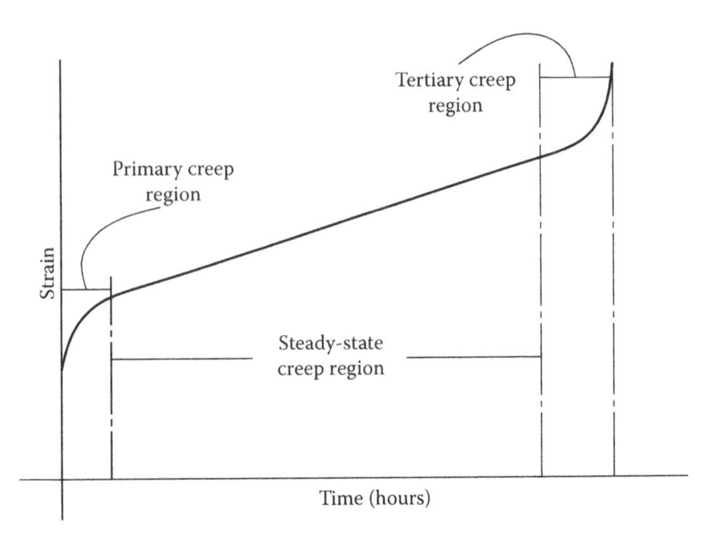

FIGURE F.1
A typical creep curve for constant load and constant temperature conditions.

Stress–Strain Relationship

To plot the stress–strain relationship in the presence of creep means that time should be frozen and then this relation plotted. As a result, we would have a different curve for each time increment, as shown in Figure F.2. However, because a constitutive relationship is needed to solve the governing equations, it is more customary to develop a stress–stain rate relationship as shown in Figure F.3.

Impact of Creep in Electronics Packaging

As heat is generated within an electronics system, its temperature rises and various components expand. Thermal stresses are developed because of a mismatch of coefficient of thermal expansions, particularly in the solder joints. Based on the foregoing discussion, these stresses, in conjunction with high temperatures, would cause creep-induced deformations, which may lead to joint failures. However, if the system operation is such that there is thermal cycling, the stresses have a tendency to oscillate between a minimum and maximum value. In these cases, often creep and fatigue work

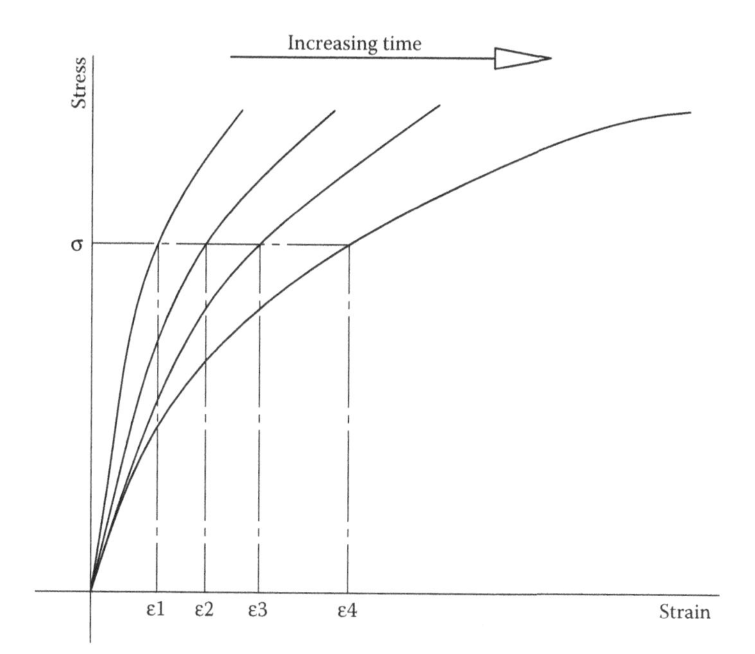

FIGURE F.2
Typical stress–strain curves for a material experiencing creep.

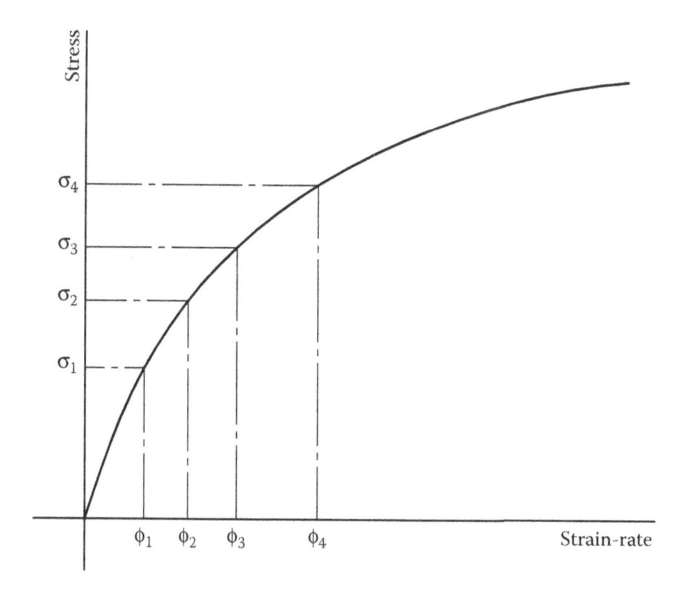

FIGURE F.3
Typical stress–strain rate curve; ϕ denotes the strain rate.

together and may create a condition called creep ratcheting (Ross and Wen 1993) and creep fatigue (Pecht 1991) that leads to solder failure. The impact of creep in a variety of solders has been studied extensively and a variety of models have been offered; as an example, see Ross and Wen (1993) and Wei et al. (2007).

Appendix G: Fatigue

Fatigue refers to a progressive failure of a material under a cyclic load with stresses generally below material yield point. The main characteristic of fatigue is that the material develops a small fracture line under the cyclic load. Initially, this fracture grows slowly; however, eventually, the material loses its load-carrying capacity and the part suddenly breaks. The cause of this process has been postulated to be affected by the relative movement or slip of the material's crystal structure. Over time, these slip bands form short cracks, which in turn join to create larger cracks and fractures.

Fatigue strength of a material may be altered by such factors as frequency of cycling, cold working of material, temperature, corrosion, residual stresses, surface finish, and mean stress, among other reasons. It may be shown that fatigue life of a material under load has an inverse relationship to the applied load, and thus the stress levels. In other words, the larger the stress values, the shorter the life of the part.

The relationship between the stress level and the number of cycles to failure is called an *S–N* curve and is generally plotted in a semilog graph. Although there have been many efforts to develop a deterministic (mathematical) relationship between stress and number of cycles to failure, fatigue has a probabilistic nature and there is a considerable scatter in the data even under the best of circumstances. See, for example, Figure G.1. The *S–N* curve usually reported for a given metal is often taken to represent a 50% probability of failure. In other words, for a given stress level, 50% of the test specimens fail before reaching the reported number of cycles to failure, and the other 50% are still functional. Thus, stress values above this line present a higher probability of failure, and values below this line present a lower probability of failure.

Based on this argument, one may estimate the life of a part if the stress level is known. For example, suppose that a cantilever beam is made with the iron alloy whose *S–N* curve is given in Figure G.1. Furthermore, suppose that this beam is subjected to a cyclic load that creates a stress level that varies between ±90 ksi. It would be reasonable to assume that this beam would survive nearly 60,000 stress reversal cycles—keeping the 50% rule in mind. By estimating the life of our cantilever beam in this manner, we have effectively made the following assumptions:

1. 90 ksi is below material yield stress value
2. Load cycling frequency is the same as the test specimen used in developing the *S–N* curve
3. Material surface condition (e.g., surface roughness) is the same as the test specimen

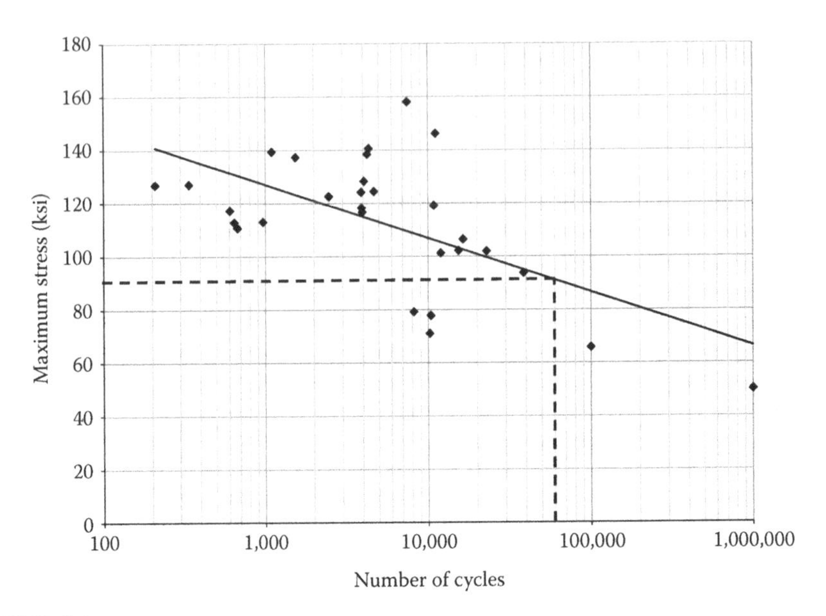

FIGURE G.1
A sample *S–N* curve of an iron alloy depicting the scatter in data as well as a theoretical trend line.

4. Material processing (e.g., work hardening) is similar to the test specimen
5. Test conditions such as temperature are similar as well

Does this mean that 60,000 cycles is wrong if the conditions are not exactly identical? The answer may be that the 60,000 must be used as an indicator and an estimate because of the probabilistic nature of fatigue. Thus, the more deviation is made from the test specimen conditions, the more inaccuracies are introduced.

Having said this, it should be pointed out that the *S–N* curve is generally used for so-called high cycle fatigue. High cycle fatigue refers to a high number of cycles to failure and is generally governed by a state of stress well below yield strength. In contrast to high cycle fatigue is low cycle fatigue. In this category, elastoplasticity and full plasticity of the material play important roles and, as such, one should not consider stress levels alone to determine the life of the material under study; rather, the state of the strain must be considered as well.

The threshold between low and high cycle fatigue has been reported from a few 1,000 to 100,000 cycles (Adams 2007; Fatigue 2007), and a value of 10,000 cycles seems to be the fairly accepted border. Figure G.2 depicts two *S–N* curves for a 300 M steel alloy for different loading conditions. It is clear that if we were to use the yield point to identify the border between high and low cycle fatigue, it would shift depending on the loading conditions.

Typically, in vibration problems, the induced stresses are well below yield and fatigue is in the high cycle region. In this case the *S–N* curve is a good

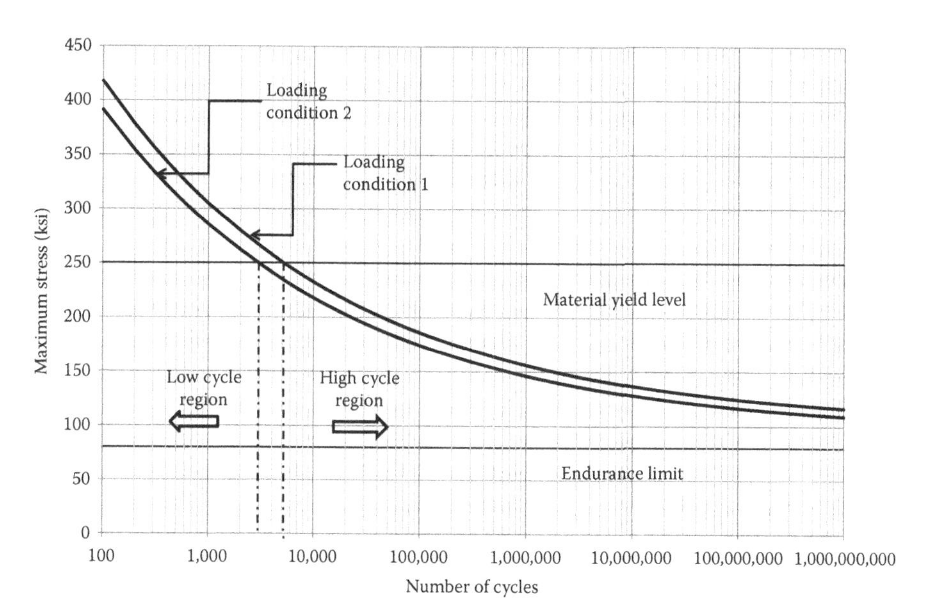

FIGURE G.2
Low and high cycle fatigue and its relationship to material yield level.

tool to determine the material's life expectancy. In contrast, temperature cycling may cause stresses that are affected by creep and exceed the yield point of the material; thus, fatigue would be in the low cycle region (Pecht 1991). Under such circumstances, *S–N* curves are inappropriate tools for life expectancy calculations. Although the boundary between low and high cycle fatigue is generally influenced by the value of yield strength, endurance limit may be influenced by proportional point on the stress–strain curve.

Some materials, particularly steel and titanium alloys, exhibit a stress value below which the test specimen may be cycled indefinitely—or at least to an extremely high number of cycles. This stress value is defined as endurance limit. An exploration of the physical reasons for this phenomenon is beyond the scope of this book; however, as design engineers, we need to be aware of this limit and try to maintain stresses below this level. A good rule of thumb is that endurance limit (if it exists), which is about 25%–30% of yield stress.

Mathematical Expressions for *S–N* Curves

Once the fatigue test data for a particular material have been collected, it is possible to use statistical tools to find a relationship between the stress level and the number of cycles to failure. For instance, MIL-HDBK-5J (2003) provides the following relationship for a 300 M alloy:

$$\log N = 14.8 - 5.2\ \log\left(S_{eq} - 94.2\right)$$

$$S_{eq} = S_{max}\left(1 - R\right)^{.38}$$

$$\text{Standard deviation of} \log\left(\text{life}\right) = 65.8\frac{1}{S_{eq}}$$

where R is the ratio of minimum stress to maximum stress in a uniaxial test. Figure G.3 provides this S–N relationship for three different values of R. One may observe that for a fully reversed loading condition ($R = -1$), fatigue strength is substantially lower.

Although MIL-HDBK-5J (2003) provides relatively complicated S–N relationships, others (Sloan 1985; Steinberg 1988) give much simpler relationships. For instance, Sloan (1985) suggests the following formula:

$$N\sigma^{b} = B$$

where b and B are material-dependent values. For example, for aluminum, $b = 9.10$ and $B = 2.307 \times 10^{47}$ psi per cycle, or for G-10 glass epoxy, $b = 11.36$ and $B = 7.253 \times 10^{55}$ psi per cycle.

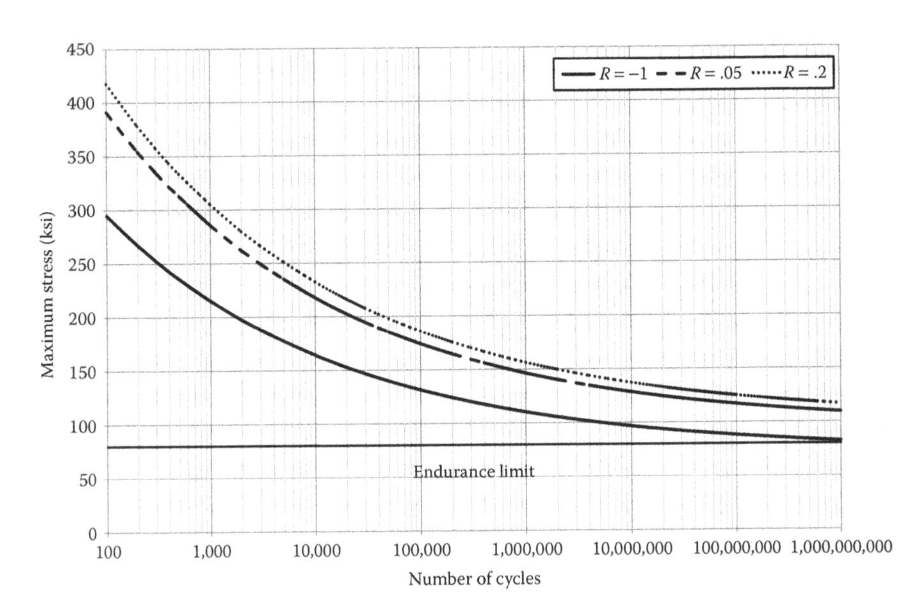

FIGURE G.3
A 300 M alloy S–N curve based on empirical formulation.

References

Adams, V., *Fatigue Analysis in Product Design—An Overview of Technologies and Techniques,* Milwaukee, WI: Impact Engineering Solutions, Inc., Retrieved on November 12, 2007, from http://www.impactengsol.com/uploads/whitepapers/wp_fatigue.pdf.

Aida, Y., Niwa, H., and Maeda, Y., Dynamic vibration absorber, US Patent 5,896,961, April 27, 1999.

Anderson, Jr., J.D., *Introduction to Flight,* New York: McGraw-Hill, 1978.

Antler, M., Contact fretting of electronic connectors. *IEICE Transactions on Electronics,* E82-C(1), 3, 1999.

ARINC Standards, *Air Transport Avionics Equipment Interfaces, ARINC Specification 600–8 and 746,* Annapolis, MD: Aeronautical Radio, Inc., 1991.

ASHRAE Handbook—Fundamentals (I-P Edition). American Society of Heating, Refrigerating and Air-Conditioning Engineers, Inc., 2013.

Asrar, N., Vancauwenberghe, O., and Prangere, S., Tin whiskers formation on electronic products: A case study, *Journal of Failure Analysis and Prevention,* 2007 7: 179–182, 2007.

Barron, R. F., *Industrial Noise Control and Acoustics,* New York: Marcel Dekker, 2003.

Bathe, K. J., *Finite Element Procedures in Engineering Analysis,* Englewood Cliffs, NJ: Prentice Hall, 1982.

Baumeister, T., Avallone, E. A., and Baumeister III, T., *Mark's Standard Handbook for Mechanical Engineers,* 8th edn, New York: McGraw-Hill, 1979.

Bazovsky, I., *Reliability Theory and Practice,* Mineola, NY: Dover, 2004.

Bejan, A., Tsatsaronis, G., and Moran, M., *Thermal Design & Optimization,* New York: Wiley Interscience, 1996.

Bellcore TR-332, *Reliability Prediction Procedure for Electronic Equipment.* Report, Issue 5, NJ, December, 1995.

Bethea, R. M. and, Rhinehart, R. R., *Applied Engineering Statistics,* New York: Marcel Dekker, 1991.

Bianchi, S., Corsini, A., and Sheard, A. G., Synergistic noise-by-flow control design of blade-tip in axial fans: Experimental investigation. In: Siano, D. (ed.) *Noise Control, Reduction and Cancellation Solutions in Engineering,* 2012, Retrieved on December 2014 from: http://www.intechopen.com/books/noise-control-reduction-and-cancellation-solutions-in-engineering/synergistic-noise-by-flow-control-design-of-blade-tip-in-axial-fans-experimental-investigation.

Bisbee, L. H., Queen, H. E., and Bezzant, R. K., *Ultrasonic Testing Metric for Creep Damage Assessment,* Retrieved on November 23, 2007, from http://www.structint.com/tekbrefs/t99014r0.pdf.

Boresi, A. P., Sidebottom, O. M., and Smith, J. O., *Advanced Mechanics of Materials,* 3rd edn, New York: John Wiley & Sons, 1978.

Boyce, W. E. and DiPrima, R. C., *Elementary Differential Equations and Boundary Value Problems,* New York: John Wiley & Sons, 1977.

Bumiller, E., Pecht, M., and Hillman, C., Electrochemical Migration on HASL Plated FR-4 Printed Circuit Boards, Presented at the *9th Pan Pacific Microelectronics Symposium Exhibits & Conference*, Kahuku, HI, February 2004.

Chomette, B., Chesné, S., Rémond, D., and Gaudiller, L., Damage reduction of on-board structures using piezoelectric components and active modal control—Application to a printed circuit board, *Mechanical Systems and Signal Processing*, 24(2), pp. 352–364, 2010.

Condra, L. W., *Reliability Improvement with Design of Experiments*, 2nd edn, Revised and Expanded, New York: Marcel Dekker, 2001.

Crede, C. E. and Ruzicka, J. E., Theory of vibration isolation, In: Harris, C. M., *Shock and Vibration Handbook*, 4th edn, New York: McGraw-Hill, 1996.

Crede, C., *Shock and Vibration Concepts in Engineering Design*, Englewood Cliffs, NJ: Prentice Hall, 1965.

Crede, C., *Vibration and Shock Isolation*, New York: John Wiley & Sons, 1951.

Culham, J. R., Teertstra, P., and Yovanovich, M. M., Natural convection modeling of heat sinks using web-based tools, *Electronics Cooling Magazine* Online, www.electronics-cooling.com, 2000.

Curtis, A. and Lust, S., Concepts in vibration data analysis, In: Harris, C. M., *Shock and Vibration Handbook*, 4th edn, New York: McGraw-Hill, 1996.

Cyganowski, A., Brusse, J., and Leidecker, H., *Nickel Cadmium Batteries: A Medium for the Study of Metal Whiskers and Dendrites*, Retrieved on November 23, 2007, from http://nepp.nasa.gov/WHISKER/dendrite/2006-cyganowski-paper.pdf.

Daniel, I. M., Wang, T. M., and Gotro, J. T., Thermomechanical behavior of multilayer structures in microelectronics, *Journal of Electronic Packaging*, 112, 11–15, 1990.

Davis, J. R., *Handbook of Materials for Medical Devices*, Materials Park, OH: ASM International, 2003.

Dodson, B., Schwab, H., *Accelerated Testing—A Practitioner's Guide to Accelerated and Reliability Testing*. Society of Automotive Engineers, Inc., (2006). Available at: http://app.knovel.com/hotlink/toc/id:kpATAPGAR3/accelerated-testing-practitioners/accelerated-testing-practitioners.

Earth Atmosphere model from Glen Research Center, Retrieved on February 4, 2008, from http://www.grc.nasa.gov/WWW/K-12/airplane/atmos.html.

Engelman, M. and Jamnia, M. A., Gray-body surface radiation coupled with conduction and convection for general geometries, *International Journal of Numerical Methods in Fluids*, 13, 1029, 1991.

Ezrin, M., *Plastics Failure Guide: Cause and Prevention*, New York: Hanser Publishers, 1996.

Fatigue (material) From Wikipedia, the free encyclopedia, Retrieved on November 12, 2007, from http://en.wikipedia.org/wiki/Fatigue_(material).

Federal Airworthiness Regulation, (FAR) Part 25, Washington, DC, 2008.

Feldstein, M. D. and Dumas, P. E., Enhanced heat transfer for electronics applications by composite electroless coatings, *Metal Finishing*, p. 10, 2000.

FIDAP *Theoretical Manual*, Evanston, IL: Fluid Dynamics International, 1989.

Gallego, N. C. and Klett, J. W. Carbon foams for thermal management, *Carbon*, 41, pp. 1461–1466, 2003.

Galvanic Compatibility, Retrieved on November 23, 2007, from http://www.engineersedge.com/galvanic_capatability.htm.

Glenn Research Center, *Earth Atmosphere Model*, Retrieved on July 18, 2007, from http://www.grc.nasa.gov/WWW/K-12/airplane/atmosmet.html.
Goodman, S. H., *Handbook of Thermoset Plastics*, 2nd edn, Westwood, NJ: Noyes Publications, 1998.
Haimovich, J., Cu-Sn intermetallic compound growth in hot-air-leveled tin at and below 100°C, *AMP Journal of Technology*, 3, 46–54, 1993.
Hall, P. M., Thermal expansivity and thermal stress in multilayered structures. In: John H. Lau (ed.) *Thermal Stress and Strain in Microelectronics Packaging*. New York: Van Nostrand Reinhold, 1993.
Harris, C. M., *Shock and Vibration Handbook*, 4th edn, New York: McGraw-Hill, 1996.
Hellström, B., *Modelling of Sound in Public Spaces*, Stockholm, Sweden: Arkitekturskolan Kungliga Tekniska Högskolan, 2001.
Hill, T. and Lind, J., Designing ventilation grilles for electronic equipment. In: *Machine Design*, San Diego, CA: Hill Engineering, 1990.
Hnatek, E. R., *Practical Reliability of Electronic Equipment and Products*, New York: Marcel Dekker, 2003.
Holdsworth, S. R., Mazza, E., and Jung, A., *Creep-Fatigue Damage Development during Service-Cycle Thermo-Mechanical Fatigue Tests of 1CrMoV Rotor Steel*, Retrieved on November 23, 2007, from http://www.zfm.ethz.ch/e/lifetime/papers/Creep_fatigue_damage_development.pdf.
Hubbard, H. H. and Houbolt, J. C., Vibration induced by accoustic waves. In: Harris, C. M. (ed.) *Shock and Vibration Handbook*, 4th edn, New York: McGraw-Hill, 1996.
ISO 24501:2010 Ergonomics—Accessible design—Sound pressure levels of auditory signals for consumer products, 2010.
ISO 3744:2010 Acoustics—Determination of sound power levels and sound energy levels of noise sources using sound pressure—Engineering methods for an essentially free field over a reflecting plane, 2010.
ISO 7779:2010 Acoustics—Measurement of airborne noise emitted by information technology and telecommunications equipment, 2010.
Jeon, M. S., Song, J. K., Lee, E. J., Kim, Y. N., Shin, H. G., and Lee, H. S., Failure analysis of thermally shocked NiCr films on Mn-Ni-Co spinel oxide substrates, *Materials Research Society Symposia Proceedings*, 875, 389–394, 2005.
Jones, Robert M., *Mechanics of Composite Materials*, Washington, DC: Scripta Book Company, 1975.
K&K Associates, *Thermal Network Modeling Handbook*, Version 97.003, 1999–2000.
Kim, J. Y., Yu, J., Lee, J. H., and Lee, T. Y., The effects of electroplating parameters on the composition and morphology of Sn-Ag solder, *Journal of Electronic Materials*, 33(12), 2004.
Klinger, D., Nakada, Y., and Menendez, M. *AT&T Reliability Manual*, New York: Van Nostrand Reinhold Company, 1990.
Krieth, F. *Principles of Heat Transfer*, New York: Harper & Row, 1973.
Krum, A., Thermal management. In: Harper, C. A. (ed.) *Electronic Packaging and Interconnection Handbook*, 3rd edn, New York: McGraw-Hill, 2000.
Krumbein, S. J., Metallic Electromigration Phenomena, Presented at 33rd *IEEE Conference on Electrical Contacts*, Chicago, IL, September 21–23, 1987.

Kuhlkamp, P., *Lead(Pb)—Free Plating for Electronics and Avoidance of Whisker Formation*, Retrieved on November 23, 2007, from http://www.atotech.com/data/publications/Pb_free_plating_for_electronics.pdf.

Lau, J. H., Thermomechanics for electronics packaging. In: Lau J. H. (ed.) *Thermal Stress and Strain in Microelectronics Packaging*, New York: Van Nostrand Reinhold, 1993.

Lau, J. H., Wong, C. P., Prince, J. L., and Nakayama, W., *Electronic Packaging: Design, Materials, Process, and Reliability*, New York: McGraw-Hill, 1998.

Lau, John H., *Thermal Stress and Strain in Microelectronics Packaging*, New York: Van Nostrand Reinhold, 1993.

Leatherman, R., Thermal analysis and heat transfer in electronics design, One-Day Presentation at *Thermal & Fluid Analysis Workshop*, Hosted by NASA Lewis Research Center and Ohio Aerospace Institute, August 1996.

Lee, S.-B., Yoo, Y.-R., Jung J.-Y., Park Y.-B., Kim Y.-S., and Joo Y.-C., Electrochemical migration characteristics of eutectic SnPb solder alloy in printed circuit board, *Thin Solid Films*, 504, 294–297, 2006.

Lieberman F. N. and Brodsky M. A., Silver dendritic growth in plastic IC packages, *Chip Scale Review*, 2004, Retrieved on November 23, 2007, from ChipScaleReview.com

Lustinger, A., Environmental stress cracking: The phenomenon and its utility. In: Brostow, W. and Corneliussen, R. D. (eds.) *Failure of Plastics*, New York: Hanser Publishers, 1989.

Lyon, R. H., *Designing for Product Sound Quality*, New York: Marcel Dekker, 2000.

MacGregor, C. W., Symond, J., Vidosic, J. P., Hawkins, H. V., Thomson, W. T., and Dodge, D. D., Strength of materials. In: Baumeister, T., Avallone E. A., and Baumeister III, T. (eds.) *Mark's Standard Handbook for Mechanical Engineers*, 8th edn, New York: McGraw-Hill, 1979.

Madhusudana, C. V., *Thermal Contact Conductance*, New York: Springer-Verlag, 1996.

Mahalingam, R., Heffington, S., Jones, L., and Williams, R. Synthetic jets for forced air cooling of electronics, *Electronics Cooling*, 13(2), 14–21, 2007.

Malfa, S. L., Laura, P. A. A., Rossit, C. A., and Alvarez, O., Use of a dynamic absorber in the case of a vibrating printed circuit board of complicated boundary shape, *Journal of Sound and Vibration*, 230(3), 721–724, 2000.

McAdams, W. H., *Heat Transmission*, 3rd edn, New York: McGraw-Hill, 1954.

McKeown, S. A., *Mechanical Analysis of Electronic Packaging Systems*, New York: Marcel Dekker, 1999.

McLean, H. W., *HALT, HASS, and HASA Explained, Accelerated Reliability Techniques*, Milwaukee, WI: ASQ Quality Press, 2009.

MIL-HDBK-1250A, *Handbook for Corrosion Prevention and Deterioration Control in Electronic Components and Assemblies*, Washington, DC: Department of Defense, 1995.

MIL-HDBK-217F, *Reliability Prediction of Electronic Equipment*, Washington, DC: Department of Defense, 1991.

MIL-HDBK-251, *Reliability/Design Thermal Applications*, Washington, DC: Department of Defense, 1978.

MIL-HDBK-338B, *Electronic Reliability Design Handbook Equipment*, Washington, DC: Department of Defense, 1988.

MIL-HDBK-338B, *Military Handbook: Electronic Reliability Design Handbook*, Washington, DC: Department of Defense, 1998.

MIL-HDBK-5J, *Metallic Materials and Elements for Aerospace Vehicle Structures*, Washington, DC: Department of Defense, 2003.

MIL-STD-810F, *Test Method Standard for Environmental Engineering Considerations and Laboratory Tests*, Washington, DC: Department of Defense, 2000.

MIL-STD-889, *Dissimilar Metals*, Washington, DC: Department of Defense, 1976.

Nelson, W. B., *Accelerated Testing, Statistical Models, Test Plans, and Data Analysis*, New York: John Wiley & Sons, 2004.

Nuventix, Enabling new electronics designs through advanced cooling technology, *Thermal Management of Electronics, 7th International Business & Technology Summit* August 22–23, Natick, MA, 2007.

O'Connor, P. D. T. and Kleyner, A., *Practical Reliability Engineering*, 5th edn, New York: John Wiley & Sons, 2012.

Peak, R. S., Fulton, R. E., and Sitaraman, S. K., "Thermomechanical CAD/CAE Integration in the TIGER PWA Toolset," *Advances in Electronic Packaging* EEP-Vol. 19 - 1, Suhir, E. et al (eds.) InterPACK'97, June 15–19, 1997, Kohala Coast, HI, 957–962.

Pecht, M., *Handbook of Electronic Package Design*, New York: Marcel Dekker, 1991.

Radio Technical Commission for Aeronautics, RTCA/DO-160C, Washington, DC, 1989.

Rao, S. S., *The Finite Element Method in Engineering*, New York: Pergamon Press, 1982.

Ross, R. G. and Wen, L.-C., Solder creep-fatigue interactions with flexible leaded surface mount components, In: Lau J. H. (ed.) *Thermal Stress and Strain in Microelectronics Packaging*, New York: Van Nostrand Reinhold, 1993.

Roubaud, P., Ng, G., Henshall G., Bulwith, R., Herber R., Prasad, S., Carson F., Kamath, S., and Garcia A., Impact of intermetallic growth on the mechanical strength of Pb-Free BGA assemblies, Presented at *APEX 2001* on January 16–18, 2001, San Diego, CA.

Sabour, M. H. and Bhat, R. B., *Lifetime Prediction in Creep-Fatigue Environment*, Retrieved on November 23, 2007, from http://www.materialsscience.pwr.wroc.pl/ap/articles/ms_2007_036.pdf.

SAE Aerospace Applied Thermodynamic Manual, SAE Standards, Document Number: AIR1168/V1.

Sang-Myeong, K., Pietrzko, S., and Brennan, M. J., Active vibration isolation using an electrical damper or an electrical dynamic absorber, *IEEE Transactions on Control Systems Technology*, 16(2), 245–254, 2008.

Shapiro, A. B., *FACET-A Radiation View Factor Computer Code for Axisymmetric, 2-D Planar, and 3-D Geometries with Shadowing*, Lawrence Livermore Laboratory Report UCID-19889, 1983.

Siegel, R. and Howell, J. R., *Thermal Radiation Heat Transfer*, 2nd edn. New York: McGraw-Hill, 1981.

Siewert, T. A., Madeni, J. C., and Liu, S., *Formation and Growth of Intermetallics at the Interface between Lead-Free Solders and Copper Substrates*, Retrieved on November 23, 2007, from http://www.boulder.nist.gov/div853/Publication%20files/NIST_Apex94_Siewert.pdf.

Silva, W. P. and Silva, C. M. D. P. S., LAB Fit Curve Fitting Software (Nonlinear Regression and Treatment of Data Program) V 7.2.39 (1999–2007), Retrieved on August 1, 2007, from www.labfit.net.

Sloan, J., *Design and Packaging of Electronic Equipment*, New York: Van Nostrand Reinhold Company, 1985.

Staley, T., Somlea, D., Tsinovkin, A., Tuan Le, Morris, J., and Coleman, C., Novel solid state air pump for forced convection electronics cooling, *Electronic Components and Technology Conference*, Lake Buena Vista, FL, May 27–30, 2008.

Steinberg, D., *Cooling Techniques for Electronic Equipment*, 2nd edn, New York: John Wiley & Sons, 1991.

Steinberg, D., *Preventing Thermal Cycling and Vibration Failures in Electronic Equipment*, New York: John Wiley & Sons, 2001.

Steinberg, D., *Vibration Analysis for Electronic Equipment*, 2nd edn, New York: John Wiley & Sons, 1988.

Suhir, E., Predicted bow of plastic packages of integrated circuit (IC) devices. In: Lau J. H. (ed.) *Thermal Stress and Strain in Microelectronics Packaging*, New York: Van Nostrand Reinhold, 1993.

Taguchi, G. and Clausing, D., Robust quality, *Harvard Business Review*, No. 90114, January–February 1990, p. 65.

Talty, J. T. *Industrial Hygiene Engineering—Recognition, Measurement, Evaluation and Control*, 2nd Edn. Park Ridge, NJ: William Andrew Publishing/Noyes, 1988.

Tasooji, A., Ghaffarian, R., and Rinaldi, A., *Design Parameters Influencing Reliability of CCGA Assembly: A Sensitivity Analysis*, Retrieved on November 23, 2007, from http://trs-new.jpl.nasa.gov/dspace/bitstream/2014/39705/1/06-0593.pdf.

Tay, F. E. H. and Li, H., Designing a fluid pump for microfluidic cooling systems in electronic components, *Electronic Packaging Technology Conference*, Piscataway, NJ, October 8–10, 1997.

Timoshenko, S. P. and Goodier, J. N., *Theory of Elasticity*, 3rd edn, New York: McGraw-Hill, 1970.

Turner, M. J., Clough, R. W., Martin, H. C., and Topp, L. C., Stiffness and deflection analysis of complex structures. *Journal of the Aeronautical Sciences*, 23, 805–882, 1956.

Van Dyke, M., *An Album of Fluid Motion*, Stanford, CA: Parabolic Press, 1982.

Viswanadham, P. and Singh, P., *Failure Modes and Mechanisms in Electronic Packages*, New York: Chapman & Hall, 1988.

Wang, F. F., Relating sinusoid to random vibration for electronic equipment testing, *COTS Journal*, 53–57, April 2003.

Wei, Y., Chow, C. L., Fang, H. E., and Neilsen, M. K., *Characteristics of Creep Damage for 60Sn-40Pn Solder Material*, Information Bridge: DOE Scientific and Technical Information, Retrieved on November 23, 2007, from http://www.osti.gov/bridge/servlets/purl/10363-4HhIa4/webviewable/10363.pdf.

Wereszczak, A. A. Ferber, M. K. and Kirkland, T. P., *Effects of Oxidation and Creep Damage Mechanisms on Creep Behavior in HIPed Silicon Nitrides*, Retrieved on November 23, 2007, from http://www.osti.gov/bridge/servlets/purl/10122404-fM4fYX/native/10122404.pdf.

White, J. R., Conduction mechanisms in contaminant layers on printed circuit boards, *IBM Journal of Research and Development*, 37(2), 1993.

Wirsching, P., Paez, T., and Ortiz, H., *Random Vibrations: Theory and Practice*, New York: John Wiley & Sons, 1995.

Xu, C., Fan, C., Vysotskaya, A., Abys, J. A., Zhang, Y., Hopkins, L., and Stevie, F., *Understanding Whisker Phenomenon—Part II: Competitive Mechanisms*, Retrieved on November 23, 2007, from http://www.hkpc.org/hkiemat/mastec03_notes/11.pdf.

Yan, Z., *Thermal Properties of Graphene and Applications for Thermal Management of High-Power Density Electronics*, UC Riverside Electronic Theses and Dissertations, Retrieved on December 29, 2014, from https://escholarship.org/uc/item/8m67d089.

Yeh, C.-P., Ume, C., Fulton, R. E., Wyatt, K. W., and Stafford, J. W., Correlation of analytical and experimental approaches to determine thermally induced PWB warpage, *IEEE Transactions on Components, Hybrids, and Manufacturing Technology*, 16(8), 986–994, 1993.

Yovanovich, M. M., Culham, J. R., and Teertstra, P., Calculating Interface Resistance, *Electronics Cooling Magazine* Online Retrieved on July 18, 2001, from www.electronics-cooling.com.

Zhang, Y., Xu, C., Fan, C., Vysotskaya A., and Abys, J. A., *Understanding Whisker Phenomenon—Part I: Growth Rates*, Retrieved on November 23, 2007, from http://www.hkpc.org/hkiemat/mastec03_notes/10.pdf.

Zribi, A., Chromik, R. R., Presthus, R., Clum, J., Zavalig, L., and Cotts, E. J., Effect of Au-intermetallic compounds on mechanical reliability of Sn-Pb/Au-Ni-Cu joints, *Advances in Electronic Packaging*, 26-2(2), 1573–1577, 1999.

Zweben, C., Advanced Thermal Management Materials, *Thermal 2012, Advancements in Thermal Management 2012*, Denver, CO, September 18, 2012.

Zweben, C., *CVD Diamond for Microelectronic and Optoelectronic Thermal Management Applications*, Retrieved on December 29, 2014, from http://usapplieddiamond.com/pdfs/Zweben.pdf.

Zweben, C., Polymer-Carbon Composite Heat Exchangers—Challenges and Opportunities, *IHTC 2010, International Heat Transfer Conference*, Washington, DC, August 9, 2010.

Zweben, C., Thermal management and electronic packaging applications, In: ASM International (ed.) *ASM Handbook, Volume 21: Composites*, Materials Park, OH: ASM International, pp. 1078–1084, 2001.

Index

Note: Page numbers followed by f and t refer to figures and tables, respectively.

D

Q

R

S

W

Y

Printed in the United States
by Baker & Taylor Publisher Services